Selected Titles in This Subseries

42 S. Tabachnikov, Editor, Differential and Symplectic Topology of Knots and Curves (TRANS2/190)

41 V. Buslaev, M. Solomyak, and D. Yafaev, Editors, Differential Operators and Spectral Theory (M. Sh. Birman's 70th anniversary collection) (TRANS2/189)

40 M. V. Karasev, Editor, Coherent Transform, Quantization, and Poisson Geometry (TRANS2/187)

39 A. Khovanskiĭ, A. Varchenko, and V. Vassiliev, Editors, Geometry of Differential Equations (TRANS2/186)

38 B. Feigin and V. Vassiliev, Editors, Topics in Quantum Groups and Finite-Type Invariants (Mathematics at the Independent University of Moscow) (TRANS2/185)

37 Peter Kuchment and Vladimir Lin, Editors, Voronezh Winter Mathematical Schools (Dedicated to Selim Krein) (TRANS2/184)

36 V. E. Zakharov, Editor, Nonlinear Waves and Weak Turbulence (TRANS2/182)

35 G. I. Olshanski, Editor, Kirillov's Seminar on Representation Theory (TRANS2/181)

34 A. Khovanskiĭ, A. Varchenko, and V. Vassiliev, Editors, Topics in Singularity Theory (TRANS2/180)

33 V. M. Buchstaber and S. P. Novikov, Editors, Solitons, Geometry, and Topology: On the Crossroad (TRANS2/179)

32 R. L. Dobrushin, R. A. Minlos, M. A. Shubin, and A. M. Vershik, Editors, Topics in Statistical and Theoretical Physics (F. A. Berezin Memorial Volume) (TRANS2/177)

31 R. L. Dobrushin, R. A. Minlos, M. A. Shubin, and A. M. Vershik, Editors, Contemporary Mathematical Physics (F. A. Berezin Memorial Volume) (TRANS2/175)

30 A. A. Bolibruch, A. S. Merkur'ev, and N. Yu. Netsvetaev, Editors, Mathematics in St. Petersburg (TRANS2/174)

29 V. Kharlamov, A. Korchagin, G. Polotovskiĭ, and O. Viro, Editors, Topology of Real Algebraic Varieties and Related Topics (TRANS2/173)

28 L. A. Bunimovich, B. M. Gurevich, and Ya. B. Pesin, Editors, Sinai's Moscow Seminar on Dynamical Systems (TRANS2/171)

27 S. P. Novikov, Editor, Topics in Topology and Mathematical Physics (TRANS2/170)

26 S. G. Gindikin and E. B. Vinberg, Editors, Lie Groups and Lie Algebras: E. B. Dynkin's Seminar (TRANS2/169)

25 V. V. Kozlov, Editor, Dynamical Systems in Classical Mechanics (TRANS2/168)

24 V. V. Lychagin, Editor, The Interplay between Differential Geometry and Differential Equations (TRANS2/167)

23 Yu. Ilyashenko and S. Yakovenko, Editors, Concerning the Hilbert 16th Problem (TRANS2/165)

22 N. N. Uraltseva, Editor, Nonlinear Evolution Equations (TRANS2/164)

Published Earlier as Advances in Soviet Mathematics

21 V. I. Arnold, Editor, Singularities and bifurcations, 1994

20 R. L. Dobrushin, Editor, Probability contributions to statistical mechanics, 1994

19 V. A. Marchenko, Editor, Spectral operator theory and related topics, 1994

18 Oleg Viro, Editor, Topology of manifolds and varieties, 1994

17 Dmitry Fuchs, Editor, Unconventional Lie algebras, 1993

16 Sergei Gelfand and Simon Gindikin, Editors, I. M. Gelfand seminar, Parts 1 and 2, 1993

15 A. T. Fomenko, Editor, Minimal surfaces, 1993

14 Yu. S. Il'yashenko, Editor, Nonlinear Stokes phenomena, 1992

(See the AMS catalog for earlier titles)

Differential and Symplectic Topology of Knots and Curves

American Mathematical Society

TRANSLATIONS

Series 2 • Volume 190

Advances in the Mathematical Sciences — 42

(*Formerly Advances in Soviet Mathematics*)

Differential and Symplectic Topology of Knots and Curves

S. Tabachnikov
Editor

American Mathematical Society
Providence, Rhode Island

ADVANCES IN THE MATHEMATICAL SCIENCES
EDITORIAL COMMITTEE

V. I. ARNOLD
S. G. GINDIKIN
V. P. MASLOV

1991 *Mathematics Subject Classification.* Primary 57Gxx; Secondary 53Cxx.

ABSTRACT. This collection of papers concerns the topology and geometry of knots and curves, especially in the presence of contact structures. The methods of study vary from contemporary knot theory to symplectic and contact topology.

Library of Congress Card Number 91-640741
ISBN 0-8218-1354-4
ISSN 0065-9290

Copying and reprinting. Material in this book may be reproduced by any means for educational and scientific purposes without fee or permission with the exception of reproduction by services that collect fees for delivery of documents and provided that the customary acknowledgment of the source is given. This consent does not extend to other kinds of copying for general distribution, for advertising or promotional purposes, or for resale. Requests for permission for commercial use of material should be addressed to the Assistant to the Publisher, American Mathematical Society, P. O. Box 6248, Providence, Rhode Island 02940-6248. Requests can also be made by e-mail to reprint-permission@ams.org.

Excluded from these provisions is material in articles for which the author holds copyright. In such cases, requests for permission to use or reprint should be addressed directly to the author(s). (Copyright ownership is indicated in the notice in the lower right-hand corner of the first page of each article.)

© 1999 by the American Mathematical Society. All rights reserved.
The American Mathematical Society retains all rights
except those granted to the United States Government.
Printed in the United States of America.

∞ The paper used in this book is acid-free and falls within the guidelines
established to ensure permanence and durability.
Visit the AMS home page at URL: http://www.ams.org/

10 9 8 7 6 5 4 3 2 1   04 03 02 01 00 99

Contents

Preface	ix
Contact Topology, Taut Immersions, and Hilbert's Fourth Problem Juan Carlos Álvarez Paiva	1
On Legendre Cobordisms Emmanuel Ferrand	23
Vassiliev Invariants of Knots in $\mathbb{R}^3$ and in a Solid Torus Victor Goryunov	37
Finite Type Invariants of Generic Immersions of M^n into $\mathbb{R}^{2n}$ are Trivial Tadeusz Januszkiewicz and Jacek Świątkowski	61
On Enumeration of Unicursal Curves Sergei K. Lando	77
Vassiliev Invariants Classify Flat Braids Alexander B. Merkov	83
New Whitney-type Formulas for Plane Curves Michael Polyak	103
Tree-Like Curves and Their Number of Inflection Points Boris Shapiro	113
Geometry of Exact Transverse Line Fields and Projective Billiards Serge Tabachnikov	131
Shadows of Wave Fronts and Arnold–Bennequin Type Invariants of Fronts on Surfaces and Orbifolds Vladimir Tchernov	153
A Unified Approach to the Four Vertex Theorems. I Masaaki Umehara	185
A Unified Approach to the Four Vertex Theorems. II Gudlaugur Thorbergsson and Masaaki Umehara	229
Topology of Two-Connected Graphs and Homology of Spaces of Knots Victor A. Vassiliev	253

Preface

In recent years we have been witnessing a revival of interest in topology of curves and knots. This interest was stimulated by progress in singularity theory, the advent of "quantum topology" and the theory of finite-type knot invariants, and also by the rapid growth of symplectic topology in which curves and knots appear as the simplest, one-dimensional Lagrangian and Legendrian manifolds. The three areas, global singularities of wave fronts and caustics, finite-type invariants of knots and knot-like objects and contact geometry and topology of Legendrian knots, define a rather wide spectrum of problems concerned in this volume.

It would be appropriate to comment on the word "curves" in the title of this collection. Immersed plane curves can be treated as knot-like objects: a generic curve has neither triple points nor self-tangencies, but both types of singularities generically occur in one-parameter families of curves (compare to curves in the three-dimensional space, where double points generically appear in one parameter families). Thus the topological theories of triple point free curves and of non-self-tangent curves are analogous to knot theory. This point of view was put forward in the recent work of the Moscow Singularity School represented in this collection by about half of the participants.

A smooth plane curve can be regarded as a Legendrian curve in the contact manifold of contact elements of the plane, and curves without self-tangencies are Legendrian knots. In this way differential geometry of curves is "embedded" in contact geometry of Legendrian curves. An example much studied in recent years is the classical 4-vertex theorem by S. Muckhopadhyaya (published in 1909): the curvature of a closed convex plane curve has at least 4 extrema; this result can be interpreted and generalized as a theorem of symplectic topology. Another example is the Möbius theorem that a noncontractible embedded curve in the projective plane has at least three distinct inflection points. These results and their numerous generalizations are closely related to works of Sturm, Hurwitz, Kellogg and others on disconjugate differential equations.

For the convenience of the reader, we give very brief summaries of the papers.

J. C. Alvarez extends the theory of taut immersions from plane curves to wave fronts in a Desarguesian surface, that is, the plane or the sphere with a Finsler metric whose geodesics are straight lines or great circles.

E. Ferrand's paper concerns several notions of Legendre cobordism; one of the applications to Legendrian knots in the standard contact 3-space is an inequality for a new contact isotopy invariant, the ribbon Legendre genus, similar to the Bennequin inequality for the self-linking of a Legendrian knot.

V. Goryunov extends the construction of the Kontsevich integral from 3-space to the solid torus and gives a description of the space of finite-type invariants of knots and framed knots therein in terms of chord diagrams.

T. Januszkiewicz and J. Swiatkowski define finite-type invariants for immersions of n-manifolds to $2n$-dimensional space and prove that, in a certain sense, these invariants are trivial.

S. Lando gives an upper bound for the number of isotopic classes of one-component immersed spherical curves; the technique is that of matrix integration.

The main result of A. Merkov's paper is that finite-type invariants classify flat braids; the latter plays a role in the theory of triple point free plane curves similar to that of the usual braids in knot theory.

M. Polyak finds a family of local formulas for the winding number of an immersed plane curve generalizing a classical result by H. Whitney; similar formulas are given for the "strangeness", a first degree invariant of triple point free curves introduced by Arnold.

B. Shapiro gives bounds for the minimal number of inflection points of smooth plane curves within a diffeomorphism class.

S. Tabachnikov studies a class of transverse line fields along hypersurfaces generalizing the field of normals in Euclidean space; the motivation for this study comes from the theory of mathematical billiards.

V. Tchernov's paper concerns knot invariants of first degree in circle fibrations and an extension of V. Turaev's shadow techniques to invariants of wave fronts on surfaces.

The two papers by M. Umehara and G. Thorbergsson provide a unified approach to many known results of the 4-vertex theorem type (one of the most recent is the theorem by E. Ghys: the Schwarzian derivative of every projective line diffeomorphism has at least 4 zeroes).

V. Vassiliev's paper concerns a promising new approach to computing homology of the space of knots (the zero-dimensional homology being the space of knot invariants).

In concluding this brief introduction, it is a pleasure to mention the truly international nature of this collection: the countries of affiliation of the authors include Belgium, France, Germany, Israel, Japan, Poland, Russia, Sweden, United Kingdom, and the U.S. All papers have been refereed, and it is a pleasure to acknowledge the valuable contributions made by the anonymous referees in the preparation of this volume.

V. Arnold
S. Tabachnikov

Contact Topology, Taut Immersions, and Hilbert's Fourth Problem

Juan Carlos Álvarez Paiva

ABSTRACT. Symplectic and contact geometry are used to study the geometry of wave fronts in Desarguesian Finsler surfaces.

1. Introduction

Let M be a smooth compact manifold immersed in an n-dimensional Euclidean space. For each point $x \in \mathbb{R}^n$ we may consider the distance-squared function $m \mapsto \|m-x\|^2$ defined on M. The theory of taut immersions studies the relation between the geometry of M and the critical points of the family of functions obtained by letting the point x range over all of $\mathbb{R}^n$.

Since Lagrangian intersection theory is a generalization of critical point theory, it is natural to try to give a *symplectization* of the theory of taut immersions. The aim of this paper is to carry out this symplectization in the simplest of cases: that of curves in the plane or the sphere.

The advantage of the symplectic approach is twofold: on one hand one may extend the theory of taut immersions to wave fronts, while on the other hand one can extend it to the class of Finsler spaces whose geodesics are straight lines or great circles. Included among this class of Finsler spaces are Banach spaces with unit spheres which are smooth and strictly convex. Moreover, the theory loses nothing of its geometric appeal.

For the reader not familiar with Finsler spaces a word of motivation is in order. The problem of constructing and studying all metric structures on $\mathbb{R}P^n$ for which the projective lines are geodesics was posed by Hilbert as the fourth of his famous list of problems. Loosely speaking, Finsler spaces are the most general sort of metric spaces that have a system of curves which deserves to be called geodesic. It is then natural to restrict oneself to the study of Finsler metrics on projective spaces, or on open convex subsets, whose geodesics are projective lines. We also remark that, while the construction of these Finsler spaces has been known for some time, this seems to be the first time that the geometry of their submanifolds has been studied.

We now describe our results using as little contact and Finsler geometry as possible.

1991 *Mathematics Subject Classification.* Primary 58F05; Secondary 58B20.

©1999 American Mathematical Society

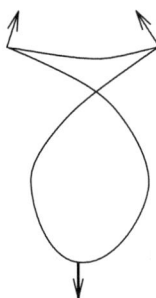

FIGURE 1. Normals of a cusped curve

An immersion $\gamma : S^1 \to \mathbb{R}^2$ is said to be taut if for every generic point $x \in \mathbb{R}^2$, the function $t \mapsto \|\gamma(t) - x\|^2$ has only two critical points.

In what follows we will favor the following equivalent definition:

DEFINITION. *An immersion $\gamma : S^1 \to \mathbb{R}^2$ is said to be taut if through every generic point in the plane there pass no more than two lines normal to γ.*

The concept of tautness and the characterization of taut curves is due to Banchoff (see [**9**]).

THEOREM (Banchoff). *A taut immersed curve is a round circle.*

Our first remark is that the concept of normal line, and hence the concept of tautness, makes sense for curves whose only singularities are self-intersections and cusps. At a cusp both branches of the curve are tangent and they share the same normal. If the curve is co-oriented we may also orient the normal lines. It is important to note that even if the curve has singularities, the field of normal lines may be a smoothly immersed curve in the manifold of oriented lines on the plane. A possibly singular curve on the plane with a smooth field of normal lines will be called a *wave front*.

REMARK. For standard and complete definitions of terms like *wave front*, *Finsler metric*, and *Legendrian curve*, the reader may refer to Sections 3 and 5.

DEFINITION. *A wave front on the plane is said to be taut if through every generic point in the plane there pass no more than two normal lines.*

A first generalization of Banchoff's theorem is the following result:

THEOREM. *A taut wave front is a round circle.*

We now discuss the other important direction in which the theory of taut immersions will be generalized. Roughly speaking, we will change the definition of normality as follows.

Let $\mathcal{C} \subset \mathbb{R}^2$ be a smooth curve which is strictly convex and symmetric with respect to the origin. If l is a line passing through the origin, we define its normal as the line which intersects $\mathcal{C}$ at those points where the tangent of $\mathcal{C}$ is parallel to l. If l is co-oriented (i.e., it is the boundary of a distinguished half-plane) we orient its normal so that it points to the interior of the half-plane defining the co-orientation.

A smooth choice of a strictly convex, centrally symmetric curve on the tangent plane at each point $x \in \mathbb{R}^2$ is classically known as a *Finsler metric* on $\mathbb{R}^2$. Note

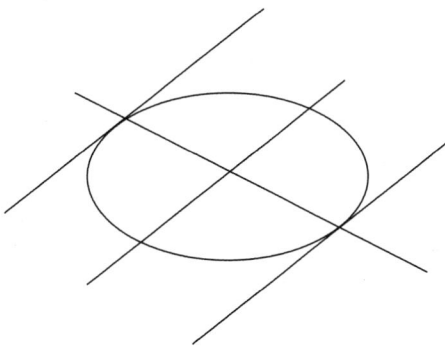

Figure 2. Normals in convex geometry

that, in this light, a Riemannian metric on $\mathbb{R}^2$ is the smooth choice of an ellipse on each tangent plane. Like a Riemannian metric, a Finsler metric defines a metric structure and geodesic lines. This will be reviewed in Section 6.

Given a point x on a cusped curve, we define its *normal geodesic* with respect to a Finsler metric to be the geodesic which passes through x in the normal direction. The *normal system* to a cusped curve is the 1-parameter family of all normal geodesics (see Figures 1 and 2).

We shall be solely interested in a very special class of Finsler metrics on the plane and the sphere.

DEFINITION. *A 2-dimensional manifold S provided with a complete Finsler metric is said to be a Desarguesian surface if there exists a diffeomorphism from S to either the plane or the sphere which maps geodesics to straight lines or great circles.*

Some geometric constructions of these metrics will be presented in Section 7. For now, the reader is advised to keep in mind the simplest example on the plane: choose a smooth, centrally symmetric, strictly convex curve on the tangent space at the origin and translate it to every other point. This corresponds to considering a Banach norm on $\mathbb{R}^2$ whose unit sphere is smooth and strictly convex.

Let us take a closed curve in the sphere or the plane such that its only singularities are cusps. At each point of the curve we have a well-defined tangent and, therefore, a well-defined normal direction. We can speak then of the normal system of a cusped curve in a Desarguesian surface.

Using geodesics instead of straight lines or great circles we can extend the notion of tautness to Desarguesian surfaces.

DEFINITION. *A wave front on a Desarguesian surface S is said to be taut if through every generic point $x \in S$, there pass no more than two normal geodesics.*

To state the generalization of Banchoff's theorem we must specify what we consider to be a round circle in a general Finsler surface.

DEFINITION. *A wave front in a Finsler surface is a Lie circle if it is made up of the endpoints of all geodesic segments of a fixed length which start at a fixed point.*

MAIN THEOREM. *A taut wave front on a Desarguesian surface is a Lie circle.*

An easy consequence of this theorem is the following amusing result.

COROLLARY. *Given a wave front on a Desarguesian surface S there exists a point $x \in S$ through which there pass no less than four normal geodesics.*

Before going into the main body of the paper, we would like to discuss two interesting ideas around the symplectization of the theory taut immersions. The first relates contact geometry and the Hopf–Rinow theorem, the second relates it to integral geometry.

Contact geometry and the theorem of Hopf–Rinow. An extension of the Hopf–Rinow theorem states that given a compact immersed submanifold N of a complete Riemannian (or Finsler) manifold M and a point $x \in M$ there exists at least one geodesic passing through x which is normal to N.

It would be interesting to extend this result to wave fronts. For instance, in Section 7 we prove the following result.

THEOREM. *Let S be a Finsler surface whose space of geodesics is a manifold and let $\mathcal{L}$ be an embedded Legendrian curve in the manifold of co-oriented contact elements of S. If $\mathcal{L}$ is contact isotopic to the Legendrian curve consisting of all co-oriented contact elements over a point, then for any point $x \in S$ there are at least two geodesics passing through x and which are normal to the front of $\mathcal{L}$.*

In particular, the result is true for Desarguesian Finsler surfaces and Riemannian metrics on the plane with nonpositive curvature (see [**18**]).

When the space of geodesics of a complete Riemannian or Finsler manifold of arbitrary dimension is itself a manifold, the Hopf–Rinow theorem and its possible extensions are easily translated into problems of Lagrangian intersection theory (some of them unsolved!). Nevertheless, it is reasonable to expect that this hypothesis is unnecessary. For example, it is likely that the answer to the following problem is in the affirmative.

PROBLEM. *Let $\mathcal{L}$ be an embedded Legendrian curve in the manifold of co-oriented contact elements of a complete Finsler surface M. If $\mathcal{L}$ is contact isotopic to the Legendrian curve consisting of all co-oriented contact elements over a point, then for any point $x \in M$ there is at least one geodesic passing through x and which is normal to the front of $\mathcal{L}$.*

Contact geometry and integral geometry. Let γ be an immersed curve on the unit sphere. For each point $x \in S^2$ let $n_\gamma(x)$ denote the number of great circles which pass through x and are normal to γ. Clearly $n_\gamma \geq 2$ and Banchoff's theorem tells us that if for almost every x, $n_\gamma(x) = 2$, then γ is a round circle. We may interpret this statement as a characterization of the absolute minima of the functional

$$S(\gamma) := \frac{1}{4} \int_{S^2} n_\gamma(x) \, \omega,$$

where ω is the standard area form on the unit sphere.

Using Crofton's formula, it is not hard to see that $S(\gamma)$ is the length of the curve of great circles normal to γ. Using a little spherical geometry, it follows that

$$S(\gamma) = \int_\gamma \sqrt{1 + \kappa_g^2}\, ds,$$

where κ_g is the geodesic curvature.

Observe that, in the original formulation, the functional S is defined for wave fronts. It must then be possible to give both an integral-geometric and a local formulation of S for Legendrian curves in the manifold of co-oriented contact elements of the sphere.

The integral-geometric formulation is obtained by noticing that if $\mathcal{L}$ is a Legendrian curve and x is a point in the sphere, then the geodesics passing through x and normal to the front of $\mathcal{L}$ are in one-to-one correspondence with the points of intersection of $\mathcal{L}$ and the torus made up of all the co-oriented contact elements which are tangent to some round circle centered at x. If we denote this torus by T_x, we write the extension of the functional S to Legendrian curves as

$$S(\mathcal{L}) := \frac{1}{4} \int_{S^2} \#(T_x \cap \mathcal{L})\, \omega.$$

To obtain the local formula for $S(\mathcal{L})$ we note that the space of co-oriented contact elements on the sphere can be naturally identified with $\mathbb{R}P^3$, and that when the front of $\mathcal{L}$ is an immersed curve on the sphere, the functional is just the length of $\mathcal{L}$ for the standard metric. This metric is normalized so that the lengths of projective lines equal 2π.

What we obtain is an interesting integral-geometric representation of the length functional for Legendrian curves in $\mathbb{R}P^3$ different from Crofton's formula.

2. Projective topology

The idea underlying the proof of the main theorem is to isolate a projective property of normal systems of wave fronts in Desarguesian surfaces. This approach has also proved fruitful in the 3-dimensional case (see [4]), where the necessary projective differential geometry is much richer.

For the moment let us consider only Finsler metrics on the sphere whose geodesics are great circles. The space of geodesics can then be easily identified with the unit sphere in $\mathbb{R}^3$ by associating to every oriented great circle the unit vector normal to the plane that contains it and oriented according to the right-hand rule. The normal system to a wave front is an immersed spherical curve (see Proposition 6.3) and we are interested in its projective properties. The one we need in this paper is the following:

PROPOSITION 2.1. *Consider a Finsler metric on S^2 such that its geodesics are great circles and let $\mathcal{L}$ be a closed Legendrian curve in the space of co-oriented contact elements on S^2. If the normal system of $\mathcal{L}$ is embedded, then it is not contained in any open hemisphere.*

SKETCH OF PROOF. We will see in Section 6 that the Finsler metric induces an area form ω on its space of geodesics (in this case the sphere) such that $a^*\omega = -\omega$, where a is the antipodal map. We will also see that if a normal system is embedded, then it divides the sphere into two pieces of equal ω-area. It follows that an embedded normal system cannot be contained in any open hemisphere. □

An interesting consequence of this theorem is a four vertex theorem in Desarguesian surfaces.

A point on a curve, or wave front, on a Desarguesian surface shall be called a *vertex* if the geodesic normal to the curve at this point is tangent to the evolute at one of its cusps. Let us remark that in Euclidean or elliptic geometry this condition is equivalent to the classical condition on the vanishing of the derivative of the geodesic curvature. By projective duality, the cusps of the evolute are in one-to-one correspondence with the inflection points of the normal system. Proposition 2.1 and a theorem of Arnold (see [**6**]), stating that an embedded spherical curve not contained in any open hemisphere must have at least four inflection points, give an immediate proof of the following result:

THEOREM 2.2. *Consider a Desarguesian surface diffeomorphic to a sphere. If the normal system of a wave front is embedded, then the wave front has at least four vertices.*

We will prove the main theorem in the case of Desarguesian surfaces diffeomorphic to the sphere assuming Proposition 2.1. The first step is to consider the case where the normal system has self-intersections.

PROPOSITION 2.3. *If $\nu \subset S^2$ is an immersed curve with self-intersections, then there exists an open set of great circles which intersect it in no less than four points.*

As a trivial consequence we have that a wave front whose normal system is not embedded cannot be taut.

Let us state the key projective result which intervenes in the proof of the theorem:

PROPOSITION 2.4. *Let $\nu \subset S^2$ be an immersed curve which intersects almost every great circle in at most two points. If ν is not a great circle, then it is contained in an open hemisphere.*

As an obvious consequence we have that an immersed spherical curve which intersects almost every great circle in exactly two points must be a great circle.

PROOF OF THE MAIN THEOREM. The previous propositions imply that if a wave front is taut, then its normal system is a great circle. In turn, this says that all the normals meet at one point and so the wave front is a round circle for the given Finsler metric. □

In the proofs of Propositions 2.3 and 2.4 we shall use a simple technical result which we will state without proof.

TECHNICAL LEMMA. *Let $\alpha \subset S^2$ be a smoothly embedded arc such that the great circle tangent to α at a point p does not osculate it to order two. There exists a small neighbourhood U of p in α such that the set of oriented circles passing through pairs of distinct points in U is an open set in the space of oriented great circles.*

PROOF OF PROPOSITION 2.4. Let us assume that the curve is not a great circle and thus it contains infinitely many points at which the tangent great circle does not osculate the curve to order two or higher. Let m be such a point on the curve ν and take two points on the curve to the right and to the left of m. For a generic choice of these points, the great circle determined by them does not cut the curve in any other point, and by taking them closer and closer to m we see that the curve ν is contained in a closed hemisphere bounded by the great circle which is tangent to ν at m.

We have just shown that the curve ν is contained in a closed hemisphere. If ν is not a great circle, we may chose a new point n such that the great circle tangent to the curve at n is different from that tangent at m. We repeat the above construction for n and obtain that ν is contained in two different closed hemispheres. This immediately implies that ν is contained in an open hemisphere. □

Using Proposition 2.4, we shall give a quick proof of Proposition 2.3.

PROOF OF PROPOSITION 2.3. We shall assume that the curve ν lies in an open hemisphere for if that is not the case, then the conclusion holds regardless of the existence of self-intersections. Now, by using the standard affinization of the projective plane we may map our curve into a plane curve and the great circles into lines. We are left with an immersed plane curve with self-intersections and which is, therefore, not convex.

To complete the proof, it is enough that we show that the set of oriented (straight) lines which intersect a nonconvex curve in four points or more contains an open set. To do this we note that if the curve is not convex, then its Gauss map, which assigns to every point its normal vector, must have an open set of regular values with two or more preimages. By cutting the curve with lines perpendicular to these normal directions we shall find an open set of lines which cuts the curve in four or more points. □

PROOF OF THE MAIN THEOREM IN THE PLANAR CASE. We begin by remarking that in the case of Finsler metrics on the plane whose geodesics are straight lines, the space of geodesics can be identified with the right circular cylinder in $\mathbb{R}^3$ (see Section 4). We will prove in Section 6 (see Proposition 6.8) that if the normal system of a wave front is embedded, then it must intersect every ellipse which is obtained by cutting the cylinder with planes passing through origin in $\mathbb{R}^3$.

Using the radial projection from the cylinder to the sphere we map curves on the cylinder to curves on the sphere in such a way that the ellipses considered in the previous paragraph are mapped to great circles. Note then that a closed immersed curve in the cylinder that cuts almost every ellipse in only two points will be mapped into a closed immersed curve on the sphere which cuts almost every great circle in only two points. Propositions 2.3 and 2.4 tell us that such a curve has to be a great circle and so the original curve has to be the intersection of the cylinder with a plane passing through the origin. We conclude that the normal system of a taut wave front in the plane, with any Desarguesian Finsler metric, is the normal system of a point and hence the normal system of a Lie circle. □

REMARK. In the preceding argument we used that if a curve on the cylinder cuts almost every plane passing through the origin in at most two points, then its image under the radial projection is a spherical curve which cuts almost every great circle in at most two points. In the Cayley–Klein model of hyperbolic geometry or in the Hilbert geometries the space of geodesics can be identified with a piece K of the cylinder. Unfortunately, it does not seem that a curve on K which intersects almost all the ellipses lying on K in exactly two points will necessarily be itself an ellipse and our arguments fail in the hyperbolic case. Note however that the definition of Desarguesian surface we have given in the introduction excludes all Finsler metrics whose geodesic are straight lines and which are defined on a proper subset of the plane.

3. Contact geometry

Let M be a smooth manifold. A contact element of M based at a point $x \in M$ is a vector subspace of codimension one in $T_x M$. A co-orientation of a contact element is a choice of one of the half-spaces bounded by it.

The manifold of co-oriented contact elements on M will be denoted by ST^*M. This manifold carries a natural geometric structure which consists of a choice of hyperplane on every tangent space. This hyperplane is called the *contact hyperplane* and is defined by the skating condition:

Let $\pi : ST^*M \to M$ be the natural projection, and let p_m be a contact element based at $m \in M$. The vector $v_{p_m} \in T_{p_m} ST^*M$ belongs to the contact hyperplane if the vector $\pi_* v_{p_m} \in T_m M$ lies in the contact element p_m.

In other words, the co-oriented contact element can move infinitesimally around its base point and along itself, but never parallel to itself.

This geometric structure is called the canonical *contact structure* on ST^*M. Let us recall that a contact structure on a manifold N of dimension $2n+1$ is a field of tangent hyperplanes with the following strong nonintegrability condition:

In a neighbourhood of every point of N there exists a 1-form α so that the field of hyperplanes is given by the kernel of α and so that the $(2n+1)$-form $\alpha \wedge d\alpha \wedge \cdots \wedge d\alpha$ never vanishes.

A submanifold $\mathcal{L} \subset ST^*M$ such that at every point $m \in \mathcal{L}$ the tangent space $T_m \mathcal{L}$ is contained in the contact hyperplane is called an *integral submanifold*. The strong nonintegrability condition implies that the maximal dimension of an integral submanifold in a $(2n+1)$-dimensional contact manifold has dimension n.

DEFINITION. *A maximal integral submanifold of ST^*M is called a Legendrian submanifold. The image under the canonical projection π of a Legendrian submanifold is called a wave front.*

EXAMPLES.
- If m is a point in M, then the fiber $\pi^{-1}(m)$ is a Legendrian submanifold diffeomorphic to a sphere.
- If $N \subset M$ is a smooth co-oriented hypersurface, then the submanifold of ST^*M which consists of the pairs $(m, T_m N)$, $m \in N$, is a Legendrian submanifold. In particular every smooth hypersurface is a wave front.

The space of co-oriented contact elements on $\mathbb{R}^2$ can be identified with the solid torus $S^1 \times \mathbb{R}^2$ and its points may be represented by coordinates (x_1, x_2, ϕ), where $(x_1, x_2) \in \mathbb{R}^2$ is the base point and $\phi \in S^1$ is the angle that the co-oriented contact element makes with respect to the positive x_1-axis.

The contact hyperplanes are also defined as the kernel of the 1-form $\alpha := \cos(\phi) dx_1 + \sin(\phi) dx_2$.

The space of co-oriented contact elements on unit sphere S^2 can be identified with the group of rotations $SO(3)$ which is diffeomorphic to the projective space $\mathbb{R}P^3$. Indeed, using the standard metric on S^2 we may identify a co-oriented contact element with the tangent vector based at the same point and normal to it. The

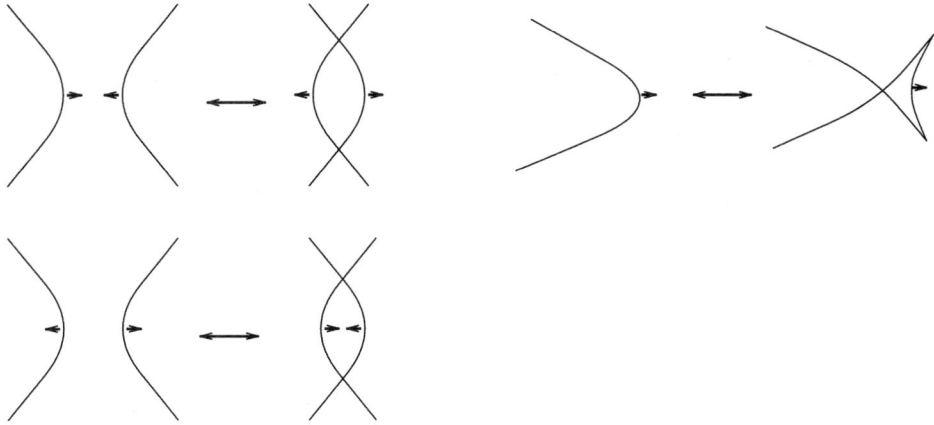

Figure 3. Some moves in Legendrian knot theory

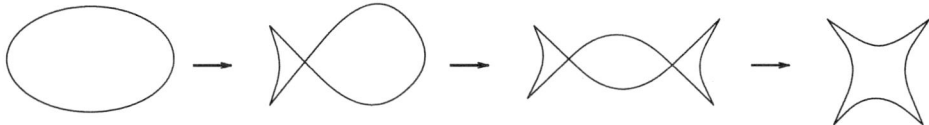

Figure 4. Some examples of quasi-circles

identification with $SO(3)$ is then defined by assigning to the contact element the unique rotation matrix whose first column vector is the base point, and the second column vector is the normal to the contact element.

If M is a 2-dimensional manifold, the manifold ST^*M is 3-dimensional and its Legendrian submanifolds are curves.

DEFINITION. *Two embedded Legendrian curves on ST^*M are said to define the same Legendrian knot if there is an isotopy of embedded Legendrian curves joining them.*

In Section 7 we will deal with only one type of Legendrian knot:

DEFINITION. *A Legendrian curve on ST^*M defining the same Legendrian knot as a fiber $\pi^{-1}(m)$ of the canonical projection is said to be a quasi-circle.*

There are various moves, analogous to the Reidemeister moves in knot theory, that can be applied to wave fronts in surfaces, and which do not change the Legendrian knot type of the associated Legendrian curve. These moves, illustrated in Figure 3, include the birth of two cusps and the birth of two double points. The small arrows denote the co-orientation of the front. Figure 4 illustrates some quasi-circles on $\mathbb{R}^2$.

4. The manifold of oriented lines

The generalization of the theory of taut immersions becomes more natural if we take the line instead of the point as the principal geometric object. Indeed, the theory then deals with intersections of curves in the space of lines.

It is important to understand the geometry of the space of lines on the plane and the space of great circles on the sphere.

Manifold of oriented lines on $\mathbb{R}^2$. The set of oriented lines normal to a wave front on $\mathbb{R}^2$ forms a curve in the space of all oriented lines on the plane. This space can be easily identified with the cylinder

$$\mathcal{C} := \{(x_1, x_2, x_3) \in \mathbb{R}^3 : x_1^2 + x_2^2 = 1\} \ .$$

An oriented line on the x_1x_2-plane is uniquely determined by the unit vector $q \in \mathbb{R}^2 \subset \mathbb{R}^3$ defining its direction, and by the vector v representing the point on the line closest to the origin. The point $q + q \times v$ is the point on the cylinder which corresponds to the oriented line.

This identification has the property that the set of all oriented lines passing through a point $x \in \mathbb{R}^2 \subset \mathbb{R}^3$ corresponds to the intersection of the cylinder with a plane passing through the origin of $\mathbb{R}^3$. If x is a point on $\mathbb{R}^2$, we shall denote the resulting ellipse by Γ_x.

In the previous section we identified the manifold of co-oriented contact elements on $\mathbb{R}^2$ with the hypersurface

$$\{(x, q) \in \mathbb{R}^2 \times \mathbb{R}^2 : \|q\| = 1\} \ .$$

With this identification, the natural projection

$$\pi_2 : ST^*\mathbb{R}^2 \longrightarrow \mathcal{C},$$

which assigns to a co-oriented contact element its normal geodesic, is given by the formula

$$(x, q) \mapsto q + q \times x \ .$$

The pullback of the area form on the cylinder under π_2 equals the exterior differential $d\alpha$ of the standard contact form $\alpha := q_1 dx_1 + q_2 dx_2$.

Manifold of oriented lines on S^2. The set of all oriented great circles on S^2 can be easily identified with the sphere itself. A point $\mathbf{q}$ on the sphere can be identified with the great circle obtained by intersecting the sphere with the plane orthogonal to $\mathbf{q}$. The orientation is then given by the right-hand rule.

The normal system to a wave front on the sphere is then a curve on the sphere. When the wave front is a point, the normal system is itself a great circle in the manifold of oriented great circles. We shall denote the set of all great circles passing through a point $\mathbf{q} \in S^2$ by $\Gamma_\mathbf{q}$.

In the previous section we identified the space of co-oriented contact elements on S^2 with the group of rotations $SO(3)$. Under this identification, the natural projection

$$\pi_2 : ST^*S^2 \longrightarrow S^2,$$

which assigns to a co-oriented contact element its normal geodesic, is given by assigning to a rotation matrix its third column vector. If α is the canonical contact form on ST^*S^2, then the pullback under π_2 of the standard area form on S^2 is $d\alpha$.

REMARK. Note that the Euclidean group acts transitively on the space of oriented lines on the plane, and that the rotation group acts transitively on the space of oriented great circles on the sphere. These actions are area preserving. In general, spaces of geodesics of symmetric rank-one spaces are themselves homogeneous manifolds and, in fact, coadjoint orbits for the group of isometries.

For the rest of this section we shall denote both the sphere and the plane by M. The cylinder, or the sphere of great circles, being the manifolds of geodesics of the plane and sphere for the canonical metrics, shall be denoted by $Geod(M, can)$.

The Huygens correspondence. The double fibration

$$M \xleftarrow{\pi_1} ST^*M \xrightarrow{\pi_2} Geod(M, can)$$

allows us to relate the metric geometry of M, the contact geometry of ST^*M, and the symplectic geometry of $Geod(M, can)$. Let us call this double fibration the *Huygens correspondence*. We shall see later that there is a Huygens correspondence associated to every Desarguesian Finsler metric and, indeed, to any Finsler metric whose space of geodesics is a manifold (see [1]).

Note that if x is a point in M, then $\pi_2(\pi_1^{-1}(x)) = \Gamma_x$. If γ is a point in $Geod(M, can)$, then $\pi_1(\pi_2^{-1}(\gamma))$ is the geodesic (as a point set) represented by γ.

The Huygens correspondence has two important properties:
1. The fibers of the projection $\pi_1 : ST^*M \to M$ are Legendrian curves.
2. The pullback of the area form on $Geod(M, can)$ under π_2 is $d\alpha$.

5. Finsler geometry

In this section we present the basic concepts of Finsler geometry in the form in which they will be used in the next two sections. A prerequisite is the notion of duality in convex geometry:

Let V be a finite dimensional vector space, and let $K \subset V$ be a smooth hypersurface which encloses the origin and is strictly convex. We shall define an embedding of K into the dual space V^* by assigning to each point $v \in K$ the unique point $k(v) \in V^*$ such that

$$\{w \in V : \langle k(v), w \rangle = 1\} = T_v K \subset V.$$

The image of K under the map k is called the *dual hypersurface*. It is easy to see that it is also strictly convex and is centrally symmetric if and only if K is centrally symmetric. However, it is not necessarily smooth.

DEFINITION. Let M be a smooth manifold and let $TM \setminus 0$ denote its tangent bundle with the zero section deleted. A Finsler metric on M is a smooth positive function

$$\varphi : TM \setminus 0 \longrightarrow \mathbb{R}$$

with the following properties:
1. $\varphi(\lambda v_m) = |\lambda| \varphi(v_m)$, for any nonzero real number λ and any nonzero tangent vector $v_m \in T_m M$.
2. For each $m \in M$,

$$\{v_m \in T_m M : \varphi(v_m) = 1\} \subset T_m M$$

is a strictly convex hypersurface with a smooth dual.

REMARKS.
- Condition (1) implies that the function φ is completely determined by the choice of the convex hypersurfaces

$$\{v_m \in T_m M : \varphi(v_m) = 1\}$$

at every tangent space $T_m M$. These hypersurfaces are classically known as the *indicatrices* of the Finsler metric. The dual hypersurfaces are known as the *figuratrices*.
- The definition of a Finsler metric given above coincides with the classical definition where the smoothness of the dual sphere is expressed in terms of the Hessian of φ with respect to the fiber variables. For more information on Finsler metrics see [**23**].

Finsler metrics induce metric structures in the same way as Riemannian metrics:

If $c : [a, b] \to M$ is an immersed curve on M, we define its length with respect to the Finsler metric φ as the integral

$$S(c) := \int_a^b \varphi(c'(t))dt.$$

Let x and y be two points in M. We define the distance between x and y as the number

$$d(x, y) := \inf\{S(c) : c \text{ joins } x \text{ and } y\}.$$

The geodesic flow of a Finsler metric. The field of figuratrices in T^*M uniquely determines a smooth positive function

$$H : T^*M \setminus 0 \longrightarrow \mathbb{R}$$

with the following properties:
1. $H(\lambda p_m) = \lambda^2 H(p_m)$, for any nonzero real number λ and any nonzero covector $p_m \in T_m^* M$.
2. For each $m \in M$, the hypersurface

$$\{p_m \in T_m^* M : H(p_m) = 1\}$$

is a figuratrix.

Note that a Finsler metric on M is completely determined by the *unit co-sphere bundle* $S_H^* M := H^{-1}(1) \subset T^*M$.

Let us denote by α_H the restriction of the canonical 1-form α_0 in T^*M to S_H^*M and let us define X_H to be the unique vector field on S_H^*M satisfying

$$d\alpha_H(X_H, \cdot) = 0, \quad \alpha_H(X_H) = 1.$$

The Finsler metric is said to be *complete* if X_H defines a flow. This flow is known as the *geodesic flow* of the Finsler metric. The projections onto M of the orbits of the geodesic flow are precisely the geodesics of the Finsler metric.

Finsler geometry and contact geometry. The kernel of the form α_H defines a field of hyperplanes on S_H^*M. Using the tautological identification of ST^*M and S_H^*M, it is immediate that these hyperplanes correspond to the contact hyperplanes in ST^*M. The following theorem is well known and—modulo terminology—dates back to Huygens.

THEOREM 5.1. *The geodesic flow of a Finsler metric preserves the contact structure on ST^*M.*

A possible viewpoint in Finsler geometry is that a Finsler metric is defined by the choice of a (special) contact form α_H on the manifold of co-oriented contact elements ST^*M.

In the rest of the paper we will freely change our point of view from the *Lagrangian formalism* (the function φ) to the *Hamiltonian formalism* (the function H) to the geometric formalism (the choice of the distinguished contact form α_H on $S_H^* M$).

An important concept in Finsler geometry is that of normality. Normal directions are defined only for hypersurfaces as was done in the introduction. The geometric construction given there is equivalent to the following:

Let $\Pi \subset T_x M$ be a co-oriented hyperplane. By duality, Π defines an oriented line or a ray in $T_x^* M$. Let us intersect that ray with the figuratrix over x and use the resulting point as an initial condition for the geodesic flow. We obtain a geodesic on M and its tangent vector over x is normal to Π.

For a proof of the equivalence see the discussion of Huygens' principle [**8**, pp. 249–252].

The preceding construction implies that we can obtain the normal system of a co-oriented hypersurface by just lifting it to a Legendrian submanifold in ST^*M as in Section 3 and using the points of the Legendrian as initial conditions for the geodesic flow.

6. Construction of Desarguesian surfaces

Let us start by recalling the definition of Desarguesian surfaces from the introduction. We mainly follow the work of Buseman ([**12**]) and Pogorelov ([**22**]).

DEFINITION. *A 2-dimensional manifold S provided with a complete Finsler metric is said to be a Desarguesian surface if there exists a diffeomorphism from S to either the plane or the sphere which maps geodesics to straight lines or great circles.*

We shall now construct Finsler metrics on $\mathbb{R}^2$ whose geodesics are straight lines. Clearly any Desarguesian surface diffeomorphic to the plane is isometric to $\mathbb{R}^2$ provided with one of these metrics. What follows is a variation of Busemann's integral-geometric construction.

Consider an area form on the cylinder $\mathcal{C} = \{(x_1, x_2, x_3) \in \mathbb{R}^3 : x_1^2 + x_2^2 = 1\}$ which is anti-invariant under the antipodal map. We associate to this area form a metric on $\mathbb{R}^2$ as follows:

Identify each point $x \in \mathbb{R}^2 \subset \mathbb{R}^3$ with the ellipse

$$\Gamma_x := \{q + q \times x : q = (q_1, q_2, 0),\ q_1^2 + q_2^2 = 1\} \subset \mathcal{C}.$$

If x and y are points in the plane the union of the ellipses Γ_x and Γ_y divide the cylinder into four connected components. Two of these components are bounded and two are unbounded. The invariance of the area under the antipodal map implies that the area of the bounded components are equal. We shall denote half of that area by $d(x,y)$.

If x, y and z are three collinear points in $\mathbb{R}^2$, then the three ellipses intersect in two antipodal points. It is easy to see that if y is between x and z, then $d(x,z) = d(x,y) + d(y,z)$.

THEOREM 6.1. *The metric d is induced by a Finsler metric on $\mathbb{R}^2$ whose geodesics are straight lines. Moreover, any such Finsler metric can be obtained in this way.*

For a proof see [**22**]. The reader should be warned that he or she will have to perform an easy translation between the integral geometric construction and that given above.

The reader may also enjoy checking that if the area form on the cylinder is the one induced by its embedding into $\mathbb{R}^3$, then the resulting metric on the plane is the Euclidean metric. Note also that if the area form is independent of the z-coordinate on the cylinder, then the resulting metric is a Banach metric on $\mathbb{R}^2$.

REMARK. It is not clear to the author which are the necessary and sufficient conditions on the area form for the metric d to be complete. An obvious necessary condition is that the area of the cylinder be infinite. A sufficient condition, that seems to be close to necessary, is that the action of the Euclidean group on the cylinder preserves the subsets of infinite area.

Let us proceed with the construction of Finsler metrics on the sphere whose geodesics are great circles. Consider an area form on the sphere $\{(x_1, x_2, x_3) \in \mathbb{R}^3 : x_1^2 + x_2^2 + x_3^2 = 1\}$ which is anti-invariant under the antipodal map. We associate to this area form a metric on S^2 as follows:

Identify each point $\mathbf{q} \in S^2 \subset \mathbb{R}^3$ with the great circle $\Gamma_\mathbf{q}$ obtained by intersecting the sphere with the plane $\{x \in \mathbb{R}^3 : \langle x, \mathbf{q} \rangle = 0\}$.

If $\mathbf{q}_1$ and $\mathbf{q}_2$ are points in the sphere, the union of the great circles $\Gamma_{\mathbf{q}_1}$ and $\Gamma_{\mathbf{q}_2}$ divide the sphere into four connected components. The invariance of the area form under the antipodal map implies that only two of the areas are distinct. We shall denote half of the smallest area by $d(x, y)$.

THEOREM 6.2. *The metric d is induced by a Finsler metric on the sphere whose geodesics are great circles. Moreover, any such Finsler metric on the sphere which is invariant under the antipodal map can be constructed in this way.*

A proof of this fact is contained in [**22**].

REMARKS.
- From Theorem 6.2 it is easy to deduce that the lengths of all geodesics in a Desarguesian Finsler metric on the sphere are of the same length if we require the Finsler metric to be invariant under the antipodal map. Another simple consequence of the integral-geometric construction is that the common length of the geodesics is half the total area of the sphere with the area form we use to define the metric. This also holds for Desarguesian metrics on the sphere which are not necessarily invariant under the antipodal map (see Proposition 6.4).
- This author does not know how to construct Desarguesian metrics on the sphere which are not invariant under the antipodal map. If we weaken the definition of Finsler metric and require the function φ to be only positively homogeneous (nonsymmetric Finsler metrics), then some very interesting constructions can be found in [**11**].
- In a beautiful three-page paper ([**20**]) Gromoll and Grove show that if all the geodesics of a Riemannian metric on the sphere are closed, then they have no self-intersections and all have the same length. A look at the proof will convince the reader that the result also holds for Finsler metrics as defined in this paper. However, the result does not hold for *nonsymmetric*

Finsler metrics; the reason their proof breaks down is that the Lusternik–Schnirelmann theorem on the existence of three simple closed geodesics is not true in the nonsymmetric case.

The Huygens correspondence and the proof of the main theorem.
One of the main properties of a Desarguesian surface is that its space of geodesics is a manifold. This manifold is diffeomorphic to the cylinder or the sphere depending on whether the surface is diffeomorphic to $\mathbb{R}^2$ or S^2.

Let $(\mathcal{S}, H)$ be a Desarguesian surface and let $Geod(\mathcal{S}, H)$ denote its manifold of geodesics. The map that assigns to a unit covector $p_x \in S_H^*\mathcal{S}$ the unique geodesic with initial value p_x defines a fibration

$$\pi_2 : S_H^*\mathcal{S} \longrightarrow Geod(\mathcal{S}, H).$$

The *Huygens correspondence* is the double fibration

$$\mathcal{S} \xleftarrow{\pi_1} S_H^*\mathcal{S} \xrightarrow{\pi_2} Geod(\mathcal{S}, H).$$

The manifold $Geod(\mathcal{S}, H)$ carries a natural area form ω defined by the equality $d\alpha_H = \pi_2^*\omega$. This is a particular case of the well-known procedure of symplectic reduction.

As in the case of the standard metrics, the Huygens correspondence relates the metric geometry of $\mathcal{S}$, the contact geometry of $ST^*\mathcal{S}$ and the symplectic geometry of $Geod(\mathcal{S}, H)$.

When the Desarguesian surface $(\mathcal{S}, H)$ is diffeomorphic to $\mathbb{R}^2$ (S^2), then the manifold of geodesics is the cylinder $\mathcal{C}$ (the sphere S^2), but with an area form which is not the standard one. We shall show below that areas are invariant under the antipodal map and the constructions given previously in this section are just the way to recover the Finsler metric H from the area form ω.

REMARK. The interplay between the metric geometry of a Finsler manifold and the symplectic geometry of its space of geodesics could be called—as Arnold has done—an *area-length duality*. Except in the 2-dimensional case, this duality is not an integral-geometric phenomenon. For more on the area-length duality and its applications see [1–4], [7], [26].

We shall now complete the arguments given in Section 2 and finish the proof of the main theorem.

PROPOSITION 6.3. *If $\mathcal{L} \subset S_H^*\mathcal{S}$ is an immersed Legendrian curve, then the curve $\pi_2(\mathcal{L})$ is an immersed curve in $Geod(\mathcal{S}, H)$.*

PROOF. The fact that the form ω defined by the equality $d\alpha_H = \pi_2^*\omega$ is an area form implies that the fibers of the projection π_2 are transversal to the contact planes given by the kernel of α_H. From this it follows immediately that immersed Legendrian curves are mapped to immersed curves in $Geod(\mathcal{S}, H)$. □

PROPOSITION 6.4. *If the common length of all geodesics of a Desarguesian Finsler metric on the sphere is equal to l, then the total area of its space of geodesics equals $2l$.*

PROOF. From the relation $d\alpha_H = \pi_2^*\omega$ and the fact that α_H is the connection 1-form of the circle bundle $\pi_2 : S_H^*\mathcal{S} \to Geod(\mathcal{S}, H)$, we have that the 2-form $\frac{\omega}{l}$ is the Chern class of this bundle. Since the Chern number of this bundle is two, we have that the integral of $\frac{\omega}{l}$ over $Geod(\mathcal{S}, H)$ equals two, and the total ω-area of the space of geodesics is twice the length of the geodesics. $\square$

For the rest of the paper we assume, by rescaling the metric if necessary, that the common length of all geodesics of a Desarguesian Finsler metric of the sphere equals 2π and that the total ω-area of its space of geodesics equals 4π.

PROPOSITION 6.5. *Let $(\mathcal{S}, H)$ be a Desarguesian surface and let $a : Geod(\mathcal{S}, H) \to Geod(\mathcal{S}, H)$ be the map which sends one oriented geodesic to the geodesic with the opposite orientation. The form ω satisfies the equality $a^*\omega = -\omega$. In particular, the area of a region $U \subset Geod(\mathcal{S}, H)$ and the area of $a(U)$ are equal.*

PROOF. Consider the map $A : T^*\mathcal{S} \to T^*\mathcal{S}$, which equals multiplication by -1 on the fibers. This map obviously induces a map in $S_H^*\mathcal{S}$, which is a lift of a and which we still denote by A. Since $A^*\alpha_H = -\alpha_H$, and since $\pi_2^*\omega = d\alpha_H$, we have that $a^*\omega = -\omega$. $\square$

PROPOSITION 6.6. *Let $(\mathcal{S}, H)$ be a Desarguesian surface which is diffeomorphic to the sphere and let $\mathcal{L} \subset S_H^*\mathcal{S}$ be a closed Legendrian curve. If the normal system of $\mathcal{L}$ is embedded, then it cuts $Geod(\mathcal{S}, H)$ into two pieces of equal ω-area.*

PROOF. The key point is to note that, for the connection form α_H on the bundle $\pi_2 : S_H^*\mathcal{S} \to Geod(\mathcal{S}, H)$, the Legendrian curve $\mathcal{L}$ is the horizontal lift of its normal system. Since the curve $\mathcal{L}$ is closed, the holonomy around $\mathcal{L}$ must be equal to $2\pi k$, for some integer k. The relation between holonomy and curvature tells us that the projection of $\mathcal{L}$ must bound a region with ω-area equal to $2\pi k$. Using that this projection is embedded and that the total ω-area equals 4π, we have that it must bound a region of area 2π and so cuts $Geod(\mathcal{S}, H)$ into two pieces of equal area. $\square$

With this last proposition we have completed the proof of Proposition 2.1 and hence the proof of the main theorem in the spherical case. We shall now consider the planar case.

PROPOSITION 6.7. *Let $(\mathcal{S}, H)$ be a Desarguesian surface which is diffeomorphic to the plane. If β is a 1-form on $Geod(\mathcal{S}, H)$ such that $\alpha_H - \pi_2^*(\beta)$ is exact and $\mathcal{L} \subset S_H^*\mathcal{S}$ is a closed Legendrian curve, then the integral of β over the curve $\pi_2(\mathcal{L})$ equals zero.*

PROOF. Since $\alpha_H - \pi_2^*(\beta)$ is exact, its integral over $\mathcal{L}$ equals zero. We have then
$$0 = \int_{\mathcal{L}} \alpha_H - \pi_2^*(\beta) = -\int_{\mathcal{L}} \pi_2^*(\beta) = -\int_{\pi_2(\mathcal{L})} \beta.$$
The second equality makes use of the fact that $\mathcal{L}$ is Legendrian and so the restriction of α_H to $\mathcal{L}$ is identically zero.

The existence of a 1-form β with the desired properties follows from the exactness of ω on $Geod(\mathcal{S}, H)$ and the fact that the projection π_2 is a homotopy equivalence between $S_H^*\mathcal{S}$ and $Geod(\mathcal{S}, H)$. $\square$

Let us now consider some easy consequences of this proposition. In what follows we assume, after applying an isometry if necessary, that the geodesics of $(\mathcal{S}, H)$ are straight lines. The space of geodesics is then the right circular cylinder, and the pencil of lines passing through a point in the plane is an ellipse obtained by intersecting the cylinder with a plane passing through the origin in $\mathbb{R}^3$.

The first consequence we draw from Proposition 6.7 is that if the curve $\pi_2(\mathcal{L})$ is embedded, then it winds once around the cylinder. Indeed, no embedded curve winds more than once around the cylinder and if it does not wind at least once, then the curve is the boundary of an embedded disc D. Since $d\beta = \omega$, the Stokes theorem says that the integral of β along $\pi_2(\mathcal{L})$ equals the ω-area of D. This area cannot be zero, and so $\pi_2(\mathcal{L})$ cannot be embedded.

The second consequence will finish the proof of the main theorem in the planar case.

PROPOSITION 6.8. *Let $(\mathcal{S}, H)$ be a Desarguesian surface which is diffeomorphic to the plane and let $\mathcal{L}$ be a closed Legendrian curve. If the normal system of $\mathcal{L}$ is embedded, then it intersects every ellipse which is obtained by cutting the cylinder with a plane passing thought the origin.*

PROOF. Let E be one of the ellipses mentioned in the statement and let γ be the normal system of $\mathcal{L}$. If E and γ do not intersect, then their union forms the boundary of an open set in the cylinder. Indeed, both curves are embedded, disjoint, and wind once around the cylinder. Now using the Stokes theorem and Proposition 6.7, we conclude that the ω-area of this open set equals zero, which is impossible. □

7. Geodesic flows of Desarguesian surfaces

In this last section we show that the geodesic flows of Desarguesian Finsler metrics on the plane or the sphere are conjugate, in a very special way, to the geodesic flows of the standard Riemannian metrics on these surfaces. We deduce from this fact the Hopf–Rinow-type theorem mentioned in the introduction.

Since it takes no additional effort to prove these results for general Finsler surfaces whose space of geodesics is a smooth manifold, we will deal with this general case. Besides Desarguesian Finsler surfaces, Riemannian metrics of nonpositive curvature in the plane and Zoll metrics on the sphere (see [**18**] and [**10**]) have spaces of geodesics which are smooth. We also mention that locally (in geodesically convex neighbourhoods) every Finsler manifold has a smooth space of geodesics.

To avoid having to rule out pathologies, we will assume at once that the Finsler surfaces are diffeomorphic to the plane or the sphere, that the spaces of geodesics are diffeomorphic to the cylinder or the sphere, and that the canonical projection from $S_H^*\mathcal{S}$ to $Geod(\mathcal{S}, H)$ is a smooth fibration.

We have seen in Section 5 that a Finsler manifold is completely determined by its unit co-sphere bundle. Recall also that the 1-form α_H determines the vector field X_H by the formulas

$$d\alpha_H(X_H, \cdot) = 0, \quad \alpha_H(X_H) = 1,$$

and that the flow generated by X_H is the geodesic flow. We have reviewed these facts in order not to surprise the reader with the following simple statement:

ASSERTION. *If there exists a diffeomorphism between the unit co-sphere bundles of two Finsler manifolds which sends the canonical 1-form of one bundle to the canonical 1-form of the other, then the geodesic flows of these Finsler manifolds are conjugate.*

THEOREM 7.1. *Let $(\mathcal{S}, H)$ be a Finsler surface diffeomorphic to S^2. If the space of geodesics of $(\mathcal{S}, H)$ is smooth, then there exists a diffeomorphism*

$$\Psi : S_H^*\mathcal{S} \longrightarrow S^*S^2$$

such that $\Psi^*\alpha_{can} = \alpha_H$. Moreover, Ψ can be taken so that it maps quasi-circles to quasi-circles.

THEOREM 7.2. *Let $(\mathcal{S}, H)$ be a Desarguesian surface diffeomorphic to $\mathbb{R}^2$. If the space of geodesics of $(\mathcal{S}, H)$ is smooth, then there exists an embedding*

$$\Psi : S_H^*\mathcal{S} \longrightarrow S^*\mathbb{R}^2$$

such that $\Psi^*\alpha_{can} = \alpha_H$. Moreover, Ψ can be taken so that it maps quasi-circles to quasi-circles.

REMARK. The diffeomorphism (embedding) Ψ induces a diffeomorphism (embedding) ψ which makes the diagram

$$(*) \quad \begin{array}{ccc} S_H^*\mathcal{S} & \xrightarrow{\Psi} & S^*M \\ \pi_2 \downarrow & & \downarrow \pi_2 \\ Geod(\mathcal{S}, H) & \xrightarrow{\psi} & Geod(M, can) \end{array}$$

commute. Here M denotes the plane or the sphere depending on whether $\mathcal{S}$ is diffeomorphic to the plane or the sphere. Notice that $\psi^*\omega_{can} = \omega$.

The proofs of Theorems 7.1 and 7.2 make heavy use of the fact that the fibration

$$\pi_2 : S_H^*\mathcal{S} \longrightarrow Geod(\mathcal{S}, H)$$

together with the 1-form α_H give a *prequantization* of the symplectic manifold $(Geod(\mathcal{S}, H), \omega)$. This means that α_H may be considered as a connection 1-form whose curvature $d\alpha_H$ is $\pi_2^*\omega$. The strategy for the proof is as follows:

First we will show that there is an area-preserving diffeomorphism (resp. embedding) from $(Geod(\mathcal{S}, H), \omega)$ to (S^2, ω_{can}) (resp. $(\mathcal{C}, \omega_{can})$). Then we show that this map can be lifted to a connection-preserving map from $S_H^*\mathcal{S}$ to S^*S^2 (resp. $S^*\mathbb{R}^2$).

PROOF OF THEOREM 7.1. Assume without loss of generality that the lengths of the closed geodesics of $(\mathcal{S}, H)$ equal 2π. This implies that the total symplectic area of its manifold of geodesics equals 4π, which is the symplectic area of $Geod(S^2, can)$. The existence of the area-preserving diffeomorphism ψ is now guaranteed by the well-known theorem of Moser ([**21**]) on the equivalence of volume forms. The lifting of ψ to a connection-preserving map follows from the fact that the sphere is simply connected and a theorem of Souriau [**24**] (see [**1**], §2.3, for a statement and proof in the spirit of spaces of geodesics). This connection-preserving map can be chosen so that it sends quasi-circles to quasi-circles stems from the proof of Moser's theorem. Namely, the area-preserving map is constructed by integrating a vector field and therefore it is isotopic to the identity. The lifts also embed in an isotopy of maps starting with the identity and, therefore, they preserve the type of Legendrian knots. □

PROOF OF THEOREM 7.2. Let us first construct the area-preserving embedding. The following technique was taken from [**16**].

As was mentioned in the previous section, $Geod(\mathcal{S}, H)$ is nothing but the cylinder $\mathcal{C}$ with a nonstandard area form ω. For $k \in \mathbb{N}$ we let

$$\mathcal{C}_k := \{(x_1, x_2, x_3) \in \mathcal{C} : -k \leq x_3 \leq k\}$$

and define ω_k to be an area form on the cylinder which agrees with ω in $\mathcal{C}_k$ and with ω_{can} outside $\mathcal{C}_{k+1}$. Note that ω_k and ω_{k+1} agree on $\mathcal{C}_k$ and outside of $\mathcal{C}_{k+2}$. We set ω_0 to be ω_{can}.

Moser's theorem guarantees the existence of a diffeomorphism ψ_k from the cylinder to itself such that $\psi_k^* \omega_k = \omega_{k+1}$. Moreover, ψ_k is the identity on $\mathcal{C}_k$ and outside $\mathcal{C}_{k+2}$.

The sequence of maps $\psi_n \cdots \psi_0$ stabilizes on every compact set and, therefore, the map

$$\psi := \cdots \psi_n \cdots \psi_0$$

is well defined. It is easy to see that ψ is a smooth embedding and that $\psi^* \omega_{can} = \omega$.

The question of lifting the area-preserving map to a connection-preserving map is also more subtle than in the case of the sphere. It is possible that the map ψ we just constructed has no such lift.

The necessary and sufficient condition for the map ψ to have a connection-preserving lift is that there exist 1-forms β and β_{can} on $\mathcal{C}$ such that $d\beta = \omega$, $d\beta_{can} = \omega_{can}$, and $\psi^* \beta_{can} - \beta$ is an exact 1-form.

The cohomology class of the form $\psi^* \beta_{can} - \beta \in H^1(\mathcal{C}, \mathbb{R}) \cong \mathbb{R}$ is known as the *Calabi invariant* of ψ and will be denoted by $C(\psi)$. We have just stated that the only area-preserving maps that have connection-preserving lifts are those with a zero Calabi invariant. This is well known, but nevertheless we refer the reader to [**1**, §2.3] for a proof.

We do not have to prove that our map ψ has a zero Calabi invariant. Rather we note that the composition of the map ψ with the map

$$T_h : \mathcal{C} \longrightarrow \mathcal{C}$$

defined by $T_h(x, y, z) := (x, y, z + h)$ satisfies $(T_h \circ \psi)^* \omega_{can} = \omega$ and its Calabi invariant equals $h + C(\psi)$.

We conclude that the map $T_{-C(\psi)} \circ \psi$ is an area-preserving embedding of $(\mathcal{C}, \omega)$ into $(\mathcal{C}, \omega_{can})$ that has a connection-preserving lift Ψ. □

We shall now use Theorems 7.1 and 7.2 and some Lagrangian intersection theorems of Chekanov and Givental to prove the Hopf–Rinow-type theorem mentioned in the Introduction.

Let $(\mathcal{S}, H)$ be a Desarguesian surface and let

$$\pi_2 : S_H^* \mathcal{S} \longrightarrow Geod(\mathcal{S}, H)$$

be the natural projection. Let us also denote by Γ_x, $x \in \mathcal{S}$, the set of all geodesics passing through the point x.

THEOREM 7.3. *Let $(\mathcal{S}, H)$ be a Finsler surface whose space of geodesics is smooth, and let $\mathcal{L} \subset S_H^*\mathcal{S}$ be a quasi-circle. For any point $x \in \mathcal{S}$, the projection of $\mathcal{L}$ to $Geod(\mathcal{S}, H)$, $\pi_2(\mathcal{L})$, and the set of all oriented geodesics passing through x, Γ_x, intersect in at least two points.*

In order to prove this theorem we need to recall some theorems of Chekanov and Givental. Their results can be summarized in the following simple statement:

THEOREM 7.4. *Let M denote either the plane or the sphere with its canonical Riemannian metric. If $\mathcal{L}_1$ and $\mathcal{L}_2$ are two quasi-circles in ST^*M, then their normal systems, $\pi_2(\mathcal{L}_1)$ and $\pi_2(\mathcal{L}_2)$ intersect in at least two points.*

For a proof the reader is referred to the papers [**14**], [**15**], [**17**], [**19**]. Note that this theorem implies Theorem 7.3 when the Finsler surface is the plane or the sphere with their standard Riemannian metrics.

PROOF OF THEOREM 7.3. Let $\mathcal{L} \subset S_H^*\mathcal{S}$ be a quasi-circle, let x be a point in $\mathcal{S}$, and let ψ and Ψ be as in the diagram $(*)$. Clearly the number of intersection points of $\pi_2(\mathcal{L})$ and Γ_x is equal to the number of intersection points of $\psi(\pi_2(\mathcal{L}))$ and $\psi(\Gamma_x)$. Note now that, because of the commutativity of $(*)$, $\psi(\pi_2(\mathcal{L})) = \pi_2(\Psi(\mathcal{L}))$ and $\psi(\Gamma_x) = \psi(\pi_2(\pi_1^{-1}(x))) = \pi_2(\Psi(\pi_1^{-1}(x)))$.

Notice that the intersections are now taking place on $Geod(M, can)$ and that both $\Psi(\mathcal{L})$ and $\Psi(\pi_1^{-1}(x))$ are quasi-circles. Theorem 7.4 guarantees that the curves intersect in at least two points. □

ACKNOWLEDGMENTS. I gladly thank A. McRae, L. Traynor, and C. Durán for some very helpful conversations and their interest in this work. To S. Tabachnikov and the anonymous referee no amount of thanks will ever be sufficient; their thorough reading, their many critical remarks, and their pointing out numerous mathematical and stylistic mistakes is greatly appreciated.

References

1. J.C. Álvarez, *The symplectic geometry of spaces of geodesics*, Ph.D. Thesis, Rutgers University (1995).
2. _____, *A symplectic construction of projective metrics on R^n*, preprint (1995).
3. J.C. Álvarez, I.M. Gelfand, and M. Smirnov, *Crofton densities, symplectic geometry, and Hilbert's fourth problem*, Arnold-Gelfand Mathematical Seminars, Geometry and Singularity Theory (V.I. Arnold, I.M. Gelfand, M. Smirnov, and V.S. Retakh, eds.), Birkhauser, Boston, 1997, pp. 77–92.
4. J.C. Álvarez and A. McRae, *The projective topology of line congruences*, preprint (1997).
5. V.I. Arnold, *On topological properties of Legendre projections in contact geometry of wave fronts*, St. Petersbourg Math J. **6** (1995), no. 3, 439–452.
6. _____, *Topological invariants of plane curves and caustics*, University Lecture Series, vol. 5, Amer. Math. Soc., Providence, RI, 1994.
7. _____, *The geometry of spherical curves and the algebra of quaternions*, Uspekhi Mat. Nauk **50** (1995), no. 1, 3–68; English transl. in Russian Math. Surveys **50** (1995).
8. _____, *Mathematical methods of classical mechanics*, Springer-Verlag, New York, 1978.
9. T. Banchoff, *The spherical two-piece property and tight surfaces on spheres*, J. Diff. Geometry **4** (1970), 193–205.
10. A. Besse, *Manifolds all of whose geodesics are closed*, Springer-Verlag, Berlin, Heidelberg, 1978.
11. R. Bryant, *Projectively flat Finsler 2-spheres of constant curvature*, preprint (1996).
12. H. Busemann, *Geometries in which planes minimize area*, Rend. Circ. Matem. Palermo **55** (1961), 171–190.

13. T. Cecil and P. Ryan, *Tight and taut immersions of manifolds*, Research Notes in Math., Pitman, Boston, London, 1985.
14. M. Chaperon, *On generating families*, The Floer's memorial volume (H. Hofer et al., eds.), Birkhauser, Basel, 1995, pp. 283–296.
15. Yu. V. Chekanov, *Critical points of quasi-functions and generating families of Legendrian manifolds*, Funktsional. Anal. i Prilozhen. **30** (1996), no. 2, 56–69; English transl., Funct. Anal. Appl. **30** (1996), no. 2, 118–128.
16. Y. Eliahsberg and M. Gromov, *Convex symplectic manifolds.* II **52** (1991), Amer. Math. Soc., Providence, RI, 135–162.
17. E. Ferrand, *On a theorem of Chekanov*, Symplectic Singularities and Geometry of Gauge Fields (Robert Budzynski et al., eds.), vol. 39, Banach Center Publications, Warszawa, 1997, pp. 39–48.
18. _____, *Sur la structure symplectique de l'espace des géodésiques d'une variété de Hadamard*, to appear in Geom. Dedicata.
19. A. Givental, *Nonlinear generalization of the Maslov index*, Adv. in Soviet Math., vol. 1, Amer. Math. Soc., Providence, RI, 1990, pp. 71–103.
20. D. Gromoll and K. Grove, *On metrics on S^2 all of whose geodesics are closed*, Invent. Math. **65** (1981), 175-177.
21. J. Moser, *On the volume elements of a manifold*, Trans. Amer. Math. Soc. **120** (1965), 286–294.
22. A.V. Pogorelov, *Hilbert's fourth problem*, Scipta Series in Mathematics, Winston and Sons, 1979.
23. H. Rund, *The differential geometry of Finsler spaces*, Springer-Verlag, Berlin, Göttingen, Heidelberg, 1959.
24. J.M. Souriau, *Structure des systemes dynamiques*, Dunod, Paris, 1970.
25. S.L. Tabachnikov, *Around four vertices*, Uspekhi Mat. Nauk **45** (1990), no. 1, 191–192; English transl., Russian Math. Surveys **45** (1990), no. 1, 229–230.
26. _____, *Outer billiards*, Uspekhi Mat. Nauk **48** (1993), no. 6, 75–102; English transl. in Russian Math. Surveys **48** (1993).

UNIVERSITÉ CATHOLIQUE DE LOUVAIN, INSTITUT DE MATHÉMATIQUES PURE ET APPL., CHEMIN DU CYCLOTRON 2, B–1348 LOUVAIN–LA–NEUVE, BELGIUM.

E-mail address: `alvarez@agel.ucl.ac.be`

On Legendre Cobordisms

Emmanuel Ferrand

ABSTRACT. Given a generic Legendre cobordism, there exists some *noncritical* values of the time function such that the reduced slice has self-intersections. We consider the classification of Legendre submanifolds up to Legendre cobordism which satisfy the constraint that all noncritical reduced slices are embedded. We also classify Legendre knots in the manifold of co-oriented contact elements of the plane up to Legendre cobordism under the constraint that the wave fronts of all reduced slices are free of J^--singularity. We analyze some relationships between Legendre cobordisms and knot concordance in the standard contact space and we show that the *slice* Bennequin inequality is not sharp.

1.

1.1. Introduction. In [**Ar2**], Arnold introduced a Vassiliev-like theory for planar immersions of the circle. In order to obtain a rough classification of curves, he classified *smooth curves* in the plane up to some version of curve cobordism where codimension-one events of a given type are forbidden. In this paper, we study a similar problem for Legendre knots in a three dimensional contact manifold: what can be said about the classification of Legendre knots modulo Legendre cobordisms under the constraint that the reduced slices must never feature some given singularity? The paper is organized as follows. In 1.2 we recall some definitions, adapt Arnold's original definition of Legendre cobordisms ([**Ar1**]) in a slightly more general setting and study its basic properties. In 2.1 we consider J^+-cobordisms, i.e., Legendre cobordisms such that all noncritical reduced slices are embedded. We prove that two Legendre knots are J^+-cobordant if and only if they are Legendre cobordant. In 2.2 we specialize to the case of the contact manifold $ST^*\mathbb{R}^2$, in which we classify Legendre links up to J^--cobordisms, i.e., Legendre cobordisms such that the fronts of all reduced slices are free of J^--singularity (see [**Ar2**]). These results allow us to define in 3.1 the *Legendre genus* of a Legendre knot l in the standard contact manifold $M = \mathbb{R}^3 = J^1(\mathbb{R}, \mathbb{R})$, the space of one-jets of functions on $\mathbb{R}$. We show that this number is not only a contact isotopy invariant, but is in fact an *ambient isotopy* invariant: it coincides with the slice genus. This result is interesting in view of the *slice Bennequin inequality* of [**Ru3**]. This leads to the definition, in 3.3, of another contact isotopy invariant, the *ribbon Legendre genus*. We prove an

---

1991 *Mathematics Subject Classification.* Primary 53C15; Secondary 57M25.

inequality involving this number, the Maslov and the Bennequin invariants. In 4.1 we discuss the sharpness of the slice Bennequin inequality from this view point.

1.2. Preliminaries. Lagrange and Legendre cobordisms were introduced in [**Ar1**]. The reader will find all necessary definitions and an introduction to contact geometry and Legendre cobordisms in [**AG**]. Conceptual view points on this theory are exposed in the books [**Au**] and [**Va**].

Arnold has defined Legendre cobordisms in several distinct settings; namely for the three contact manifolds which are naturally associated to a given manifold B (see [**AG**]). The manifold $J^1(B, \mathbb{R}) = T^*B \times \mathbb{R}$ of one-jets of functions on B, the manifold of (resp. co-oriented) contact elements of B, which is the projectivized cotangent bundle PT^*B, (resp. the spherized cotangent bundle ST^*B). We consider here a version of Legendre cobordism which is defined for all contact manifolds. In the case when the contact manifold is $J^1(B, \mathbb{R})$, we recover what Arnold calls the *cylindrical cobordism of Legendre submanifolds in $J^1(B, \mathbb{R})$* (a theory which is isomorphic to the theory of cylindrical cobordism of *exact Lagrange* submanifolds in T^*B). We point out that in the case when the contact manifold is PT^*M or ST^*M, Arnold calls his construction *cylindrical cobordisms of wave front in M*, and that our general construction is different. In paragraph 2.3, we observe that the classification of Legendre submanifolds in ST^*B modulo our definition of Legendre cobordism is a refinement of Arnold's classification of wave fronts in B modulo cylindrical cobordism of wave fronts.

Consider a contact manifold M equipped with a globally defined contact form α. The manifold $M \times T^*[0,1]$ carries a natural contact structure, defined by the following construction. Denote by λ the Liouville form of $T^*[0,1]$. The one-form $\alpha - \lambda$ on $M \times T^*[0,1]$ is contact. A simple way to check this is to observe that the product of a Darboux chart of M with $T^*[0,1]$ is a Darboux chart for $M \times T^*[0,1]$. Choose another contact form $f \cdot \alpha$, where $f : M \to \mathbb{R}$ is a nonvanishing function. Denote by ϕ the diffeomorphism of $M \times T^*[0,1]$ which consists in multiplying by f in the fiber of $T^*[0,1]$. Since $\phi^*(f \cdot \alpha - \lambda) = f \cdot (\alpha - \lambda)$, the contact structure induced by $\alpha - \lambda$ does not depend on the choice of α. This manifold will be referred to as the *big contact manifold* and its Legendre submanifolds as *big Legendre submanifolds*.

Denote by (t, w) an element of $T^*[0,1] = [0,1] \times \mathbb{R}$. Given an *oriented* big Legendre submanifold $\mathcal{B} \subset M \times T^*[0,1]$ and a time $s \in [0,1]$, one can construct a Legendre submanifold $l_s \subset M$, the *reduced slice at time s*, by *reduction* along the hypersurface $\{t = s\}$:

$$l_s = \{x \in M, \ \exists \ w \in \mathbb{R} \text{ such that } (x, s, w) \in \mathcal{B}\}.$$

For a generic time s, l_s is immersed. It inherits an orientation from $\mathcal{B}$. If $\mathcal{B}$ is generic, l_s is embedded for a generic time s. The big Legendre submanifold $\mathcal{B}$ and the collection l_t, $t \in [0,1]$ of its reduced slices are called a *Legendre cobordism between l_0 and l_1*. It can be shown that this cobordism relation is an equivalence relation between embedded oriented Legendre submanifolds of M. The union of submanifolds induces a semi-group structure on the set of Legendre cobordism classes.

PROPOSITION 1.1. *This semi-group is always a group. The inverse of the Legendre cobordism class of l is the class of $-l$, the Legendre submanifold obtained from l by reversing its orientation.*

PROOF. We first prove the theorem for l embedded in $M = J^1(l, \mathbb{R})$ as the zero section. The choice of a coordinate system q on l induces a natural coordinate system (q, u, p) on $J^1(l, \mathbb{R})$ in which its standard contact form is $du - pdq$. In $J^1(l, \mathbb{R}) \times T^*[0, 1]$, consider the big Legendre submanifold defined by

$$(q, u = \varepsilon \cdot (s/2 + \sin(2s)/4),\ p = 0,\ t = \varepsilon \cdot \sin(s),\ w = \cos(s)), \quad q \in l,\ s \in [0, \pi],$$

where ε is a small positive real number. The reduced slice at time $\{t = 1\}$ is empty. At $\{t = 0\}$, it consists in the union of l and of a manifold l^ε which is contact-isotopic to l modulo orientation. This shows that the Legendre cobordism class $l \amalg (-l)$ is trivial. This particular case implies the general statement via the following classical result (see [**AG**]). Given any Legendre submanifold l in any contact manifold M, there exists a tubular neighborhood of l which is contactomorphic to a neighborhood of the zero section in $J^1(l, \mathbb{R})$. □

EXAMPLE 1.2. If $M = ST^*\mathbb{R}^2 = J^1(S^1, \mathbb{R})$ the group of Legendre cobordism classes is isomorphic to $\mathbb{Z} \times \mathbb{Z}$ and the isomorphism is induced by the Maslov index and the projection in $H_1(M) = \mathbb{Z}$. This follows from the computations of paragraph 2.2. The techniques used here are nothing but generalizations of the elementary constructions used by Arnold in [**Ar1**] to show that the Maslov class induces an isomorphism between the group of Legendre cobordism classes of $M = J^1(\mathbb{R}, \mathbb{R})$ and $\mathbb{Z}$.

The above construction is limited to manifolds equipped with a co-orientable contact structure. For example, this excludes the case when $M = PT^*B$. However one can adapt the preceding construction to the non-co-orientable case. The contact structure can be viewed as a line sub-bundle $\mathcal{L}$ of T^*M. Consider the two-fold cover $\widetilde{M} \to M$, where $\widetilde{M}$ is the spherization of $\mathcal{L}$. The pull back of the contact structure is co-orientable. Extend the $\mathbb{Z}_2$-action on $\widetilde{M}$ in a $\mathbb{Z}_2$-action on $\widetilde{M} \times T^*[0, 1]$ by the $\mathbb{Z}_2$-action on $T^*[0, 1]$ which consists in changing the sign of the covector component. The quotient is a line bundle over $M \times [0, 1]$ with a non-co-orientable contact structure, which will be, by definition, the big contact manifold associated to M. The remainder of the construction is identical to the co-orientable case.

2.

2.1. J^+-cobordisms. Consider a *generic* Legendre cobordism l_t between two embedded Legendre submanifolds l_0 and l_1 in a contact manifold M. Denote by t_i the critical values of the restriction to the big Legendre submanifold of the time function. At finitely many instants t_j, different from these critical values, l_{t_j} will have a transverse self-intersection. The next result shows that one can avoid these self-intersections (i.e., realize a J^+-*cobordism*) if one replaces the initial cobordism by a more complicated one.

DEFINITION 2.1. Two Legendre submanifolds l_0 and l_1 of a contact manifold M are said to be J^+-cobordant if there exists a Legendre cobordism l_t, $t \in [0, 1]$ between them such that the Legendre submanifold l_t is embedded for $t \neq t_i$, and embedded outside the image of the critical point for $t = t_i$.

THEOREM 2.2. *Two Legendre submanifolds in a three dimensional contact manifold M are J^+-cobordant if and only if they are Legendre cobordant.*

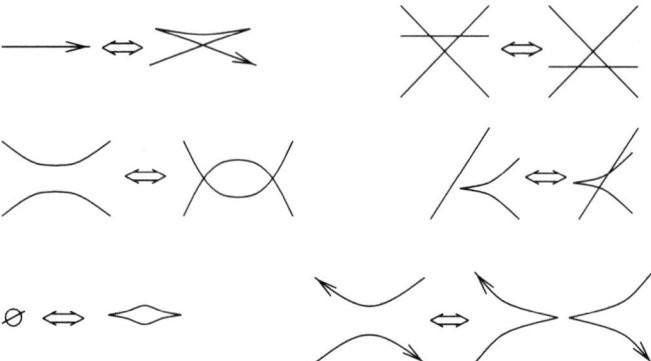

FIGURE 1. Elementary moves of Legendre cobordisms viewed in $J^0(\mathbb{R},\mathbb{R})$

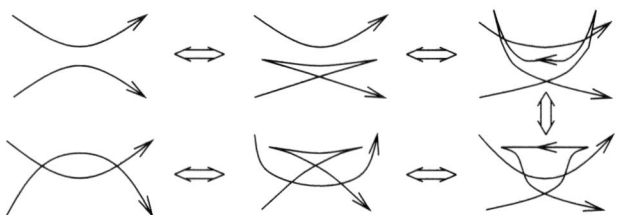

FIGURE 2. Elimination of self-tangencies at which the orientations of the two branches coincide

PROOF. We first observe that it is enough to prove the theorem for $M = J^1(\mathbb{R},\mathbb{R}) = \mathbb{R}^3$. We will modify a given Legendre cobordism only in a small neighborhood of the singularity we want to remove. We may assume that this neighborhood is contained in $U \times T^*[0,1]$, where U is a Darboux chart of M. Recall that the *wave front* of a Legendre submanifold in $J^1(\mathbb{R}^p,\mathbb{R})$ is the image of l by the projection

$$J^1(\mathbb{R}^p,\mathbb{R}) \to J^0(\mathbb{R}^p,\mathbb{R}) = \mathbb{R}^p \times \mathbb{R}.$$

A generic wave front completely determines the Legendre submanifold which is above. A transverse self-intersection of l corresponds to a quadratic self-tangency of the wave front.

From the structure of generic wave fronts in $J^0(\mathbb{R}^2,\mathbb{R})$ (see [**AG**]), it follows that the wave fronts of two Legendre cobordant Legendre submanifolds in $J^1(\mathbb{R},\mathbb{R})$ might be deduced one from the other by fibered diffeomorphisms of $J^0(\mathbb{R},\mathbb{R}) \to \mathbb{R}$, and the elementary front reconstructions shown in Figure 1.

We distinguish two types of self-tangencies, according to the fact that the orientations of the two tangent branches of the front coincide or differ. As shown in Figure 2, we can modify a given Legendre cobordism so that at a self-tangency of the front, the orientations of the two branches differ.

The sequence of moves of wave fronts of Figure 3 shows how to avoid such a self-tangency. □

REMARK 2.3. There is a higher-dimensional version of this theorem. We hope to come back to this elsewhere.

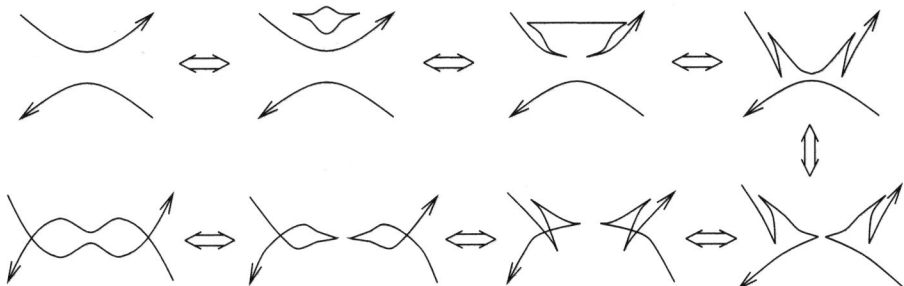

FIGURE 3. Avoiding a self-tangency

2.2. Wave fronts in the plane and J^--cobordisms. In this section, M is the contact manifold $ST^*\mathbb{R}^2$. We also call *wave front* the image of an oriented Legendre link $l \subset ST^*\mathbb{R}^2$ by the projection $ST^*\mathbb{R}^2 \to \mathbb{R}^2$. A generic wave front is an oriented and co-oriented plane curve which has only transverse self-intersections and an even number of semi-cubic cusps as singularities. A cusp is called *positive* (resp. *negative*) if the value of the covector which defines the co-orientation of the wave front at the cusp point on the velocity vector is nonnegative (resp. nonpositive) in a neighborhood of the cusp point. The *Maslov index* μ of the front is the algebraic number of cusps. The *index* i of the front is the degree of the Gauss map of the front, i.e., the rotation number of the co-orienting covector. It follows from [**Ar1**] that these two integers are the only invariants of Legendre cobordism in $ST^*\mathbb{R}^2$, and it follows from the preceding section that they are also the only invariants of J^+-cobordisms.

Denote by l^a the Legendre submanifold obtained from l by antipody in the fibers of $ST^*\mathbb{R}^2 \to \mathbb{R}^2$. The wave front of l^a coincides with the one of l, but has the opposite co-orientation. By a J^--singularity, we mean a transverse intersection between l and l^a. This happens generically in a one parameter family of Legendre immersions. On the wave front, this corresponds to a self-tangency at which the two co-orienting covectors are in opposite directions.

DEFINITION 2.4. A Legendre cobordism l_t, $t \in [0,1]$ between two Legendre links l_0 and l_1 in $ST^*\mathbb{R}^2$ is called a J^--cobordism if none of the l_t ($t \in [0,1]$) features a J^--singularity.

The addition of two J^--cobordism classes is defined as follows. Choose a representative in each of the classes such that their front belongs to two different half-planes and take the class of the disjoint union of these two representatives.

THEOREM 2.5. *The semi-group of J^--cobordism classes of Legendre links in $ST^*\mathbb{R}^2$ is a group which is isomorphic to the group $\mathbb{Z}^3$ via three additive maps μ, i, and j^-. Furthermore, any Legendre link l with trivial Maslov class is J^--cobordant to l^a and is J^--cobordant to a Legendre link with an immersed front.*

PROOF. The Maslov index μ and the index i are well-defined additive functions on the set of J^--cobordism classes. In order to define the invariant j^-, we first need to recall some basic observations. Consider a *generic* oriented and co-oriented wave front on a surface. In a neighborhood of a double point, assign an arbitrary order on the two branches which intersect at the double point. The double point is called *positive* (resp. *negative*) if the local orientation of the surface induced by

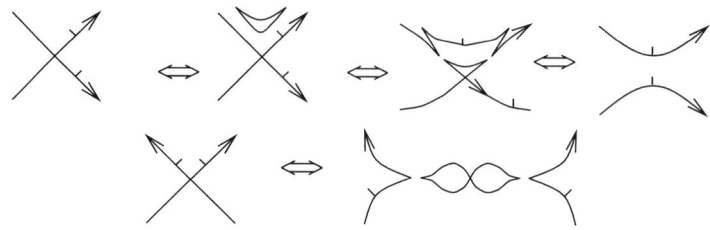

FIGURE 4. Elimination of double points

the frame made of the orienting vector of the first branch followed by the orienting vector of the second branch coincides (resp. differs) with the orientation induced by the co-orienting covector of the first branch followed by the co-orienting covector of the second branch. The sign of a double point does not depend on the choice of the ordering of the two branches.

The invariant j^- associated to a Legendre link $l \subset ST^*\mathbb{R}^2$ is some kind of linking number between l and l^a. However, such a linking number is not well defined in the solid torus, and we will define j^- in a combinatorial way. Its invariance under J^--cobordisms will be a direct consequence of this description. Choose an arbitrary point N in the plane. Call this point "north pole", and identify the plane with the north hemisphere of the round sphere S^2. The geodesic flow of S^2 can be seen as a contact flow Φ_t of ST^*S^2 via the Legendre transform which identifies ST^*S^2 with the unitary tangent bundle of S^2. We define $l^\vee = \Phi_{\frac{\pi}{2}}(l)$. By construction, $W^\vee$, the (spherical) wave front of $l^\vee$, has no "vertical" tangents, i.e., tangents which go through the poles (in particular $W^\vee$ does not meet the poles). Consider two Legendre links l_1 and l_2 in general position in $ST^*\mathbb{R}^2$. Denote by $W_1^\vee$ and $W_2^\vee$ the wave fronts of $l_1^\vee$ and $l_2^\vee$. The linking number of l_1 and l_2 is by definition

$$lk(l_1, l_2) = \frac{1}{2} \cdot \sum_{X \in \{W_1^\vee \cap W_2^\vee\}} \epsilon(X).$$

In this formula, $\epsilon(X)$ is the sign of an intersection X of $W_1^\vee$ and $W_2^\vee$. Observe that $l^\vee$ avoids a section (the direction of the poles) of $ST^*(S^2 \setminus \{\text{poles}\})$. Hence this formula coincides with the standard definition of the linking number of $l_1^\vee$ and $l_2^\vee$ via the projection on the base of a line bundle. It is invariant under contact isotopy in $ST^*\mathbb{R}^2$. It does not depend on the choice of the pole N in the plane. The invariant j^-, evaluated on some Legendre knot l in $ST^*\mathbb{R}^2$, is defined by the formula

$$j^-(l) = lk(l, l^a) - (i(l))^2.$$

If l_0 and l_1 are Legendre cobordant in $ST^*\mathbb{R}^2$, then $l_0^\vee$ and $l_1^\vee$ are also Legendre cobordant in $ST^*(S^2 \setminus \{\text{poles}\})$. By construction, j^- jumps if and only if $l^\vee$ intersects $(l^a)^\vee$, that is when l has a J^--singularity. Hence j^- is invariant under J^--cobordisms. We will check that j^- is additive later in the proof.

Now we construct the inverse of a J^--cobordism class in which we choose a representative l. Near each double point of the front of l, we perform the procedure described in Figure 4. As a result, positive double points are smoothed, whereas negative double points give rise to *bow-tie* components in the front. Furthermore, as shown in Figure 5, we can eliminate all the cusps of this modified front *modulo bow-ties*.

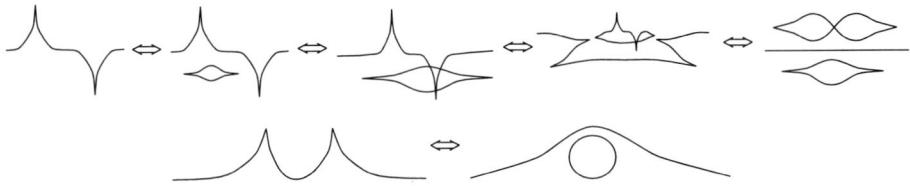

FIGURE 5. Elimination of cusps

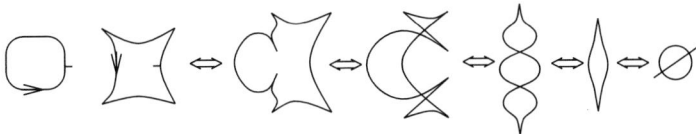

FIGURE 6. The inverse of a "circle" is a "square"

FIGURE 7. How to perform a mitosis

Observe that bow-tie components can move freely through the other components of the front. We can concentrate all of them away from the remainder of the front. Observe also that a bow-tie has an inverse, i.e., the bow-tie with the opposite orientation (and hence the opposite Maslov index). This shows that any J^--cobordism class can be represented by a front consisting in a collection of smooth nested circles and a multiple of a bow-tie, which represents exactly its Maslov class.

We now explain how to build the inverse of such a collection of nested circles. The inverse of a single circle is shown in Figure 6, where it is called a *square*. A collection of nested circles being given, one constructs a collection of nested squares by replacing all the circles by their inverses. Applying recursively the trick of Figure 6, we see that we have built the needed inverse.

Now we need to show that the invariant j^- classifies J^--cobordism classes having a given index i and which are represented by collections of oriented and co-oriented circles. All the 2^n Legendre links corresponding to the 2^n possible ways to co-orient the n circles belong to the same J^--cobordism class. Indeed, it follows from the construction of the inverse that all these Legendre links have the same inverse. Hence we can forget about the co-orientation. Using recursively the sequence of moves described in Figure 7, we see that each class can be represented as a linear combination of some elementary fronts which are made of some concentric circles, possibly with one square in the center. Two neighboring concentric circles which turn in opposite directions cancel each other out. Hence we can suppose that all the circles turn in the same direction. If there is no square inside a family of n concentric circles which turn in the positive direction, we call this elementary front A_n. If there are $n-1$ circles turning in the positive direction with a square inside, we call this front B_n if the square turns in the same direction as the circles, and

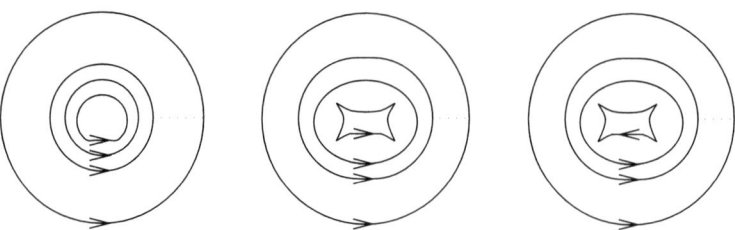

FIGURE 8. Elementary fronts of type A_n, B_n and C_n, $(n > 0)$

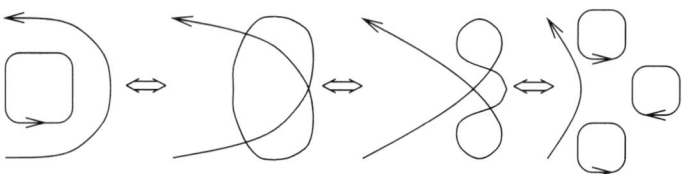

FIGURE 9. A move involved in the simplification of A_n

C_n otherwise (see Figure 8). We call A_{-n}, B_{-n}, C_{-n} the fronts obtained from A_n, B_n, C_n by reversing the orientation of all components. The index of C_k is k. The index of B_k is $k-2$ if $k > 0$, and $k+2$ if $k < 0$ ($A_0 = B_0 = C_0 = \emptyset$). If $k > 0$, a calculation shows that $A_k + B_k = 2 \cdot A_{k-1}$ and $A_{k-2} + C_k = 2 \cdot A_{k-1}$. If $k < 0$, the calculation yields $A_k + B_k = 2 \cdot A_{k+1}$ and $A_{k+2} + C_k = 2 \cdot A_{k+1}$. In other words the group is generated by A_k, $k \in \mathbb{Z}$. Another computation, based on Figure 9, shows that
$$A_k = \frac{k(k+1)}{2} \cdot A_1 + \frac{k(k-1)}{2} \cdot A_{-1}.$$
Hence we have shown that the group is generated by the classes of the bow-tie and of the fronts A_1 and A_{-1}. There are no relations between these three classes because of our three invariants. The computation of j^- on such simplified fronts shows that it is additive (a way to check this is to use the formulas of the next proposition). □

Once the front is reduced to a collection $\mathcal{C}$ of nested circles, there exists a simple way to compute j^-. It is a generalization of a formula discovered by O. Viro ([**Vi**]) in the context of smooth curves. We have already noticed that we can forget about the co-orientation. A pair of circles picked in $\mathcal{C}$ is called an *injective pair* if one of the circles lies inside the other. Otherwise, it is called a *free pair*. By definition, the sign of a pair is *positive* if the pair has the orientation of a boundary, and is *negative* otherwise. We denote by $\langle \circ \circ \mid \mathcal{C} \rangle$ (resp. $\langle \odot \mid \mathcal{C} \rangle$) the number of free pairs (resp. injective pairs), counted with signs.

PROPOSITION 2.6. *The invariant j^- of a collection $\mathcal{C}$ of n circles is given by the formula*
$$j^- = 2 \cdot \langle \odot \mid \mathcal{C} \rangle - n = 2 \cdot \langle \circ \circ \mid \mathcal{C} \rangle - i^2.$$

PROOF. The proof of the Viro formula given in [**Fe**] also yields the above formulas. □

REMARK 2.7. The invariant J^- of *plane curves without "safe" self-tangencies* defined in [**Ar2**] is in fact $J^- = j^- + 1$.

2.3. Cobordism of wave fronts.
Here we consider Arnold's version ([**Ar1**]) of Legendre cobordism in $ST^*\mathbb{R}^2$, i.e., *cobordism of wave fronts*.

DEFINITION 2.8 ([**Ar1**]). Two generic wave fronts w_0 and w_1 in $\mathbb{R}^2$ are said to be *cobordant* if they are the slices at $t = 0$ and $t = 1$ of a generic big wave front W in $\mathbb{R}^2 \times [0,1]$; i.e., W is the projection of a generic big Legendre manifold lying in $ST^*(\mathbb{R}^2 \times [0,1])$. Furthermore, they are said to be J^--cobordant if none of the slices w_t ($t \in [0,1]$) features a J^--singularity.

The wave fronts of two Legendre cobordant Legendre links of $ST^*\mathbb{R}^2$ are cobordant, but the converse does not hold. The difference consists in the Morse *perestroikas* (reconstructions) shown in Figure 10, which change the index, but not the Maslov index.

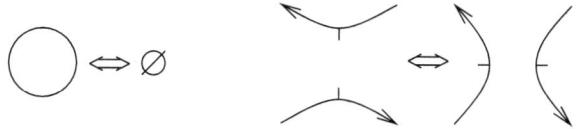

FIGURE 10. Additional reconstructions

PROPOSITION 2.9. *Two wave fronts in $\mathbb{R}^2$ are J^--cobordant if and only if they are cobordant.*

PROOF. This follows from the proof of the preceding theorem and from the fact that a collection of circles is cobordant to zero. □

3.

3.1. $g_{\text{leg}} = g_s$.
From now on M is the contact three dimensional manifold $J^1(\mathbb{R}, \mathbb{R})$. It follows from the preceding section that the semi-group of J^+-cobordism classes of Legendre links is isomorphic to its nonrestricted version, i.e., to the group $\mathbb{Z}$, the isomorphism being induced by the Maslov class. As shown in [**Ar1**], we see that any Legendre knot in $J^1(\mathbb{R}, \mathbb{R})$ is Legendre cobordant to a sum of (lifts of) bow-tie fronts. Denote by X_μ a standard (connected) representative in each class (see Figure 11).

DEFINITION 3.1. The Legendre genus g_{leg} of a Legendre *knot* l with Maslov class μ is the least integer of the form $-\chi/2$, where χ is the Euler characteristic of a big Legendre submanifold which realizes a J^+-cobordism between l and the corresponding standard representative X_μ.

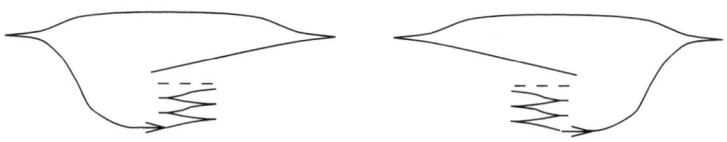

FIGURE 11. The fronts (in $J^0(\mathbb{R}, \mathbb{R})$) of X_μ (for $\mu \geq 0$ and $\mu \leq 0$)

Section 2.1 shows that $g_{\text{leg}} < \infty$. Recall that the *slice genus* g_s of a knot l in M is the minimum of all genera of the surfaces S with boundary $\partial S = S^1$ which can be smoothly embedded in $M \times [0,1)$ such that $\partial S = l \subset M \times \{0\}$.

THEOREM 3.2. *The Legendre genus is a topological invariant, $g_{\text{leg}} = g_s$.*

3.2. Proof. Denote by S the image of the big Legendre submanifold by the projection $M \times [0,1] \times \mathbb{R} \to M \times [0,1]$. It is a nonsingular and embedded surface S except near the image of the critical points of the time function. But these singularities can be removed by making S generic: a smooth embedded surface $\widetilde{S}$ is obtained from S by perturbing the Legendre knots $S \cap \{t\}$ into knots transverse to the contact structure $\widetilde{S} \cap \{t\}$, for $\{t\}$ near a critical value t_i. A surface whose genus must be at least the slice genus is obtained from $\widetilde{S}$ by smoothly pasting a disk along the component of $\partial \widetilde{S}$ lying in $\{t=1\}$. This proves that $g_{\text{leg}} \geq g_s$.

By an Ω-move, we mean the local J^+-cobordism move depicted in Figure 12. We say that two Legendre knots are Ω-equivalent if one can be obtained from the other by a sequence of contact isotopies and Ω-moves.

FIGURE 12. The Ω-move

We use now a theorem by D. Fuchs and S. Tabachnikov, reformulated as follows.

THEOREM 3.3 ([**FT**], Theorem 4.4). *Two Legendre knots are Ω-equivalent if and only if they are ambient isotopic and homotopic through Legendre immersions.*

Denote by l_t ($t \in [0,1]$) a differentiable knot cobordism having genus $-\chi_s/2 = g_s$ such that $l_0 = l$ is the Legendre knot under consideration and $l_1 = X_\mu$, the standard representative of the Legendre homotopy class of l, which is trivial as a differentiable knot. Still denote by t_i the critical values of the time function. Applying the theorem of Fuchs and Tabachnikov for $l_t, t \in (t_i, t_{i+1})$, and realizing the elementary cobordisms near the critical values t_i by their Legendrian versions, we obtain a J^+-cobordism between l and X_μ, having genus g_s. This proves that $g_{\text{leg}} = g_s$. □

3.3. Ribbon Legendre genus.

DEFINITION 3.4. A ribbon J^+-cobordism is a J^+-cobordism such that the time function, when restricted to the big Legendre manifold, has no minimum.

LEMMA 3.5. *Two Legendre knots are ribbon J^+-cobordant if and only if they are Legendre-cobordant.*

PROOF. Local minima of the time function correspond to the appearance of *lips* components in the front. One can create them from a newborn swallowtail and a saddle cobordism near one of the two cusps of the swallowtail (see Figure 13). By a swallowtail birth-death, one can choose the desired orientation of the lips. This shows that any J^+-cobordism can be modified into a ribbon J^+-cobordism (but having a higher genus). □

FIGURE 13. Creation of a "lips" component

This means that the *ribbon Legendre genus* of a Legendre knot is well defined.

The standard contact structure of $\mathbb{R}^3$ is co-oriented. Hence any Legendre knot l has a natural framing. The *Bennequin invariant* $tb(l)$ is the self-linking of this framing. If the front of l features n^+ (resp. n^-) positive (resp. negative) crossings, and c cusps, the following formula holds:

$$tb(l) = n^+ - n^- - \frac{c}{2}.$$

PROPOSITION 3.6. *Denote by $\mu(l)$ the Maslov class of a Legendre knot l, by $tb(l)$ its Bennequin invariant, and by χ the Euler characteristic of a ribbon J^+-cobordism connecting l to X_μ. The following inequality holds:*

$$\chi + tb(l) + |\mu(l)| < 0.$$

PROOF. Still denote by t_i the critical values of the time function, and by l_t the reduced slice at time t (hence $l_0 = l$ and $l_1 = X_\mu$).

By a *positive saddle* (resp. *negative saddle*) we mean an elementary saddle Legendre cobordism during which the number of cusps in the front *decreases* (resp. *increases*).

Denote by m^+ (resp. m^-, s^+, s^-) the number of maxima (resp. minima, positive saddles, negative saddles) of the time function. One can check that

$$\chi = m^+ + m^- - s^+ - s^-,$$
$$tb(l_0) - tb(l_1) = tb(l) - tb(X_\mu) = tb(l) - (-1 - |\mu|) = m^- + s^- - m^+ - s^+.$$

Notice that $tb(l)$ is a *twisted Euler characteristic*. This shows that

$$\chi + tb(l) + |\mu| + 1 = 2 \cdot (m^- - s^+).$$

The result follows from the fact that, by hypothesis, $m^- = 0$. □

4.

4.1. The slice Bennequin inequality is not sharp. Let K be an ambient isotopy class of knots in $\mathbb{R}^3$. Denote by $[K]$ the smooth concordance class of K. To a Legendre knot l we associate the number $\mathcal{S}(l) = tb(l) + |\mu(l)| + 1$. By $\mathcal{S}(K)$ is meant $sup\{\mathcal{S}(l), l \in K\}$, and $\mathcal{S}([K]) = sup\{\mathcal{S}(L), L \in [K]\}$. The Bennequin inequality ([**Be**]) (resp. slice Bennequin inequality, ([**Ru3**]) is

$$\mathcal{S}(K) \leq 2 \cdot g \quad (\text{resp. } \mathcal{S}([K]) \leq 2 \cdot g_s),$$

where g (resp. g_s) is the genus of K (resp. the slice genus of $[K]$). This implies Proposition 3.6, but it is a much deeper result. It also implies, using the notations of the previous section, that on a J^+-cobordism, not necessarily ribbon, between l and X_μ, we have $m^- - s^+ \leq 0$.

Several authors ([**CG**], [**FT**], [**Ka**] and [**Ru1**], [**Ru2**], [**Ru4**]) have remarked that the Bennequin inequality is not sharp (for some nontrivial K). We shall see that the slice Bennequin inequality contains its own nonsharpness.

PROPOSITION 4.1. $\mathcal{S}$ is over-additive:
$$\mathcal{S}([K_1]) + \mathcal{S}([K_2]) \leq \mathcal{S}([K_1] + [K_2]).$$

COROLLARY 4.2. *The slice Bennequin inequality is not sharp in the following sense: there exists concordance classes $[K]$ such that $\mathcal{S}([K]) < 2 \cdot g_s([K])$.*

4.2. Proof. Consider two concordance classes K_i, $i \in \{1, 2\}$. Recall that $[K_1] + [K_2] = [K_1 \sharp K_2]$, where $\sharp$ denotes the connected sum of knots. Let l_i be Legendre knots such that $l_i \in [K_i]$ and $\mathcal{S}(l_i) = \mathcal{S}([K_i])$. Denote by u, p, q the standard coordinates of $J^1(\mathbb{R}, \mathbb{R}) = \mathbb{R}^3$, in which the natural contact form is $du - pdq$. Denote by Φ the contactomorphism $(u, p, q) \to (-u, -p, q)$, which is isotopic to the identity map, (but not through contactomorphisms). Remark that $\mu(\Phi(l)) = -\mu(l)$. Since Φ does not change the knot type, we can assume that $\mu(l_i) \geq 0$. Putting the wave fronts of l_1 (resp. l_2) to the left-hand side (resp. right-hand side) of the axis $\{q = 0\}$, we are going to consider an element of $[K_1 \sharp K_2]$ obtained by building a bridge between the right-most cusp of the front of l_1 and the left-most cusp of the front of l_2. Using, if necessary, the birth of two cusps (by a swallowtail *perestroika*) near one of these two cusps, we can assume that the local orientation of l_1 and l_2 near these two extreme cusps allow a saddle Legendre cobordism, which will provide our bridge. The resulting connected sum l belongs to $[K_1] + [K_2]$ and verifies $|\mu(l)| = |\mu(l_1)| + |\mu(l_2)|$ and $tb(l) = tb(l_1) + tb(l_2) + 1$. In other words $\mathcal{S}(l) = \mathcal{S}(l_1) + \mathcal{S}(l_2)$. □

The function g_s is sub-additive and not additive. Hence $\mathcal{S}([\cdot]) \neq g_s$. This proves the corollary. □

4.3. Illustration. If $[K]$ has finite order, this implies that $\mathcal{S}([K]) \leq 0$. For instance *the slice Bennequin inequality is not sharp for the figure-eight knot* (whose slice genus is 1 and whose concordance class has order two).

Conversely, if one can find a Legendre representative l of K such that $\mathcal{S}(l) > 0$, then $[K]$ has infinite order. In many cases, this is a very quick way to prove that the order is infinite.

Denote by $-K$ the knot obtained from the mirror image of K by changing the orientation. Since $[-K] = -[K]$, the inequality $\mathcal{S}([K]) + \mathcal{S}([-K]) \leq 0$ holds. One can construct explicitly a Legendre version l of the left trefoil knot T such that $tb(l) = 1$ and $\mu(l) = 0$ (see Figure 14). This shows that $\mathcal{S}([T]) = 2 = 2 \cdot g_s([T])$ and hence *the slice Bennequin inequality is not sharp* for the right trefoil $-T$: $\mathcal{S}([-T]) \leq -2$. We do not know if this upper bound is sharp. The authors cited in subsection 4.1 were able to show that $\mathcal{S}(-T) = -4$.

FIGURE 14. A Legendre left trefoil, with $tb = 1$ and $\mu = 0$

4.4. There exist slice knots K such that $\mathcal{S}(K) < 0$. The connected sum of the left and right trefoil knots is such an example. Following [**FT**], $\mathcal{S}(K)$ does not exceed $e(K)$, the lower degree of the framing variable in the HOMFLY polynomial of K. A computation shows that this invariant takes the value -2 on the slice knot $T\sharp(-T)$. The estimate $\mathcal{S}(K) \leq e(K)$ is also proved in [**CG**], via a different method. Note that a related estimate is proved in [**Ru1**], [**Ru2**]: $q(K) \leq e(K)$, where $q(K)$ is the *modulus of quasi-positivity* of K, and that it is proved in [**Ru4**] that $q(K) = TB(K) = sup(tb(l), l \in K)$.

References

[AG] V. Arnold and A. Givental, *Symplectic geometry and its applications*, Encyclopaedia Math. Sci., vol. 4, Springer-Verlag, Berlin, Heidelberg, New York, 1990, pp. 1–136.

[Ar1] V. Arnold, *Lagrange and Legendre cobordisms* I, II, Funct. Anal. Appl. **14** (1980), 167–177, 252–260.

[Ar2] _____, *Singularities and curves*, Adv. in Soviet Math. vol. 21, Amer. Math. Soc., Providence, RI, 1994, pp. 33–91.

[Au] M. Audin, *Cobordismes d'immersions lagrangiennes et legendriennes*, Hermann, Paris, 1987.

[Be] D. Bennequin, *Entrelacements et équations de Pfaff*, Astérisque, **106–107** (1983), 87–161.

[CG] S. Chmutov and V. Goryunov, *Polynomial invariants of Legendrian links and wave fronts*, preprint, 1995.

[Fe] E. Ferrand, *On the Bennequin invariant and the geometry of wave fronts*, Geom. Dedicata, **65** (1997), no. 2, 219–245.

[FT] D. Fuchs and S. Tabachnikov, *Invariants of Legendrian and transverse knots in the standard contact space*, Topology, **36** (1997), no. 5, 1025–1053.

[Ka] Y. Kanda, *On the Thurston-Bennequin invariant of Legendrian knots and non exactness of the Bennequin inequality*, preprint, 1995.

[Ru1] L. Rudolph, *A congruence between link polynomials*, Math. Proc. Camb. Phil. Soc. **107** (1990), 319–327.

[Ru2] _____, *Quasi-positive annuli (construction of quasi-positive knots and links*, IV), J. Knot Theory and Ramifications, **1** (1992), no. 4, 451–466.

[Ru3] _____, *Quasi-positivity as an obstruction to sliceness*, Bull. Amer. Math. Soc. **29** (1993), no. 1, 51–59.

[Ru4] _____, *An obstruction to sliceness via contact geometry and "classical" gauge theory*, Invent. Math. **119** (1995), no. 1, 155-163.

[Va] V. A. Vassiliev, *Lagrange and Legendre characteristic classes*, Gordon and Breach, New York, 1988.

[Vi] O. Viro, *Generic Immersions of circle to surfaces and complex topology of real algebraic curves*, Transl. Math. Monographs, vol. 173, Amer. Math. Soc., Providence, RI, 1996.

CENTRE DE MATHÉMATIQUES, U.R.A. 169 DU C.N.R.S. ECOLE POLYTECHNIQUE, 91128, PALAISEAU CEDEX, FRANCE

E-mail address: `ferrand@math.polytechnique.fr`

Vassiliev Invariants of Knots in $\mathbb{R}^3$ and in a Solid Torus

Victor Goryunov

ABSTRACT. We give a chord-diagram description of finite type invariants of framed and unframed knots in a solid torus. The relation is established via appropriate versions of the universal Vassiliev–Kontsevich invariant. The framed case is treated from the singularity theory point of view that involves knots with degenerate framings.

1. Introduction

The main goal of this paper is to introduce a basis for a part of the theory of Vassiliev type invariants of regular plane curves.

Consideration of such invariants was started recently by Arnold. In [1, 2] he defined three order 1 invariants which are dual to the three generic bifurcations in families of regular plane curves. While singularities of a generic curve are only transverse double points, in 1-parameter families there appear triple points and two types of self-tangencies: direct (when the two velocity vectors at the self-tangency point have the same direction) and inverse (when the directions are opposite).

Considering invariants that do not change under triple-point and inverse self-tangency transformations, one immediately arrives at invariants of framed knots in a solid torus. Indeed, a regular plane curve with no direct self-tangencies lifts to a Legendrian knot in the solid torus $ST^*\mathbb{R}^2$. This knot has a natural framing. The appearance of direct self-tangencies corresponds to singular knots, with double points. So the theory of Vassiliev type invariants of regular plane curves without direct self-tangencies seems to be parallel to that of framed knots in a solid torus. Thus a chord-diagram interpretation (similar to [7, 12]) of the latter theory is very desirable.

Obviously, such interpretation should be constructed from two pieces: that for framed knots in $\mathbb{R}^3$ and that for unframed knots in a solid torus.

i) The chord-diagram description of finite type invariants of framed knots in 3-space has already been given. In [12] Kontsevich mentioned that the space of such real-valued invariants is dual to the real linear space spanned by circular chord diagrams modulo the 4-term relation. Later this was proved by Lê and Murakami ([13, 14]) who adjusted the method used by Kontsevich in the unframed case.

1991 *Mathematics Subject Classification.* Primary 57M25, Secondary 57R45.
Supported by an RDF grant of The University of Liverpool.

©1999 American Mathematical Society

The approach by Lê and Murakami (their regularisation of the Kontsevich integral) works only for the blackboard framing. Unfortunately, this is not sufficient for the study of plane curves. Indeed, the canonical framing of the Legendrian lift of a regular plane curve is blackboard only with respect to the projection which is not very convenient to consider if one wants to construct Vassiliev type theory for plane curves.

In the present paper we fill this gap by modifying the definition of the universal Vassiliev–Kontsevich invariant so that it serves knots with any framing.

ii) The first attempt to construct a chord-diagram interpretation for Vassiliev type theory for unframed knots in a 3-manifold was done by Kalfagiani in [11]. But, since the case she considered was rather general, there was no obvious way to complete the theory by, say, a definition of a corresponding Kontsevich integral. In our special case of a solid torus the integral is defined straightforwardly: almost the only difference with Kontsevich's original idea ([12]) is that now we use the decomposition $\mathbb{C} \times S^1$ of the solid torus instead of the decomposition $\mathbb{C} \times \mathbb{R}^1$ of $\mathbb{R}^3$.

Those are two basic constructions of this paper. The third is the singularity theory approach to invariants of framed knots. This is very close to Vassiliev's original idea of considering singular knots instead of nonsingular ones passing from embeddings of a circle into 3-space to arbitrary smooth mappings. Any of numerous equivalent definitions of a framing leaves obvious room to make a framing singular. We introduce one more definition of our own and trace the framing degenerations. The bifurcations considered show, for example, what happens with the 1-term framing-independence relation of invariants of unframed knots: it does not disappear but gets new terms that reflect framing degenerations.

The contents of the paper is as follows.

In Section 2, we introduce knots with degenerate framings and the extension of invariants of nonsingular framed knots to those with elementary singularities. We construct the diagram theory for framed knots in 3-space which basicly coincides with that of Kontsevich [12, 13, 14] (some difference appears only for $\mathbb{Z}_2$-valued invariants).

In Section 3, we define the universal Vassiliev–Kontsevich invariant for knots in $\mathbb{R}^3$ that have arbitrary framings and reprove the result of Lê–Murakami in this general setting.

In Section 4, we consider Vassiliev–Kontsevich type theory for unframed knots in a solid torus. We show that the graded space of complex-valued finite order invariants in this case is dual to the graded linear space generated by marked chord diagrams modulo marked 1- and 4-term relations. The marking is defined by the fundamental group of the solid torus.

In Section 5, we obtain the similar result for framed knots in a solid torus. We also show that all the coefficients of the version of the HOMFLY polynomial for framed knots in a solid torus are in fact Vassiliev invariants of finite order (see [7]).

REMARK 1.1. The paper [10] establishes the isomorphism between the theory of Vassiliev invariants for framed knots in a solid torus and that for regular plane curves with no direct self-tangencies.

REMARK 1.2. The constructions in the present paper are very convenient to construct spectral sequences (similar to Vassiliev's one in [17]) to calculate cohomology of spaces of framed knots in $\mathbb{R}^3$ and unframed and framed knots in a solid torus. This will be the topic of another paper.

ACKNOWLEDGMENTS. I am very thankful to Sergei Chmutov for extremely useful discussions. The main result of this paper was reported at the meeting on Geometry and Physics held in Aarhus in July 1995 ([**9**]). Since then there has been a paper published by Suetsugu ([**15**]) containing an independent construction of the universal invariant of framed links in the solid torus based on the approach by Lê and Murakami.

2. Framed knots from the singularity point of view

2.1. Framed knots as mappings. A smooth unframed knot in 3-space is the image of a smooth embedding of a circle into $\mathbb{R}^3$. So, in singularity theory, an oriented knot is treated as an element of the set Ω of all C^∞-mappings of an oriented circle into $\mathbb{R}^3$. Formally, *a nonsingular unframed oriented knot* in 3-space is a connected component of the subset of Ω that consists of all the embeddings.

In the theory of Vassiliev invariants of knots in $\mathbb{R}^3$ the key role is played by so-called singular knots. Namely, consider the subset of Ω of all the immersions whose images have only n double points with nontangent branches and no other singularities. An *unframed oriented knot with n singular points* is a connected component of this subset. *A singular unframed knot* is a knot with a finite number of singular points.

Now we introduce similar notions for framed knots.

Let $S^1 \subset \mathbb{R}^2$ be an oriented C^∞-embedded circle and $U \subset \mathbb{R}^2$ an open (annular) neighbourhood of S^1. Consider a C^∞-mapping $g : U \to \mathbb{R}^3$. It defines the mapping Tg from the restriction $T_{S^1}\mathbb{R}^2$ of the tangent bundle $T\mathbb{R}^2$ to the tangent bundle $T\mathbb{R}^3$.

DEFINITION 2.1. Two mappings $g_i : U_i \to \mathbb{R}^3$, $i = 1, 2$, are *equivalent* if the mappings $Tg_i : T_{S^1}\mathbb{R}^2 \to T\mathbb{R}^3$, $i = 1, 2$, coincide.

We denote by Ω_f the set of all C^∞-mappings $g : U \to \mathbb{R}^3$ modulo this equivalence. The class of a mapping g in Ω_f will be denoted by g as well.

DEFINITION 2.2. Consider the set of all the equivalence classes of mappings $g \in \Omega_f$ such that the restriction $Tg : T_{S^1}\mathbb{R}^2 \to T\mathbb{R}^3$ is an embedding. *A nonsingular oriented framed knot* in $\mathbb{R}^3$ is a connected component of this set.

The $Tg : T_{S^1}\mathbb{R}^2 \to T\mathbb{R}^3$ being an embedding guarantees the mapping $g : S^1 \to \mathbb{R}^3$ being an embedding too.

The image $g(S^1)$ will be called the *core* of the mapping.

Consider a subset of Ω_f which consists of all the equivalence classes of mappings g such that:

i) the mapping $g : S^1 \to \mathbb{R}^3$ represents an unframed knot with $n \geq 0$ singular points;
ii) for all $s \in S^1$, except $k \geq 0$ points none of which is mapped to a double point of the core $g(S^1)$, the mapping Tg has rank 2 on $T_s\mathbb{R}^2$;
iii) for the remaining k points $s \in S^1$, the mapping Tg has rank 1 on $T_s\mathbb{R}^2$.

DEFINITION 2.3. An *oriented framed knot* in $\mathbb{R}^3$ with $n + k$ singularities is a connected component of the above subset. *A singular oriented framed knot* is an oriented framed knot with a finite number of singularities.

Condition i) implies that the kernel of Tg, which appears in iii), is not tangent to S^1.

2.2. The framed equivalence.
The set Ω_f of framed curves in $\mathbb{R}^3$ splits into orbits of the natural equivalence group which we denote by $\mathcal{F}$ (for "framed"). This is an analog of the group of left-right equivalence of mappings (see [5, 4]). Namely, we consider a representative $g : U \to \mathbb{R}^3$ of an element of Ω_f modulo:

i) orientation-preserving diffeomorphisms of the target $\mathbb{R}^3$;
ii) diffeomorphisms of the source pair (U, S^1) preserving the orientations of the circle and its neighbourhood;
iii) terms of order greater than 1 in the direction in U which is transversal to S^1.

There is an obvious local version of the $\mathcal{F}$-equivalence for germs of mappings $g : (\mathbb{R}^2, \mathbb{R}^1, 0) \to (\mathbb{R}^3, 0)$. This has the following coordinate description.

Let x and y be coordinates on the source plane with $\mathbb{R}^1$ being the x-axis. We use the following notation:

$\mathcal{O}_{x,y}$ is the space of all real-valued C^∞-function-germs on $(\mathbb{R}^2, 0)$;
$\mathcal{O}^3_{x,y}$ is the space of all C^∞-map-germs of $(\mathbb{R}^2, 0)$ to $\mathbb{R}^3$;
$\mathcal{O}^3_g$ is the space of C^∞-map-germs from the target copy of $(\mathbb{R}^3, 0)$ to $\mathbb{R}^3$ pulled back to $(\mathbb{R}^2, 0)$ by the germ g.

The tangent space to the $\mathcal{F}$-orbit of a map-germ $g \in \mathcal{O}^3_{x,y}$ is

$$T_g(\mathcal{F}g) = \mathcal{O}^3_g + \mathcal{O}_{x,y}\langle \partial g/\partial x, y\partial g/\partial y\rangle + y^2 \mathcal{O}^3_{x,y} ,$$

where the middle summand is a module on the two generators.

EXAMPLE 2.4. A nonsingular germ $(\mathbb{R}^2, \mathbb{R}^1, 0) \to \mathbb{R}^3$ can be reduced to

$$(x, y) \mapsto (x, y, 0).$$

The tangent space to its $\mathcal{F}$-orbit is the whole of $\mathcal{O}^3_{x,y}$ (so, the germ is stable).

2.3. Bifurcation diagrams of framed curve-germs.
As usual, an $\mathcal{F}$-miniversal deformation of g (see [5, 4]) is a minimal transversal to its $\mathcal{F}$-orbit so long as the tangent space $T_g(\mathcal{F}g)$ has a finite codimension in $\mathcal{O}^3_{x,y}$.

The base of an $\mathcal{F}$-miniversal deformation of a map-germ g contains *the bifurcation diagram* $\Sigma_\mathcal{F}(g)$; i.e., the set of the values of the deformation parameters λ for which the corresponding perturbed mappings $Tg_\lambda : T_{\mathbb{R}^1}\mathbb{R}^2 \to T\mathbb{R}^3$ are not embeddings. There are two options to achieve a degeneration:

1) either the mapping $g_\lambda : \mathbb{R}^1 \to \mathbb{R}^3$ is not an embedding; or
2) for some point $s \in \mathbb{R}^1$, the differential Tg_λ is not of rank 2 on $T_s\mathbb{R}^2$.

Thus the diagram $\Sigma_\mathcal{F}(g)$ has two components which we denote by $\Sigma'_\mathcal{F}(g)$ and $\Sigma''_\mathcal{F}(g)$, respectively. Both are hypersurfaces. Forgetting the framing and considering g as a mapping of the line alone, we stay with the component $\Sigma'_\mathcal{F}(g)$ only.

EXAMPLE 2.5. A local normal form for the simplest singular framing on a smooth curve in $\mathbb{R}^3$ is

$$h_0 : (x, y) \mapsto (x, yx, 0).$$

The differential of this mapping has rank 1 at the origin.

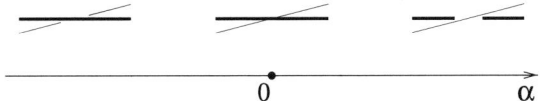

FIGURE 1. The simplest framing degeneration

FIGURE 2. Positive crossing of the stratum of nonembedded curves

For a one-parameter miniversal deformation one can take
$$h_\alpha : (x,y) \mapsto (x, yx, \alpha y),$$
where α is the deformation parameter. In Figure 1, above the parameter line, we show the corresponding framed curves. The bold line there is the core, that is the image of the x-axis. The thin line is the framing. It represents the Th_α-image of a section of $T_{\mathbb{R}^1}\mathbb{R}^2$ which contains the generator ∂_y of the kernel of $T_0 h_0$. The bifurcation diagram is $\Sigma_\mathcal{F}(h_0) = \Sigma''_\mathcal{F}(h_0) = \{\alpha = 0\}$.

The bifurcation diagram of a framed curve-germ is co-orientable in the base of an $\mathcal{F}$-versal deformation at its regular points. Namely, to co-orient the component $\Sigma''_\mathcal{F}$ we say that the local bifurcation of Figure 1, for *decreasing* α is done in the *positive direction*. To co-orient $\Sigma'_\mathcal{F}$ we say, as usual, that the bifurcation of Figure 2 is *positive*. We assume here and further on that the right orientation of $\mathbb{R}^3$ is fixed. In both cases the positive move increases the writhe of the framed curve.

EXAMPLE 2.6. The simplest local singularity of a mapping $\mathbb{R}^1 \to \mathbb{R}^3$ has a normal form
$$x \mapsto (x^2, \ x^3, \ 0).$$
Equipping this map-germ with a generic framing we arrive at normal forms
$$(x,y) \mapsto (x^2, \ x^3 \pm yx, \ y)$$
and 2-parameter miniversal deformations
$$(x^2, \ x^3 \pm yx + \alpha x, \ y + \beta x),$$
with the parameters α, β. Bifurcations in the (+)-family are shown in Figure 3. The co-oriented bifurcation diagram of the (−)-family is absolutely the same.

EXAMPLE 2.7. Another particular case useful for our further consideration is the generic degeneration of the framing at a double point of the core. Its $\mathcal{F}$-miniversal deformation is the 3-parameter family of bigerms
$$(x_1, y_1) \mapsto (x_1, y_1, 0),$$
$$(x_2, y_2) \mapsto (\alpha y_2, x_2 y_2 + \beta + \gamma y_2, x_2).$$
A co-oriented bifurcation diagram of this family, within the assumption that each of the two curve-germs is oriented by the increase of the corresponding x-coordinate, is shown in Figure 4.

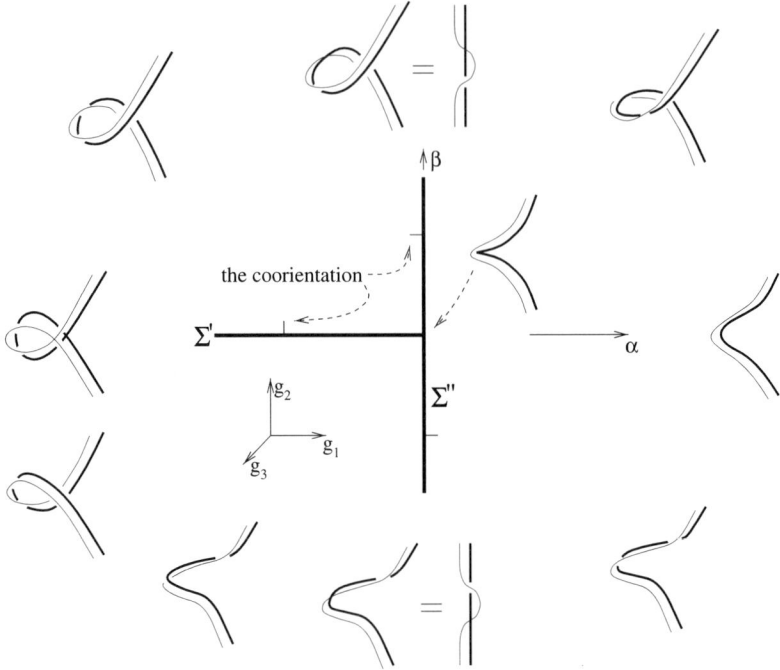

FIGURE 3. A miniversal deformation of a framed curve with a generic singular point

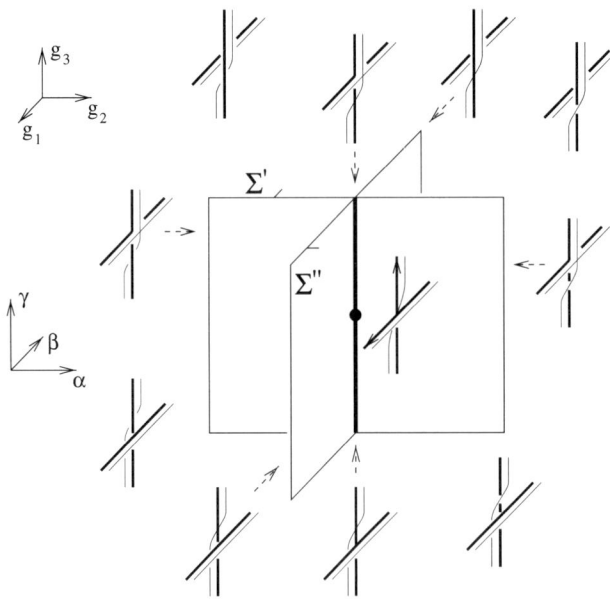

FIGURE 4. Bifurcations of a double point with a degenerate framing

2.4. Extended invariants. The obvious global version, in the space Ω_f, of the bifurcation diagram will be called the *discriminant* of Ω_f and denoted by Σ_f. This is the union of the two hypersurfaces Σ'_f and Σ''_f. We co-orient the discriminant by local means, using the above co-orientation of bifurcation diagrams.

An *invariant of oriented framed knots* is an element of the group $H^0(\Omega_f \setminus \Sigma_f)$ (with any coefficients).

A regular point of m-tuple self-intersection of Σ_f represents a singular oriented framed knot. We extend an invariant v of nonsingular framed knots to the singular ones by the following recursive setting.

$$v(\!\times\!) = v(\!\times\!) - v(\!\times\!)$$
$$v(\!=\!) = v(\!=\!) - v(\!=\!)$$

Here and below we assume that all framed curves which enter one and the same equality coincide modulo the curve and framing fragments shown. The lower line uses the bifurcation in the normal form of Example 2.5.

Application of the recursive definition to certain special degenerations of framed knots implies the following result.

PROPOSITION 2.8. *The values of an invariant on singular framed knots are subject to the 4-term, 3-term and commutativity relations:*

$$v(\!\times\!) - v(\!\times\!) + v(\!\times\!) - v(\!\times\!) = 0$$

$$v(\!\circ\!\!\!<\!) = v(\!|\!) + v(\!|\!)$$

$$v(\!\times\!) = v(\!\times\!)$$

PROOF. The 4-term relation here is in fact the one that is induced from the Vassiliev theory of invariants of unframed knots by omitting the framing. It follows from the bifurcations of a generic triple point of the core. To prove the 4-term relation one can follow [8] to resolve the double points by the definition (see Figure 5).

The two other relations are easily read from the bifurcations of Examples 2.6 and 2.7 in a similar way (see Figure 6). □

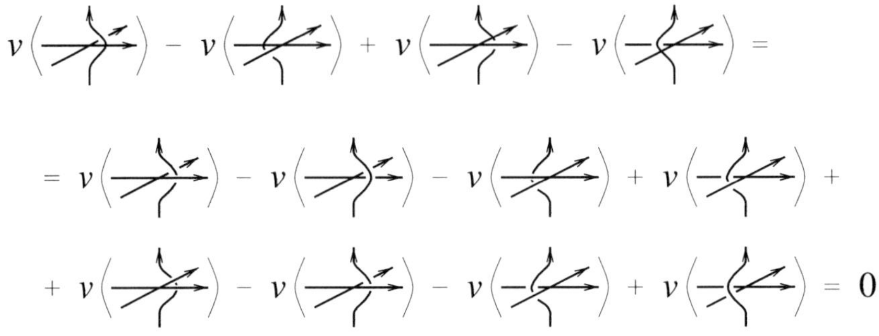

$$v\left(\overset{\uparrow}{\underset{\downarrow}{\rightarrow\!\!\!\!\!\nearrow}}\right) - v\left(\underset{\downarrow}{\nearrow\!\!\!\!\!\nwarrow}\right) + v\left(\overset{\uparrow}{\rightarrow\!\!\!\!\!\nearrow}\right) - v\left(-\!\!\!\!\nearrow\!\!\!\!\!\nwarrow\right) =$$

$$= v\left(\overset{\uparrow}{\underset{\downarrow}{\rightarrow\!\!\!\!\!\nearrow}}\right) - v\left(\overset{\uparrow}{\underset{\downarrow}{\rightarrow\!\!\!\!\!\nearrow}}\right) - v\left(\underset{\downarrow}{\nearrow\!\!\!\!\!\nwarrow}\right) + v\left(-\!\!\!\!\nearrow\!\!\!\!\!\nwarrow\right) +$$

$$+ v\left(\overset{\uparrow}{\underset{\downarrow}{\rightarrow\!\!\!\!\!\nearrow}}\right) - v\left(\overset{\uparrow}{\underset{\downarrow}{\rightarrow\!\!\!\!\!\nearrow}}\right) - v\left(-\!\!\!\!\nearrow\!\!\!\!\!\nwarrow\right) + v\left(-\!\!\!\!\nearrow\!\!\!\!\!\nwarrow\right) = 0$$

FIGURE 5. Proof of the 4-term relation

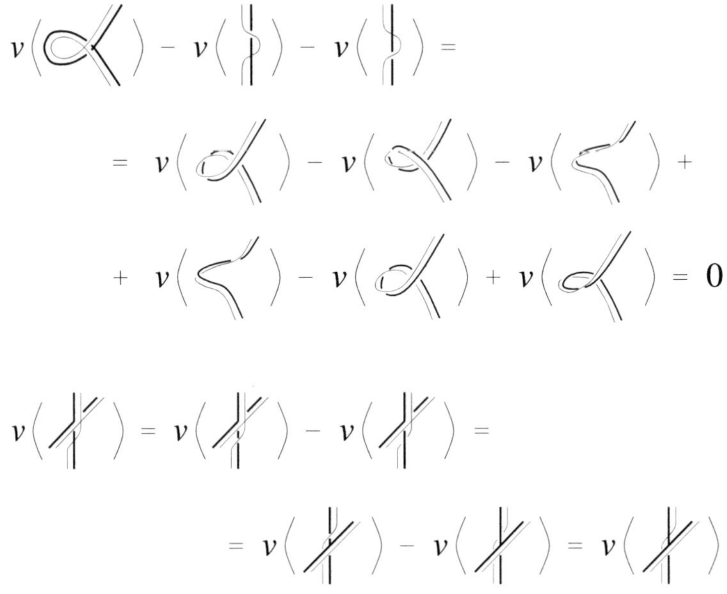

FIGURE 6. Proof of the 3-term and commutativity relations

2.5. Chord diagrams with distinguished points and invariants of finite order. Consider $2n + k$ distinct points on an oriented circle. Join $2n$ of them in n nonordered pairs. Consider such objects up to diffeomorphisms of the circle preserving its orientation. Each equivalence class will be called an *n-chord diagram with k distinguished points* or, shortly, an *(n, k)-diagram*.

We associate an (n, k)-diagram to a singular framed oriented knot $g : U \to \mathbb{R}^3$:
  i) the circle is the source $S^1 \subset U \subset \mathbb{R}^2$ (we take it to be a standard counterclockwise oriented circle on a plane and never mention this orientation in our figures);
  ii) a pair of points is the inverse image of a double point;
  iii) at a distinguished point the rank of the differential Tg is 1.

DEFINITION 2.9. An invariant of framed oriented knots in $\mathbb{R}^3$ is a (*Vassiliev*) *invariant of order less than* m if it vanishes on all oriented framed knots with m singularities.

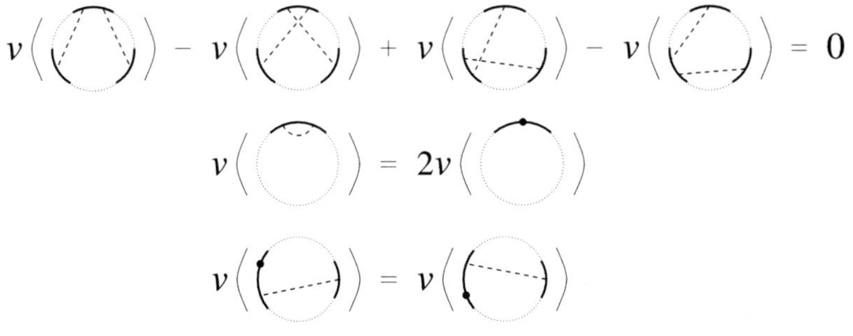

$$v\left(\vcenter{\hbox{⊗}}\right) - v\left(\vcenter{\hbox{⊗}}\right) + v\left(\vcenter{\hbox{⊗}}\right) - v\left(\vcenter{\hbox{⊗}}\right) = 0$$

$$v\left(\vcenter{\hbox{○}}\right) = 2v\left(\vcenter{\hbox{○}}\right)$$

$$v\left(\vcenter{\hbox{○}}\right) = v\left(\vcenter{\hbox{○}}\right)$$

FIGURE 7. 4-term, 2-term and floating-point relations for symbols

We denote the linear space of all invariants of order less than m by V_{m-1}^f.

Take an invariant of order m, that is, an element $v \in V_m^f \setminus V_{m-1}^f$. Its restriction to the set of all oriented framed knots with m singularities is called the *symbol* of v.

PROPOSITION 2.10. *The symbol of an order m invariant is a well-defined function on the set of all (n, k)-diagrams, $m = n + k$.*

PROOF. Consider the set $\Omega(D) \subset \Omega_f$ of all parametrizations $g : U \to \mathbb{R}^3$ of singular framed knots with the same (n, k)-diagram D. We need to show that any two elements $g_1, g_2 \in \Omega(D)$ can be deformed one into another without changing the value of the symbol.

On the first step we deform g_1 into $g_2' \in \Omega(D)$ that has the same core as g_2. This can be done by a homotopy that stays almost all the time in $\Omega(D)$ and, at the remaining finitely many instances, passes transversally through the set of parametrizations with $n + 1$ double points on the core.

Up to homotopies in $\Omega(D)$, we can assume that g_2' and g_2 have the same k points of generic framing degeneration and coincide on small neighbourhoods of these distinguished points. So, the two parametrizations differ only by the rotation of the framings along the intervals between distinguished points. Now we deform the framing of g_2' to that of g_2 by a homotopy that stays almost all the time in $\Omega(D)$ and, at the remaining finitely many instances, passes transversally through the set of parametrizations with $k + 1$ points of framing degeneration.

During the constructed homotopy the value of the symbol could change only on the two sets of the above mentioned finitely many instances. The increments are the values of the invariant on framed knots with $n + k + 1 = m + 1$ singularities, which are zeros. $\square$

The relations of Proposition 2.8 immediately imply

PROPOSITION 2.11. *The values of a symbol are subject to the 4-term, 2-term and floating-point relations of Figure 7.*

In Figure 7 and other figures we show all distinguished points and chords based on the solid arcs and none of those based on the dotted arcs. All the diagrams entering the same relation are assumed to differ only by their parts based on the solid arcs.

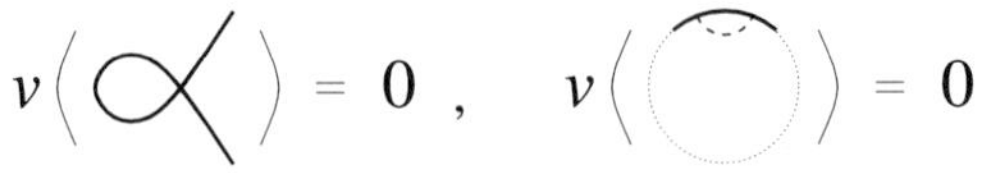

$$v\left(\begin{array}{c}\includegraphics\end{array}\right) = 0 \,, \quad v\left(\begin{array}{c}\includegraphics\end{array}\right) = 0$$

FIGURE 8. The 1-term relations for invariants of unframed knots and their symbols

REMARK 2.12. In the Vassiliev theory of unframed knots, invariants and their symbols are subject to the 1-term relations ([**17, 3, 7, 12, 6**]); see Figure 8. These relations follow from Figure 3 with all the framings omitted. Propositions 2.8 and 2.11 show what happens with the 1-term relations when we pass to the framed setting. For example, for symbols of $\mathbb{Z}_2$-valued invariants the 1-term relation still holds in the framed case.

REMARK 2.13. Twice the floating-point relation follows from the two others of Proposition 2.11. Indeed, consider the 4-term relation in the case when all its 7-through-1-o'clock arcs are solid. The two middle terms cancel one another. Expressing each of the two remaining terms by means of the 2-term relation we obtain what has been promised.

Let $\dot{\mathcal{A}}_m$ be the linear space spanned by all (n,k)-diagrams, with $m = n + k$, modulo the three relations of Proposition 2.11 considered now as relations on the diagrams rather than functions on them. Proposition 2.11 embeds the space V_m^f/V_{m-1}^f of symbols of order m invariants into the space $\dot{\mathcal{A}}_m^*$ dual to $\dot{\mathcal{A}}_m$. Set

$$\dot{\mathcal{A}} = \bigoplus_{m \geq 0} \dot{\mathcal{A}}_m.$$

Similar to [**12, 6**], the 4-term and floating-point relations imply that the connected sum of two chord diagrams with distinguished points is a well-defined element in $\dot{\mathcal{A}}$; that is, the sum does not depend on the location of the connecting surgery.

PROPOSITION 2.14. $\dot{\mathcal{A}}$ *is an algebra with respect to the connected summation of diagrams.*

If the ground field is not of *char* $= 2$, the graded algebra $\dot{\mathcal{A}}$ is isomorphic to the graded algebra $\mathcal{A} = \bigoplus_{m \geq 0} \mathcal{A}_m$ whose mth direct summand is spanned by all m-chord diagrams (with no distinguished points at all) modulo the 4-term relation only. The operation on $\mathcal{A}$ is the connected summation as well.

As it was mentioned in [**12, 6**], the 4-term relation on chord diagrams with no distinguished points turns out to be the only relation for real-valued invariants of framed knots in $\mathbb{R}^3$. Namely, indicating the real versions of the spaces by the subscript $\mathbb{R}$, we have

THEOREM 2.15. $V_{m,\mathbb{R}}^f/V_{m-1,\mathbb{R}}^f = \mathcal{A}_{m,\mathbb{R}}^*$.

This fact was proved in [**13, 14**] by introducing a version of the universal Vassiliev–Kontsevich invariant for knots with the blackboard framing. In the next section we define the universal invariant that serves knots with arbitrary framings and reprove Theorem 2.15. Our approach is distinct from that of [**13, 14**].

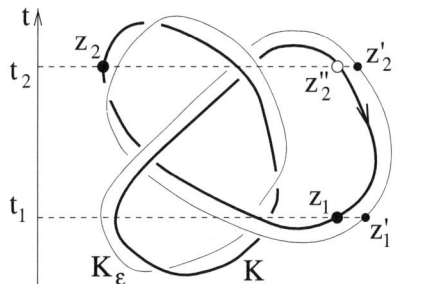
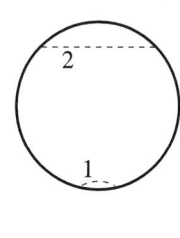

FIGURE 9. A pairing on a framed knot and the chord diagram of the pairing

3. The universal Vassiliev–Kontsevich invariant for framed knots in $\mathbb{R}^3$

3.1. The invariant of Morse knots. We represent the Euclidean 3-space as a direct product $\mathbb{C} \times \mathbb{R}$ with the complex coordinate z and the real coordinate t.

An unframed knot in $\mathbb{C} \times \mathbb{R}$ is called a *Morse knot* if t is a Morse function on it. A *framed knot* in $\mathbb{C} \times \mathbb{R}$ is called *Morse* if its core is a Morse knot.

A. Consider a nonsingular oriented framed Morse knot K^f parametrized by a mapping $g : (U, S^1) \to \mathbb{R}^3$. Let K be its core $g(S^1)$. Fix a decomposition $(U, S^1) = S^1 \times (\mathbb{R}, 0)$ of the annulus. Let y be the coordinate along the second factor. Denote by u the vector field $Tg(\partial_y)$ on K. For small $\varepsilon > 0$, we shift the core K in the direction of u:

$$(z, t) \mapsto (z, t) + \varepsilon u(z, t).$$

For each sufficiently small ε, the result K_ε of the shift is a Morse knot that does not intersect K. We orient K_ε by the orientation inherited from K.

B. In order to have a good definition of a chord diagram later on, we slightly adjust the Morse link $K \cup K_\varepsilon$. Near a local maximum of the function t on K, t has the local maximum on K_ε as well. We take the lowest of the two critical levels and remove the small arc of $K \cup K_\varepsilon$ that is locally above this level. In a similar way, we remove the small arc that is locally below the highest of the two critical levels near a local minimum of t on K. After the surgery at all the local extrema, we remain with the subsets $\widehat{K} \subseteq K$ and $\widehat{K}_\varepsilon \subseteq K_\varepsilon$.

The shift along the framing field u provides the one-to-one correspondence between the sets of intervals of monotonicity of the function t on K and K_ε. For each noncritical point $(z', t) \in \widehat{K}_\varepsilon$ this correspondence correctly defines its unique *neighbour* $(z'', t) \in \widehat{K}$ on the same t-level.

C. Now we take m different noncritical levels $t_{\min} < t_1 < t_2 < \cdots < t_m < t_{\max}$, where $t_{\min}$ and $t_{\max}$ are the global extreme values of t on $\widehat{K} \cup \widehat{K}_\varepsilon$. In each section $t = t_j$ of $\widehat{K} \cup \widehat{K}_\varepsilon$, we choose an *ordered* pair of points $(z_j, z'_j) = (z_j, z'_j)(t_j) \in \widehat{K} \times \widehat{K}_\varepsilon$. Let P be a set of m such pairs, one pair per level. The set P defines the m-chord diagram $D(P)$ as follows (see Figure 9).

In each pair we substitute $z'_j \in \widehat{K}_\varepsilon$ by its neighbour $z''_j \in \widehat{K}$. The core K is the image of the embedding of the oriented circle S^1 that we again take to be a standard counterclockwise oriented circle on the plane. If $z_j \neq z''_j$, we join the preimages of the points z_j and z''_j on the source circle by the chord. If $z_j = z''_j$, we draw a small chord between two arbitrary points on the circle which are very close to the preimage of z_j (so that on the small arc subtended by this chord there are no endpoints of any of the $m-1$ chords corresponding to the other distinguished t-levels).

D. Let $\mathcal{A}_{m,\mathbb{C}}$ be the complex linear space generated by all m-chord diagrams modulo the 4-term relation.

We introduce

DEFINITION 3.1.

$$\widehat{Z}_m(K, K_\varepsilon) = \frac{1}{(2\pi i)^m} \int_{t_{min}<t_1<t_2<\cdots<t_m<t_{max}} \sum_{P=\{(z_j,z_j')(t_j)\}} (-1)^{P_\downarrow} \bigwedge_{j=1}^m \frac{dz_j - dz_j'}{z_j - z_j'} \mathcal{D}(P)$$
$$\in \mathcal{A}_{m,\mathbb{C}},$$

where P runs through all possible pairings on $\widehat{K} \cup \widehat{K}_\varepsilon$, $P_\downarrow$ is the number of points in the m pairs at which the function t is decreasing along the oriented link $K \cup K_\varepsilon$, and $\mathcal{D}(P)$ is the class of the diagram $D(P)$ in $\mathcal{A}_{m,\mathbb{C}}$.

DEFINITION 3.2. $Z_m(K^f) = \lim_{\varepsilon \to 0} \widehat{Z}_m(K, K_\varepsilon)$.

THEOREM 3.3 (see [**12**]). i) *The limit that defines $Z_m(K^f)$ is finite.*
ii) *$Z_m(K^f)$ does not depend on the decomposition $(U, S^1) = S^1 \times (\mathbb{R}, 0)$ of the annulus used in the definition.*
iii) *$Z_m(K^f)$ is invariant under the homotopy in the class of framed Morse knots.*
iv) *$Z_m(K^f)$ is an invariant of order less than $m + 1$.*

Statement iv) concerns the extension (in the sense of subsection 2.4) of the invariant to singular framed Morse knots, none of whose singular points is a local extremum of t.

The proof of the theorem occupies the next two subsections.

3.2. Proof of the convergence. The divergence of the limit could arise from the two *dangerous* types of pairs (chords):

- *Infinitesimal pairs* that correspond to the case $z_j = z_j''$ in the definition of $D(P)$;
- *Short pairs* whose elements $z_j \neq z_j''$ are lying on two successive intervals of monotonicity of the function t on K, so that no chord connects the two semicircles into which z_j and z_j'' cut K.

In both cases, the diagram $D(P)$ is obtained from a diagram with fewer chords by insertion of an isolated chord. Due to the 4-term relation, the corresponding generator $\mathcal{D}(P)$ of $\mathcal{A}_{m,\mathbb{C}}$ does not depend on the location of the insertion. We are going to exploit this independence and group together the terms which correspond to one and the same generator of such type so that their individual divergences kill one another. The grouping is mainly based on the following example.

EXAMPLE 3.4. Consider the family of all possible pairings on $\widehat{K} \cup \widehat{K}_\varepsilon$ which have only two pairs involving points from a neighbourhood of some local maximum of the function t on K (see Figure 10), with the upper pair being dangerous and the lower one not.

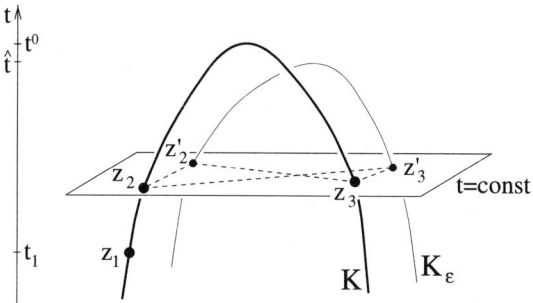

FIGURE 10. Cancellation of divergencies

Let $\hat{t}$ be the local maximum value of t on $\widehat{K} \cup \widehat{K}_\varepsilon$. Let $t_1 < \hat{t}$ be the level of a *long* pair (z_1, z_1') that joins a point inside our neighbourhood with a point outside.

Pairings in the family have only one infinitesimal or short pair in the slice $t_1 < t < \hat{t}$ of the neighbourhood. Integration, within these limits, of the sum of the four 1-forms corresponding to the four dangerous pairs gives the logarithm of the cross-ratio

$$\ln \frac{(z_2 - z_2')(z_3 - z_3')}{(z_2 - z_3')(z_3 - z_2')} \bigg|_{t_1}^{\hat{t}}.$$

The upper bound evaluation gives zero.

The divergence of the lower bound terms $\ln(z_i - z_i')$, $i = 2, 3$, for $\varepsilon \to 0$, is cancelled by the integration along the pairings in which the pair (z_i, z_i') dives under the level $t = t_1$.

In the remaining lower bound term $\ln((z_2 - z_3')(z_3 - z_2'))$, z_i' tends to z_i as $\varepsilon \to 0$, $i = 2, 3$. At the same time, $\hat{t}$ tends to the local maximum value t^0 of the function t on K (unless $\hat{t} = t^0$ from the very beginning). Now the integral

$$\int\limits_{\text{const}}^{t^0} \ln(z_2(t_1) - z_3(t_1)) \, \frac{dz_1(t_1) - dz_1''(t_1)}{z_1(t_1) - z_1''(t_1)}$$

along the levels of the long pair converges at $t_1 = t^0$.

In a general case we follow the above order of integrations and passing to the limit. We integrate first in the prelimit setting along the t-levels of infinitesimal and short chords for all the other fixed levels.

Consider two intervals, $I \subset \widehat{K}$ and $I_\varepsilon \subset \widehat{K}_\varepsilon$, of monotonicity of the function t that correspond to each other by the shift along the framing field u. We say that an *infinitesimal pair* (z_j, z_j') *is based on* $I \cup I_\varepsilon$ if the points z_j and z_j' are lying on these intervals. We say that a *short pair* (z_j, z_j') *is based on* $I \cup I_\varepsilon$ if that of the points z_j and z_j' which lies on $I \cup I_\varepsilon$ cannot be moved continuously along $\widehat{K} \cup \widehat{K}_\varepsilon$ to the neighbouring couple of intervals of monotonicity on which the other member of the pair lives. This applies in the case where the surgery that adjusted $K \cup K_\varepsilon$ to $\widehat{K} \cup \widehat{K}_\varepsilon$ was nontrivial in the neighbourhood of the local extremum of t on K. If, on the contrary, the local surgery was trivial (that is, the framing vector u at the extremum has zero t-component) the base-interval assignment is arbitrary (but should be fixed before we start the integration).

EXAMPLE 3.5. In Figure 10, the pairs (z_2, z_3') and (z_3, z_2') are based on the left and right couples of the intervals of monotonicity, respectively.

Now, assume that in the expression for $\widehat{Z}_m(K, K_\varepsilon)$ we are considering an m-pairing P that has exactly r dangerous pairs based on a certain couple $I \cup I_\varepsilon$ of intervals. Consider this particular pairing as a member of the entire family of all the pairings that have exactly r dangerous pairs based on $I \cup I_\varepsilon$ and whose remaining $m - r$ pairs $\widetilde{P}$ are exactly the same as in P. The chord diagrams of all these m-pairings define the same element in $\mathcal{A}_{m,\mathbb{C}}$.

In this family, the t-levels of the short pairs, which are based on $I \cup I_\varepsilon$ and join the points of this couple of intervals with the points of the other couple of intervals that is adjacent to $I \cup I_\varepsilon$ via the two local maxima of t on $K \cup K_\varepsilon$, are bounded from below by the level $t = t_a$ of the corresponding long pair. There is also the similar bound $t < t_b$ for the levels of the short pairs located by the two local minima of the function t on $K \cup K_\varepsilon$.

A small exercise in elementary calculus and combinatorics shows that integration in $\widehat{Z}_m(K, K_\varepsilon)$ along the described family provides the form

$$\frac{1}{(2\pi i)^m} (-1)^{\widetilde{P}_\downarrow} \bigwedge_{j=1}^{m-r} \frac{dz_j - dz_j'}{z_j - z_j'} \frac{1}{r!} \ln^r \frac{z(t_a) - z'(t_a)}{z(t_b) - z'(t_b)} \mathcal{D}(P),$$

where after the reodering of the pairings of the family we set $\widetilde{P} = \{(z_1, z_1'), \ldots, (z_{m-r}, z_{m-r}')\}$. The evaluations of the differences under the logarithm are done on the two short pairs based on $I \cup I_\varepsilon$ and lying on the corresponding t-levels. The form obtained is to be integrated along the various t-levels of the pairings $\widetilde{P}$.

After similar integration is done for all the couples of intervals, we do not finish the integration in $\widehat{Z}_m(K, K_\varepsilon)$ but immediately pass to the limit for $\varepsilon \to 0$ to get $Z_m(K^f)$. This means the substitution of z_j'' (and z'') for z_j' (and z') everywhere in the above $(m - r)$-form or in the lower degree form that has emerged from it.

So the limiting integral obtained is absolutely convergent since its only singularities (see [12, 6]) are estimated by constant multiples of the integrals like

$$\int_0^{\text{const}} \Big(\int_0^{x_s} \ldots \Big(\int_0^{x_2} \Big(\int_0^{x_1} \ln^r x_0 dx_0 \Big) dx_1 \Big) \ldots dx_{s-1} \Big) dx_s,$$

which are convergent at 0.

3.3. Invariance of the limit under horizontal moves. The prelimit integrals $\widehat{Z}_m(K, K_\varepsilon)$ do not need to be invariant under horizontal perturbations (that is, when each point stays in its t-level) of the link $K \cup K_\varepsilon$. On the contrary, the limit $Z_m(K^f)$ is invariant under horizontal isotopies of the framed link. Let us show this assuming, at first, that none of the critical levels of the function t on K moves.

From the previous subsection we see that it is enough to assume that the framing is such that, out of sufficiently small neighbourhoods of critical points of t on K, the framing field u is lying in the levels $t = const$ and is of length 1. So the distance between a point $z_j' \in \widehat{K}_\varepsilon$ and its neighbour $z_j'' \in \widehat{K}$ is ε:

$$z_j' - z_j'' = \varepsilon e^{i\varphi_j}, \quad \varphi_j \in \mathbb{R}/2\pi\mathbb{Z}.$$

Consider a slice $a < t < b$ such that the closed interval $[a, b]$ contains no critical values of t on K. The part of K in this slice consists of, say, r branches going upward and r branches going downward. Consider a horizontal isotopy of our framed knot which is nontrivial only in $a < t < b$. Absolutely similar to [**12, 6**], the invariance of the elements $Z_m(K^f) \in \mathcal{A}_{m,\mathbb{C}}$, $m \geq 0$, under such isotopy is implied, according to Stokes' formula, by flatness of the following Knizhnik–Zamolodchikov type connection.

This formal connection, which we denote by $\Theta^f_{r,r}$, is defined on the direct product of two spaces. One of them is the set of all ordered $2r$-tuples $(z_1, \ldots, z_{2r})$ of pairwise distinct complex numbers. The other is the $2r$-dimensional torus with the coordinates $\varphi_p \in \mathbb{R}/2\pi\mathbb{Z}$.

We set
$$\Theta^f_{r,r} = \sum_{1 \leq p,q \leq 2r} s_p s_q \Theta_{pq} \omega_{pq},$$
where
s_p is 1 for $p \leq r$ and -1 otherwise;
$\omega_{pq} = \frac{dz_p - dz_q}{z_p - z_q}$ when $p \neq q$;
$\omega_{pp} = d\varphi_p$;
Θ_{pq} and Θ_{pp} are the 1-chord diagrams based on $2r$ ordered parallel arrows, first r of which point upward and the others downward:

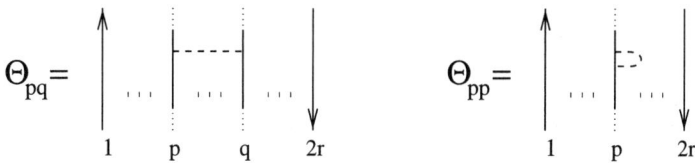

The 1-form $\Theta^f_{r,r}$ is closed.

Now, the product of two chord diagrams based on our $2r$ ordered arrows is, as usual, drawing the first of them below the second. Formal linear combinations of chord diagrams are considered modulo the 4-term relation. For example, the special "diagonal" case of the 4-term relation says that Θ_{pp} commutes with any Θ_{pq} (this is actually twice the floating-point relation). Within this understanding the fact $\Theta^f_{r,r} \wedge \Theta^f_{r,r} = 0$ is obvious (see [**6**]). Thus connection $\Theta^f_{r,r}$ is flat.

The proof of invariance of $Z_m(K^f)$ under all other moves which preserve the class of Morse framed knots is almost a word-for-word repetition of subsection 4.3.3 of [**6**].

Part ii) of Theorem 3.3 is a particular case of part iii).

Finally, as in [**12**], part iv) of Theorem 3.3 is obvious.

3.4. The universal invariant. Similarly to the unframed case ([**12, 6**]), the integrals $Z_m(K^f)$ are not invariant under the move that cancels two neighbouring local extrema of t on K. We fix the problem exactly in the same way as it was done in [**12, 6**].

Set
$$Z(K^f) = \sum_{m \geq 0} Z_m(K^f) \in \overline{\mathcal{A}}_{\mathbb{C}},$$
where $\overline{\mathcal{A}}_{\mathbb{C}} = \prod_{m \geq 0} \mathcal{A}_{m,\mathbb{C}}$. Let $\mathcal{U}^f$ be the curve of Figure 11 lying in the plane $\mathrm{Im}\, z = 0$ and equipped with the trivial framing $i\partial_z$.

The series $Z(\mathcal{U}^f) \in \overline{\mathcal{A}}_{\mathbb{C}}$ is invertible since it starts with $1 \in \mathcal{A}_{0,\mathbb{C}}$. Let c be the number of critical points of the function t on the core of a framed Morse knot K^f.

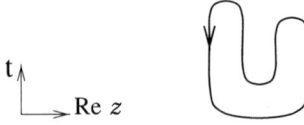

FIGURE 11. The curve $\mathcal{U}$

DEFINITION 3.6. The element
$$\widetilde{Z}(K^f) = Z(K^f) \times Z(\mathcal{U}^f)^{1-\frac{c}{2}} \in \overline{\mathcal{A}_\mathbb{C}}$$
is called the *universal Vassiliev–Kontsevich invariant* of the framed Morse knot K^f.

EXAMPLE 3.7. Let $\omega \in \mathcal{A}_{1,\mathbb{C}}$ be the 1-chord diagram. Consider an unknot with the framing that makes one positive rotation around it. The value of $\widetilde{Z}$ on this unknot is $\exp(\omega)$.

THEOREM 3.8 (see [**12**, **6**]). *For any framed Morse knot K^f, $\widetilde{Z}(K^f)$ depends only on the topological type of K^f. The degree m component $\widetilde{Z}_m(K^f) \in \mathcal{A}_{m,\mathbb{C}}$ of $\widetilde{Z}(K^f)$ is an invariant of order less than $m + 1$.*

Since the proof completely repeats that in the unframed case ([**12**, **6**]), we omit it here.

The lowest order term of $\widetilde{Z}(K^f)$ for a singular framed knot with n double points and k points of degeneration of the framing is easily seen to be $2^n \mathcal{D}(K)$, where $\mathcal{D}(K) \in \mathcal{A}_{n+k,\mathbb{C}}$ is the chord diagram of K^f (here we use the 2-term relation to treat an (n,k)-diagram as the $(n+k)$-chord diagram). As in [**12**, **6**], this fact implies the claim $V_{m,\mathbb{R}}^f / V_{m-1,\mathbb{R}}^f = \mathcal{A}_{m,\mathbb{R}}^*$ of Theorem 2.15 on the description of the space of symbols of order m real-valued invariants of framed knots in $\mathbb{R}^3$.

REMARK 3.9. Similar to Exercise 4.5 of [**6**], it is easy to see that the series $\widetilde{Z}(K^f)$ is real.

4. Unframed knots in a solid torus

4.1. Marked chord diagrams. Starting with the space Ω_{ST} of C^∞-mappings of an oriented circle to a solid torus (ST), we construct the theory of Vassiliev type invariants of oriented unframed knots in ST in the obvious way. We get notions of nonsingular and singular knots, extended invariants, chord diagrams, etc. The main new feature here is that the chord diagram of a singular knot possesses a natural integer marking (see [**11**]).

Namely, let us fix a generator of the fundamental group $\pi_1(ST) = \mathbb{Z}$. A double point of an oriented singular knot in ST cuts the knot into two subloops each of which has its class in $\pi_1(ST)$. We write the corresponding fundamental integer on the side of the corresponding chord in the diagram that faces the preimage of the corresponding subloop (see Figure 12). For convenience we also mark the circle of the diagram with the fundamental class of the whole knot. The sum of the two markings on each chord is equal to the marking of the circle.

Let $\Delta_n \subset \Omega_{ST}$ be the set of parametrizations of knots with exactly n generic double points. We say that two elements of Δ_n are *related* (see [**17**]) if they can be joined by a C^∞-homotopy that stays almost all the time in Δ_n and, at the

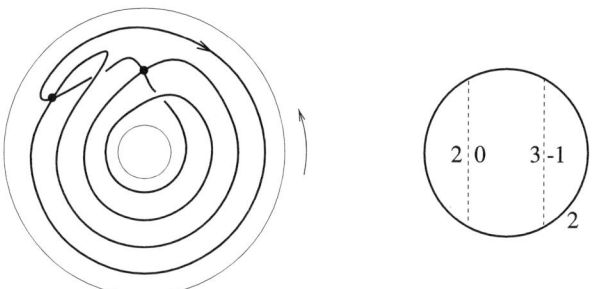

FIGURE 12. The marked chord diagram of a singular knot in a solid torus

remaining finitely many instants, crosses Δ_{n+1} transversally. We also say that *two singular knots are related* if their representatives are related.

We have the following evident result.

PROPOSITION 4.1. *Two singular knots in a solid torus are related if and only if their marked chord diagrams coincide.*

Recall that we consider chord diagrams up to diffeomorphisms of the circle which preserve the orientation. In the marked case diffeomorphisms should preserve the markings as well.

4.2. Marked relations. Of course, the values of invariants on oriented singular knots in a solid torus are subject to the 4-term relation of Proposition 2.8 and the framing-independence 1-term relation of Figure 8. On the other hand, due to Proposition 4.1, the values of a symbol of order n invariant on singular knots in ST are in fact functions on marked n-chord diagrams. This implies

PROPOSITION 4.2. *The values of a symbol v on marked chord diagrams are subject to the marked 1- and 4-term relations:*

$$v\left(\begin{smallmatrix}0\\w\end{smallmatrix}\right) = 0$$

$$v\left(\begin{smallmatrix}j&i\\&w\end{smallmatrix}\right) - v\left(\begin{smallmatrix}j&i\\&w\end{smallmatrix}\right) + v\left(\begin{smallmatrix}j&\\i+j\\w\end{smallmatrix}\right) - v\left(\begin{smallmatrix}j&\\i+j\\w\end{smallmatrix}\right) = 0$$

Here we give only partial markings that allow us to restore the complete ones.

PROOF. Indeed, the marked 1-term relation corresponds to the contraction of a small subloop of a singular knot which is unlinked with the remaining part of the knot (see Figure 3 and omit all the framings there). The fundamental class of a contractible subloop is 0.

The marked 4-term relation comes from the bifurcation of a triple point. So the partial markings are inherited from that of the marked chord diagram of a triple point:

□

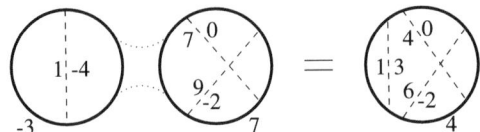

FIGURE 13. Connected sum of marked chord diagrams

4.3. Module of marked chord diagrams. An *abstract marked n-chord diagram* is an n-chord diagram with integer two-side marking of its chords and integer marking of the circle such that the sum of the two markings of each chord is the marking of the circle.

Let $\mathcal{M}^0$ be the linear space of finite linear combinations (with some fixed coefficients) of all abstract marked chord diagrams modulo the marked 1- and 4-term relations (that is, once again, the relations of Proposition 4.2 for diagrams themselves rather than for functions on them). We denote by $\mathcal{M}_n^0$ the order n part of $\mathcal{M}^0$ generated by n-chord diagrams. Unlike the nonmarked case, $\mathcal{M}_n^0$ is infinite-dimensional for any $n \geq 0$.

In the nonmarked case the linear space $\mathcal{A}^0$ generated by all chord diagrams modulo the 1- and 4-term relations is an algebra with respect to the connected sum operation [**12, 6**] ($\mathcal{A}^0$ is the quotient-algebra of the algebra $\mathcal{A} = \oplus_{m \geq 0} \mathcal{A}_m$ considered above). It is pretty obvious that $\mathcal{M}^0$ does not have similar algebra structure: in general, the markings of chords of the connected sum of two marked chord diagrams depend on the arcs on which the connecting surgery is done. The marking on the side of a chord that faces the surgery increases by the marking of the circle of the added diagram (see Figure 13) as this should be for the connected sum of singular knots in a solid torus. Nevertheless the connected sum is a well-defined operation on $\mathcal{M}^0$ in the following special case.

Consider the embedding $\zeta : \mathcal{A}^0 \to \mathcal{M}^0$ that assigns identically zero marking to a nonmarked diagram.

THEOREM 4.3. *The connected summation defines on $\mathcal{M}^0$ the structure of a module over $\zeta(\mathcal{A}^0)$.*

The proof, based on the marked 4-term relation, repeats the proof of the fact that $\mathcal{A}^0$ is an algebra with respect to the same operation (Lemma 2.1 of [**12**]). One needs only to be slightly attentive to the markings.

4.4. The universal invariant for unframed knots in a solid torus. We consider a solid torus (ST) as a direct product $\mathbb{C} \times S^1$ with the complex coordinate z and circular coordinate θ mod 2π. We take a generator of $\pi_1(\mathrm{ST})$ being a loop that runs once around the torus in the direction of increase of θ.

A knot in ST is a *Morse knot* if θ is a Morse function on it.

Let K be an oriented nonsingular Morse knot in ST.

Take n different noncritical levels $0 < \theta_1 < \theta_2 < \cdots < \theta_n < 2\pi$. In each section $\theta = \theta_j$ of K, choose an *unordered* pair of points $(z_j, z_j') = (z_j, z_j')(\theta_j)$. The set P of n such pairs, one pair per level, defines the n-chord diagram in the obvious way. The diagram is marked. Its circle is marked by the class of K in $\pi_1(\mathrm{ST})$. The marking on a chord is given by the fundamental classes of the two loops obtained by a homotopy of K in ST that glues together the points of the corresponding pair and is the identity outside a small neighbourhood of the θ-level of the pair.

We denote by $\mathcal{D}(P) \in \mathcal{M}^0_{n,\mathbb{C}}$ the class of the obtained marked diagram. Using the obvious notations we introduce the following definition.

DEFINITION 4.4.
$$Z_n^{ST}(K) = \frac{1}{(2\pi i)^n} \int_{0<\theta_1<\theta_2<\cdots<\theta_n<2\pi} \sum_{P=\{(z_j,z'_j)(\theta_j)\}} (-1)^{P_\downarrow} \bigwedge_{j=1}^n \frac{dz_j - dz'_j}{z_j - z'_j} \mathcal{D}(P)$$
$$\in \mathcal{M}^0_{n,\mathbb{C}}.$$

In exactly the same way as in [12, 6], adding only the locality of the marked 4- and 1-term relations to trace the markings in homotopies, one gets

THEOREM 4.5. i) The integral that defines $Z_n^{ST}(K)$ is absolutely convergent.
ii) $Z_n^{ST}(K)$ does not depend on the choice of the zero-level of the circular coordinate θ on the solid torus.
iii) $Z_n^{ST}(K)$ is invariant under the homotopy in the class of Morse knots in the solid torus.
iv) $Z_n^{ST}(K)$ is an invariant of Morse knots of order less than $n+1$.

We set $Z^{ST}(K) = \sum_{n\geq 0} Z_n^{ST}(K) \in \overline{\mathcal{M}^0_\mathbb{C}}$, where $\overline{\mathcal{M}^0_\mathbb{C}} = \prod_{n\geq 0} \mathcal{M}^0_{n,\mathbb{C}}$.

Consider the curve $\mathcal{U}$ of subsection 3.4, this time *unframed* and lying in a sector of our solid torus (the coordinate t is replaced by θ). The series $Z^{ST}(\mathcal{U})$ belongs to the subspace $\zeta(\overline{\mathcal{A}^0_\mathbb{C}}) \subset \overline{\mathcal{M}^0_\mathbb{C}}$ spanned by chord diagrams with identically zero markings. As we have already mentioned, $\overline{\mathcal{M}^0_\mathbb{C}}$ is a module over the algebra $\zeta(\overline{\mathcal{A}^0})$. The element $Z^{ST}(\mathcal{U})$ is invertible in $\zeta(\overline{\mathcal{A}^0})$.

DEFINITION 4.6. Let c be the number of critical points of the function θ on an oriented nonsingular Morse knot K in the solid torus. The element
$$\widetilde{Z}^{ST}(K) = Z^{ST}(K) \times (Z^{ST}(\mathcal{U}))^{1-\frac{c}{2}} \in \overline{\mathcal{M}^0_\mathbb{C}}$$
is called the *universal Vassiliev–Kontsevich invariant* of K.

THEOREM 4.7. *The element $\widetilde{Z}^{ST}(K)$ depends only on the topological type of the Morse knot K in the solid torus. The degree n component $\widetilde{Z}_n^{ST}(K) \in \mathcal{M}^0_{n,\mathbb{C}}$ of $\widetilde{Z}^{ST}(K)$ is an invariant of order less than $n+1$.*

The proof of the statement completely repeats that for knots in $\mathbb{R}^3$ given in [12, 6].

For any abstract marked chord diagram D one can find a singular knot in ST whose marked diagram is exactly D. Calculation based on the definition of the Vassiliev type extension of an invariant of nonsingular knots shows that the lowest degree term of the series $\widetilde{Z}^{ST}$ for such a knot is exactly D (see subsection 4.4.2 of [6]). This provides

THEOREM 4.8. *The space of symbols of complex-valued order n Vassiliev invariants of oriented unframed knots in a solid torus coincides with the space of all complex-valued functions on the set of all marked n-chord diagrams subject to the marked 1- and 4-term relations.*

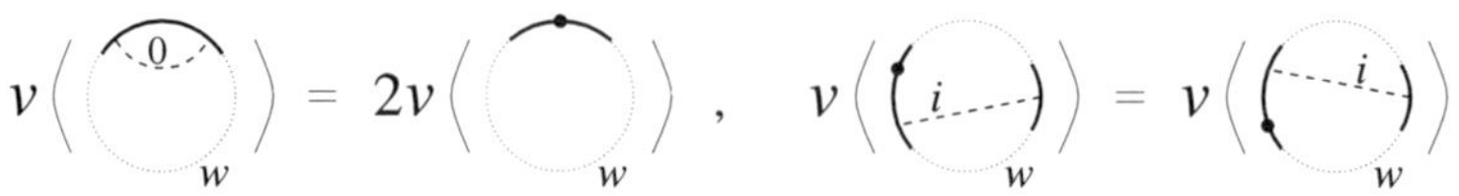

$$v\left(\begin{array}{c}\overset{0}{\frown}\\ w\end{array}\right) = 2v\left(\begin{array}{c}\overset{\bullet}{\frown}\\ w\end{array}\right), \quad v\left(\begin{array}{c}(\,i\,\cdots)\\ w\end{array}\right) = v\left(\begin{array}{c}(\cdots i\,)\\ w\end{array}\right)$$

FIGURE 14. The marked 2-term and floating-point relations

5. Framed knots in a solid torus

5.1. Finite type invariants in terms of marked chord diagrams. This case is the obvious symbiosis of the two cases above of framed knots in $\mathbb{R}^3$ and unframed knots in a solid torus. We give a brief description.

We extend invariants of nonsingular oriented framed knots in ST to those of singular knots following the two recursive settings of subsection 2.4.

Marking the mixture of Definitions 3.1 and 4.4, for a nonsingular oriented Morse framed knot K^f in ST we obtain the elements $\widehat{Z}_m^{ST}(K, K_\varepsilon) \in \mathcal{M}_{m,\mathbb{C}}$. The $\mathcal{M}_{m,\mathbb{C}}$ are the degree m components of the $\mathbb{C}$-linear space $\mathcal{M}_\mathbb{C}$ generated by all marked chord diagrams modulo the marked 4-term relation. Passing to the limit for $\varepsilon \to 0$, we define the elements $Z_m^{ST}(K^f)$. Similar to Theorems 3.3 and 4.5, we have

THEOREM 5.1. i) *The limit element $Z_m^{ST}(K^f)$ is a finite element of $\mathcal{M}_{m,\mathbb{C}}$.*
ii) *$Z_m^{ST}(K^f)$ is invariant under homotopies in the class of framed Morse knots in the solid torus.*
iii) *$Z_m^{ST}(K^f)$ is an invariant of framed Morse knots of order less than $m+1$.*

Marking the mixture of Definitions 3.6 and 4.6, we define the element $\widetilde{Z}^{ST}(K^f) \in \overline{\mathcal{M}_\mathbb{C}}$. This time we take the curve $\mathcal{U}^f$ of Figure 11 lying in a sector of the annulus $\mathrm{Im}\, z = 0$ in the solid torus and equipped with the framing $i\partial_z$. We also use the fact that $\overline{\mathcal{M}_\mathbb{C}}$ is a module over its subspace generated by chord diagrams with all the markings zero.

We have (see Theorems 3.8 and 4.7)

THEOREM 5.2. *The element $\widetilde{Z}^{ST}(K^f) \in \overline{\mathcal{M}_\mathbb{C}}$ depends only on the topological type of the Morse framed knot K^f in the solid torus.*

The final classification result now is

THEOREM 5.3. *The space of symbols of complex-valued order m Vassiliev invariants of oriented framed knots in a solid torus coincides with the space of all complex-valued functions on the set of all marked m-chord diagrams subject to the marked 4-term relation.*

REMARK 5.4. The space of symbols of arbitrarily valued order m Vassiliev invariants of oriented framed knots in ST embeds into the space of functions (with the same values as invariants) on the set of all marked (n,k)-diagrams. Symbols satisfy the marked 4-term relation of Proposition 4.2 as well as the marked 2-term and floating-point relations of Figure 14.

5.2. Coefficients of the polynomial invariants as invariants of finite order. As in the case of knots in 3-space ([7]), coefficients of the polynomial invariants of knots in a solid torus, when properly understood, turn out to be invariants of finite order. To illustrate this we consider in detail the framed version of the HOMFLY polynomial.

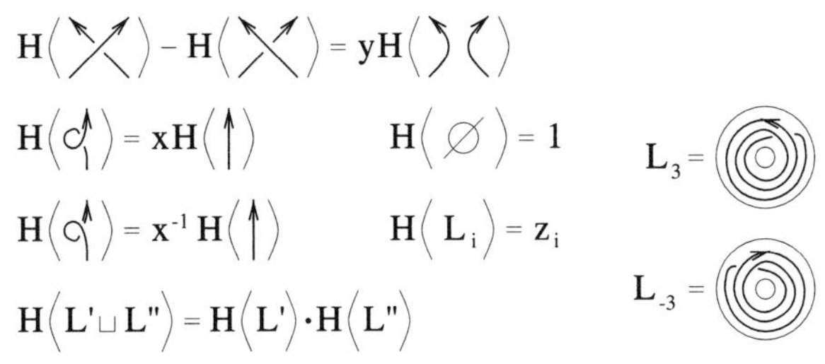

FIGURE 15. Definition of the framed version of the HOMFLY polynomial for links with the blackboard framing in a solid torus

In the definition of the HOMFLY polynomial of framed knots and links in ST, we follow the definiton of [**16**] given for the unframed setting. This time we consider the solid torus $I \times I \times S^1$, where I is the interval $[0, 1]$. The polynomial is defined on representatives of framed knots and links whose framing is blackboard with respect to the projection of the solid torus that forgets the first factor I.

DEFINITION 5.5. For a nonsingular link $L \subset I^2 \times S^1$, the polynomial $H(L) \in \mathbb{Z}[x, x^{-1}, y, y^{-1}, z_1, z_{-1}, z_2, z_{-2}, \ldots]$ is defined by the recursive and initial data of Figure 15.

In the last relation of Figure 15 the links L' and L'' are lying in the solid tori $[0, 1/2) \times I \times S^1$ and $(1/2, 1] \times I \times S^1$, respectively. The curves $L_{\pm 3}$ show the pattern for the whole of the basic series $\{L_j\}_{j=\pm 1, \pm 2, \ldots}$.

The results of [**16**] imply

THEOREM 5.6. *Function H is a well-defined function on the isotopy classes of framed links in ST.*

EXAMPLE 5.7. The value of H on an unknot with the blackboard framing is $(x - x^{-1})/y$. Participation in this fraction is the only way for y^{-1} to enter the polynomial of any link.

Fix an integer $n \neq -1$ and set
$$x = e^{(n+1)t/2}, \qquad y = e^{t/2} - e^{-t/2}$$
in $H(L)$. Since y^{-1} enters $H(L)$ only in the combination $(x - x^{-1})y^{-1}$, the result $W_n(L)$ of the substitution is an element of $\mathbf{Q}[z_{\pm 1}, z_{\pm 2}, \ldots]\{t\}$. Consider the expansion in powers of t
$$W_n(L) = \sum_{m=0}^{\infty} w_{n,m}(L) t^m,$$
where the $w_{n,m}(L)$ are polynomials in the variables z_i.

THEOREM 5.8. *For a knot K each polynomial $w_{n,m}(K)$ is a framed knot invariant of order not greater than m.*

PROOF. (see [**7, 12**]). Extend the function H to the set of singular framed knots in ST via the two recursive relations of subsection 2.4. Consider the value of H on a framed knot K_s with a double points and b points of framing degeneration. We can assume that one of the 2^{a+b} nonsingular resolutions of K_s used for the

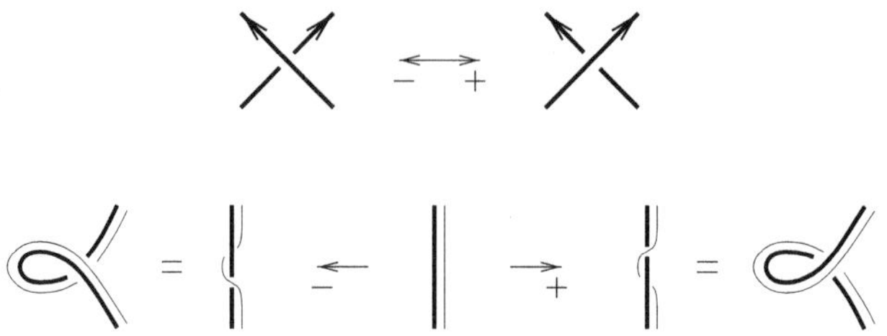

FIGURE 16. Resolutions of local singularities of a singular framed knot in the blackboard setting

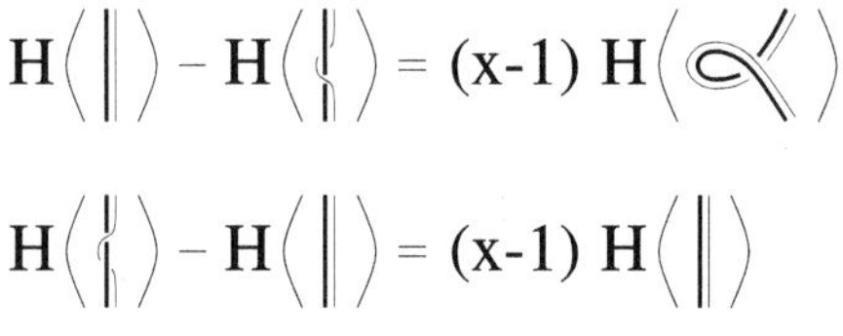

$$H\left\langle \|\right\rangle - H\left\langle \stackrel{\downarrow}{\uparrow}\right\rangle = (x-1)\, H\left\langle \text{\reflectbox{$\mathcal{Q}$}}\right\rangle$$

$$H\left\langle \stackrel{\downarrow}{\uparrow}\right\rangle - H\left\langle \|\right\rangle = (x-1)\, H\left\langle \|\right\rangle$$

FIGURE 17. Calculation of the polynomial of a knot with a degenerate framing

calculation of $H(K_s)$ is a knot with the blackboard framing and all the others are obtained from that one by the local moves of Figure 16.

The differences of the values of H on the resolutions that define the value $H(K_s)$ are given by the first line of Figure 15 and by Figure 17.

Thus the value $H(K_s)$ is $y^a(x-1)^b$ times a polynomial in x, x^{-1}, y, $(x - x^{-1})/y, z_{\pm 1}, z_{\pm 2}, \ldots$. Since both y and $x - 1$ vanish at $t = 0$, the series $W_n(K_s)$ is divisible by t^{a+b}. □

EXAMPLE 5.9. Assume that the class of the knot K in $\pi_1(ST)$ coincides with that of the basic loop L_i. Then the t-free term $w_{n,0}(K)$ is $z_i = H(L_i)$. For a contractible K the t-free term is $n + 1$.

The statements analogous to Theorem 5.8 hold for the Kauffman polynomial and the unframed version of the HOMFLY polynomial ([**16**]).

References

[1] V. I. Arnold, *Plane curves, their invariants, perestroikas and classifications*, Singularities and curves, Adv. in Soviet. Math., vol. 21, Amer. Math. Soc., Providence, RI, 1994, pp. 33–91.

[2] _____, *Topological invariants of plane curves and caustics*, Univ. Lecture Series, vol. 5, Amer. Math. Soc., Providence, RI, 1994.

[3] _____, *Vassiliev's theory of discriminants and knots*, First European Congress of Mathematicians, Paris, July 1992, vol. 1, Birkhäuser, Basel, 1993, pp. 3–28.

[4] V. I. Arnold, V. V. Goryunov, O. V. Lyashko, and V. A. Vassiliev, *Singularities II*, Encyclopaedia Math. Sciences, vol. 39, Springer-Verlag, Berlin, 1993.

[5] V. I. Arnold, S. M. Gusein-Zade, and A. V. Varchenko, *Singularities of differentiable mappings*, vol. I, Birkhäuser, Basel, 1985.

[6] D. Bar-Natan, *On the Vassiliev knot invariants*, Topology **34** (1995), 423–472.

[7] J. Birman and X.-S. Lin, *Knot polynomials and Vassiliev type invariants*, Invent. Math. **111** (1993), 225–270.

[8] S. V. Chmutov and S. V. Duzhin, *An upper bound for the number of Vassiliev knot invariants*, J. Knot Theory and Ramifications, **3** (1994), 141–151.

[9] V. V. Goryunov, *Finite order invariants of framed knots in a solid torus and in Arnold's J^+-theory of plane curves*, (1997) Geometry and Physics (J. E. Andersen, J. Dupont, H. Pedersen, and A. Swann, eds.), Lecture Notes Pure Appl. Math., vol. 184, Marcel Dekker, New York–Basel–Hong Kong, 1997, pp. 549–556.

[10] _____, *Vassiliev type invariants in Arnold's J^+-theory of plane curves without direct self-tangencies*, Topology **37** (1998), 603–620.

[11] E. Kalfagiani, *Finite type invariants for knots in three-manifolds*, Columbia University, New York, 1993, preprint.

[12] M. Kontsevich, *Vassiliev's knot invariants*, Adv. in Soviet Math., vol. 16, Part 2, Amer. Math. Soc., Providence, RI, 1993, pp. 137–150.

[13] T.Q.T. Lê and J. Murakami, *Representation of the category of tangles by Kontsevich's iterated integral*, Comm. Math. Phys. **168** (1995), 535–562.

[14] _____, *The universal Vassiliev–Kontsevich invariant for framed oriented links*, Max-Planck-Institut für Mathematik, Bonn, 1994, preprint.

[15] Y. Suetsugu, *Kontsevich invariant for links in a donut and links of satellite form*, Osaka J. Math. **33** (1996), 823–828.

[16] V. G. Turaev, *The Conway and Kauffman modules of the solid torus with an appendix on the operator invariants of a tangle*, LOMI preprint E-6-88, Leningrad, 1988.

[17] V. A. Vassiliev, *Cohomology of knot spaces*, Theory of Singularities and its Applications (V. I. Arnold, ed.), Adv. in Soviet. Math., vol. 1, Amer. Math. Soc., Providence, RI, 1990, pp. 23–69.

DEPARTMENT OF PURE MATHEMATICS, THE UNIVERSITY OF LIVERPOOL, LIVERPOOL L69 3BX, UK

E-mail address: goryunov@liv.ac.uk

Finite Type Invariants of Generic Immersions of M^n into $\mathbb{R}^{2n}$ are Trivial

Tadeusz Januszkiewicz and Jacek Świątkowski

The approach to knot invariants initiated by Vassiliev resulted in a great success. The recent activity related to invariants of plane curves also produced several valuable insights. In both of these developments the general abstract method of Vassiliev spectral sequence was reformulated into a very simple combinatorial and readily computable (though CPU time consuming) axiomatic version. This axiomatic approach was taken for a definition of Vassiliev or, more precisely, "finite type" invariants. In view of the simplicity of this approach, it is tempting to use the same approach in other situations and, in particular, to examine the limitations of the method. The first obvious candidate is the study of embedded surfaces in $\mathbb{R}^4$. However, it turned out that the arguments used in this case work (or even become simpler) for maps of n-dimensional manifolds into $\mathbb{R}^{2n}$.

To apply Vassiliev's approach, we have to consider the full space $C^\infty(M^n, \mathbb{R}^{2n})$ of smooth mappings. The regular objects are the stable maps $f : M^n \to \mathbb{R}^{2n}$. These are well known to be immersions with at most double transversal self-intersections, and to form an open and dense subset in the space $C^\infty(M^n, \mathbb{R}^{2n})$ (see [**G–G**, Chapter III, §3]). Embeddings, which correspond to knotted manifolds, are special cases of stable immersions. This is a slight departure from the viewpoint of classical knot theory, which seems very reasonable for us. (Remark 1.5 below gives a dramatic explanation of the necessity for the above setting.)

We will study invariants of stable immersions up to (a variant of) ambient isotopy, which we call $\mathcal{A}_0$-equivalence. More precisely, we say that two maps $f, f' : M^n \to \mathbb{R}^{2n}$ are $\mathcal{A}_0$-*equivalent* if there exist diffeomorphisms $\varphi \in \mathrm{Diff}_0 M^n$ and $\psi \in \mathrm{Diff}_0 \mathbb{R}^{2n}$ such that $f = \psi \circ f' \circ \varphi$, where Diff_0 is the group of diffeomorphisms isotopic to identity. The set of all maps $g \in C^\infty(M^n, \mathbb{R}^{2n})$ which are not stable is denoted by D and called the *discriminant*. Then two stable maps $f, f' : M^n \to \mathbb{R}^{2n}$ are $\mathcal{A}_0$-equivalent if and only if f and f' are in the same connected component in the complement $C^\infty(M^n, \mathbb{R}^{2n}) \setminus D$ of the discriminant. This follows easily from openness of the complement and from stability of all its maps, by recalling that all maps from a small neighbourhood of a stable map are $\mathcal{A}_0$-equivalent with each other. Thus *any* invariant of $\mathcal{A}_0$-equivalence of stable immersions $M^n \to \mathbb{R}^{2n}$ can be thought of as a locally constant function on the complement $C^\infty(M^n, \mathbb{R}^{2n}) \setminus D$.

1991 *Mathematics Subject Classification.* Primary 57N35.

In this paper we study invariants which are, moreover, of finite type (see Definition 1.7 for details).

Assume that M^n is a closed connected manifold. Let σ be the invariant which counts the number of self-intersection points of stable immersions $M^n \to \mathbb{R}^{2n}$. If the dimension n is even, let ϵ be the Euler number of the normal bundles of stable immersions (or twisted Euler number if M^n is nonorientable). The main results of the paper follow.

THEOREM 1. *Let M^n be a closed connected manifold with dimension n even. Then each finite type invariant of a stable immersion $M^n \to \mathbb{R}2n$ is a polynomial in invariants σ and ϵ.*

THEOREM 2. *Let M^n be a closed connected manifold with dimension n odd. Then each finite type invariant of a stable immersion $M^n \to \mathbb{R}^{2n}$ is a polynomial in σ.*

The paper is organized as follows. In Section 1 we start with the description of singularities appearing in generic 1-parameter families of immersions. We distinguish types of such singularities and show that immersions with finitely many typical singularities have well-defined (up to $\mathcal{A}_0$-equivalence) resolutions. This allows us to extend any invariant of $\mathcal{A}_0$-equivalence of stable immersions to immersions with finitely many typical singularities, a fact which is crucial for defining finite type invariants.

In Section 2 we study the subspace X of the discriminant consisting of all immersions with finitely many typical singularities. We show that any two such maps with the same numbers of singularities of corresponding types can be connected by a path in X preserving all those singularities and their types. This result implies that the value of a finite type invariant of order k for an immersion with k typical singularities depends only on the numbers of singularities of corresponding types (among those k singularities). Thus the tuples of nonnegative integers (representing numbers of singularities of corresponding types) in our context play the same role as chord diagrams in the case of knots.

In Section 3 we show an analogue of the 4-term relation for Vassiliev weight systems. We then use relations of this form to estimate the dimension of the space of finite type invariants of given order. Finally we show that this dimension bound is achieved by invariants appearing in the statements of Theorems 1 and 2, respectively, which proves these theorems.

In Sections 4 and 5 we present additional results: we calculate finite type invariants for nonconnected manifolds as well as invariants of the regular homotopy of immersions.

We consider our results as being of negative type. Basically these results say that a method which is a very rich source of invariants for knots in $\mathbb{R}^3$ and plane curves fails in the situations we consider. The invariants obtained are clearly not very interesting and contain little information about immersions. In our opinion this is due not to the failure of the general Vassiliev method but rather to the failure of the axiomatization of the method of finite type invariants, which was adequate for studying maps with one dimensional sources.

True Vassiliev invariants of knotted surfaces in $\mathbb{R}^4$ are yet to be discovered.

1. Typical singularities, resolutions, and invariants of finite type

This section is devoted to setting the scene. We introduce the notion of a (singular) immersion with typical singulaties and show that resolutions of such singular maps are well defined. This allows us to extend any $\mathcal{A}_0$-equivalence invariant of stable immersions to immersions with finitely many typical singularities, and consequently to define invariants of finite type. At the end of the section we give examples of finite type invariants of order 1.

Generic 1-parameter families of maps. Let M^n be an n-dimensional closed manifold and let $I = [0, 1]$ be the unit interval.

DEFINITION 1.1. A one-parameter family $H : M^n \times I \to \mathbb{R}^{2n}$ of smooth maps (which is smooth as a map on the product) is called *typical* if the induced map $\widetilde{H} : M^n \times I \to \mathbb{R}^{2n} \times I$ given by $\widetilde{H}(x,t) = (H(x,t),t)$ satisfies the following conditions:
  (a) $\widetilde{H}$ is stable, that is, its only singularities are cross-cups (having the form $(x, t_1, \ldots, t_n) \mapsto (x^2, xt_1, \ldots, xt_n, t_1, \ldots, t_n)$ in some local coordinates), it is an immersion outside these singularities, and the self-intersections of $\widetilde{H}$ are at most double and transversal (see [**G–G,** Chapter VII, Theorem 4.6] for the description of stable maps in the relevant dimensions);
  (b) the projection $\pi : \mathbb{R}^{2n} \times I \to I$ restricted to the 1-dimensional self-intersection submanifold $Q \subset \mathbb{R}^{2n} \times I$ is a Morse function with distinct critical values, regular at endpoints (corresponding to singularities mentioned in (a)), with the values at all such endpoints and critical values distinct from each other.

It is clear from the definition that typical 1-parameter families of maps are generic among all smooth 1-parameter families.

A point $t \in I$ is called *critical* if it is a critical value of the projection π restricted to the submanifold Q or if it is a value of $\pi \circ \widetilde{H}$ at a singular point of $\widetilde{H}$. A point $t \in I$ which is not critical is called *regular*. The following properties of maps in typical 1-parameter families follow immediately from the definition.

1.2. Properties of maps in typical 1-parameter families. (i) Any map $H(\cdot, t)$ for a regular $t \in I$, including $t = 0$ and $t = 1$, is nonsingular (i.e., stable).

(ii) Any map $H(\cdot, t)$ for a critical $t \in I$ is not stable infinitesimally only at one point. Consequently, it is stable outside any neighbourhood of this point. (See [**G–G**] for the definition of infinitesimal stability and the proof that it is equivalent to stability.)

Singularities and their types. A point $q \in \mathbb{R}^{2n}$ is called a *singularity* of a map $f : M^n \to \mathbb{R}^{2n}$ if for some $p \in M^n$ with $f(p) = q$, the map f is not regular at p, or if q is a self-intersection of f which is not double and transversal. A singularity q of a map f is *typical* if it has locally a form appearing in a map $H(\cdot, t)$ for some typical H and critical t. More precisely, this means that there exists a neighbourhood $U \subset \mathbb{R}^{2n}$ of q and diffeomorphisms $\varphi \in \mathrm{Diff}_0 M^n$ and $\psi \in \mathrm{Diff}_0 \mathbb{R}^{2n}$ such that
$$\psi \circ H(\cdot, t) \circ \varphi|_{f^{-1}(U)} = f|_{f^{-1}(U)}.$$
Roughly speaking, singularities appearing in generic one-parameter families are the local models for typical singularities.

We will view typical singularities as representing deformations of local pieces of maps. This means that the actual form of a singularity is not so important to us, and we define properties of typical singularities by refering to local properties of generic 1-parameter deformations in which the singularities appear. If H is typical and t is the image of a cross-cup singularity of $\widetilde{H}$ by $\pi \circ \widetilde{H}$, then one of the maps $H(\cdot, t \pm \epsilon)$, for small ϵ, has exactly one self-intersection close to the singularity of $H(\cdot, t)$ (and the other has none). We will say that this kind of typical singularity is of *type I*. If t is a critical value of π restricted to Q, then one of the maps $H(\cdot, t\pm \epsilon)$ has two self-intersections close to the singularity of $H(\cdot, t)$ (and the other has none). We will say that this kind of typical singularity is of *type II*. Our symbols I and II reflect the number of self-intersections appearing (or disappearing) after crossing a singularity of corresponding type. Equivalently, this is the number of sheets of the manifold involved in a singularity.

Since each manifold M^n is locally orientable, for n even the index of a double point appearing after crossing any singularity of type I is well defined, even if the manifold is nonorientable. Thus, we distinguish subtypes I^+ and I^-, according to a sign of the above index. We shall also refer to *moves of type I, I^+, I^-* and *II*, respectively, according to the type of singularity we are crossing in a typical 1-parameter family of maps.

We finish the subsection with the following result concerning moves.

LEMMA 1.3. *Any two stable immersions $f, f' : M^n \to \mathbb{R}^{2n}$ can be connected up to $\mathcal{A}_0$-equivalences by a finite sequence of moves corresponding to crossing typical singularities.*

Lemma 1.3 follows from contractibility of $\mathbb{R}^{2n}$ by deforming a homotopy between f and f' into a typical one.

Resolutions. In this subsection we define specific deformations which turn maps with finitely many typical singularities into stable maps. We study the uniqueness question for stable maps resulting from such deformations.

Denote by $X^{(k)}$ the subset of the discriminant consisting of maps with exactly k singularities all of which are typical. Let $f \in X^{(k)}$ and assume that each singularity q of f is determined locally by a model map $H_q(\cdot, t)$ for some typical 1-parameter family H_q and some critical t. This means that $f|_{f^{-1}(U_q)} = \psi \circ H_q(\cdot, t) \circ \varphi|_{f^{-1}(U_q)}$ for some diffeomorphisms $\psi \in \mathrm{Diff}_0 M^n$ and $\varphi \in \mathrm{Diff}_0 \mathbb{R}^{2n}$ and for some neighbourhood $U_q \subset \mathbb{R}^{2n}$ of q. We assume that H_q together with t, U_q, ψ and φ is a part of data of the typical singularity q.

A smooth family $r_f(\cdot, \theta)$ of maps from M^n to $\mathbb{R}^{2n}$ parametrized with parameter $\theta \in [0, \epsilon)$ is called a *resolving deformation* of f if it satisfies the following properties:
  (i) $r_f(\cdot, 0) = f$;
  (ii) $r_f(\cdot, \theta)$ is a stable immersion for $\theta > 0$;
  (iii) for each singularity q of f we have $r_f(\cdot, \theta)|_{f^{-1}(W_q)} = \psi \circ H_q(\cdot, t \pm \theta) \circ \varphi|_{f^{-1}(W_q)}$ for some neighbourhood W_q of q contained in U_q, where the sign in $\pm$ depends only on q.

Each map $r_f(\cdot, \theta)$ with $\theta > 0$ in a resolving deformation r_f of f is called a *resolution* of f. A resolution $r_f(\cdot, \theta)$ is called *positive* at a singularity q of f if for $\theta > 0$ new self-intersections appear in r_f close to q; it is called *negative* at q if no such self-intersections appear. Accordingly, we speak also about signs of resolving deformations of f at singularities of f. A crucial property of resolutions which allows the definition of finite type invariants is the following.

LEMMA 1.4. *For each singularity of a map $f \in X^{(k)}$ fix a sign $\pm$. Then there exists a resolving deformation r_f of f having a prescribed sign at each singularity. Moreover, any two resolutions of f with prescribed signs are $\mathcal{A}_0$-equivalent.*

PROOF. Choose neighbourhoods $W_q \subset U_q$ of singularities q and a real number $\epsilon > 0$ so that
(a) W_q are all disjoint;
(b) for each singularity q and for any $\theta \in (0, \epsilon)$ the map $\psi \circ H_q(\cdot, t \pm \theta) \circ \varphi|_{f^{-1}(W_q)}$ has no singularities, where the $\pm$ sign in the expression above is taken appropriately with respect to the desired sign of r_f at q.

Let $\tilde{r}_f$ be the 1-parameter family of maps defined on the sum $\cup_q W_q$ by the formulas
$$\tilde{r}_f(x, \theta) = \psi \circ H_q(\cdot, t \pm \theta) \circ \varphi(x) \text{ for } x \in W_q,$$
where the $\pm$ sign in each formula is taken as before. Then for small θ the maps $\tilde{r}_f(\cdot, \theta)$ are close to f near the boundary of each W_q. Since f is stable outside the set $\cup_q W_q$, it is possible to use maps close to f to extend $\tilde{r}_f$ to a resolving deformation r_f of f, choosing ϵ smaller if necessary. This proves the existence part of the lemma.

The uniqueness up to $\mathcal{A}_0$-equivalence follows by a similar argument, noting that any two resolutions with required signs at the singularities of f coincide on some neighbourhood of the set of these singularities. This finishes the proof.

REMARK 1.5. Denote by $X^{(\Sigma)}$ the sum of all spaces $X^{(k)}$, including $X^{(0)}$—the space of stable immersions. There is a striking difference between the situation we are considering and that of knots in $\mathbb{R}^3$. In our case closures of the components of embeddings in the space $X^{(\Sigma)}$ are disjoint. (This is clear since any move introduces a double point.) Thus, to apply Vassiliev's approach to the problem of constructing knot invariants in dimensions $n \to 2n$, we are forced to consider all stable maps.

Finite type invariants. Let V be an invariant of the $\mathcal{A}_0$-equivalence of stable immersions $M^n \to \mathbb{R}^{2n}$. We extend V to maps with finitely many typical singularities, i.e., to maps from the sets $X^{(k)}$ for all integers $k \geq 1$. This extension is given inductively with respect to the number of singularities by the formula

$$V(\text{map with singularity}) = V(\text{positive resolution}) - V(\text{negative resolution}).$$

For the maps with several singularities, the value of V clearly does not depend on the order in which the singularities enter the calculation process.

DEFINITION 1.6. We say that an invariant V is a *finite type invariant of order k*, if $V|_{X^{(k+1)}} \equiv 0$.

We denote by $V^{(k)}$ the space of all finite type invariants of order k.

REMARK 1.7. Note that $V^{(k)}$ is the linear space, and that
(a) $V^{(0)}$ is the set of constant invariants;
(b) $V^{(k)} + V^{(l)} \subset V^{(\max(k,l))}$;
(c) $V^{(k)} \cdot V^{(l)} \subset V^{(k+l)}$.

Remark 1.7(a) follows from Lemma 1.3, while the others are the direct consequences of the definition of finite type invariants. Compare with [**BN**], where the Vassiliev invariants for knots are introduced in the way we follow above.

Examples of finite type invariants of order one. The number of double points of a stable immersion, denoted by σ, is clearly a finite type invariant of order 1.

For immersions of even dimensional manifolds we have another slightly more sophisticated invariant. Assume first that M is orientable. Orient M^n and $\mathbb{R}^{2n}$ and consider the induced difference orientation on the normal bundle ν. Then ν has its Euler class which we can integrate against M^n. This gives the Euler number ϵ which is clearly independent of the orientation on M^n (but does depend on the orientation of $\mathbb{R}^{2n}$); see [**M–S**], page 95 onward. If M^n is nonorientable we proceed similarly: the normal bundle ν is nonorientable and its orientation homomorphism is the same as that of the tangent bundle. Hence we can integrate the *twisted* Euler class against the twisted fundamental cycle to again get the Euler number (compare [**Sp**], exercises for Chapter 5).

For those who want to avoid machinery, the Euler number admits a very convenient intersection theoretic interpretation. Let ξ be an n-dimensional vector bundle over M^n, such that its orientation homomorphism is equal to an orientation homomorphism of M^n. Take a generic section s of ξ and at each zero, count the intersection number of a piece of the zero section locally oriented in an arbitrary way with a piece of s oriented locally with orientation induced from a projection onto a zero section. This intersection number (equal ± 1) is clearly independent of local orientations chosen, and the sum of all local intersection numbers is the Euler number.

The Euler number clearly does not change under move II. It changes by ± 1 under move $I^{\pm}$, as it follows from:

(a) the possibility of interpreting move $I^{\pm}$ as a connected sum with the immersed sphere with one double point of the corresponding sign;
(b) the additivity of an Euler number under connected sums;
(c) the explicit computation of Euler numbers of the corresponding spheres with one double point.

All this is essentially due to Whitney [**Wh**].

2. Strata of the discriminant

The goal of this section is to study the structure of the space of immersions with finitely many typical singularities. In the case of knots such structure is described by means of chord diagrams. We show that in our context the role of chord diagrams is played by tuples of nonnegative integers which represent numbers of singularities of corresponding types (see Proposition 2.5 for the precise statement).

We start the section with a discussion of the necessary technicalities.

Recall from Section 1 that a typical singularity of a map $f : M^n \to \mathbb{R}^{2n}$ is a point $q \in \mathbb{R}^{2n}$ together with a neighbourhood U_q and a typical 1-parameter family H_q, such that

$$f|_{f^{-1}(U_q)} = \psi \circ H_q(\cdot, t) \circ \varphi|_{f^{-1}(U_q)}$$

for some critical t and some diffeomorphisms $\psi \in \mathrm{Diff}_0\mathbb{R}^{2n}$ and $\varphi \in \mathrm{Diff}_0 M^n$. For any $\theta \neq 0$ the map $r_q(\cdot, \theta) : f^{-1}(U_q) \to \mathbb{R}^{2n}$ defined by

$$r_q(\cdot, \theta) = \psi \circ H_q(\cdot, t \pm \theta) \circ \varphi|_{f^{-1}(U_q)}$$

will be called a *local resolution* of f at q. Local resolutions clearly have their signs defined as before.

DEFINITION 2.1. A *normal neighbourhood* of a typical singularity q of f is an open set $B_q \subset \mathbb{R}^{2n}$ satisfying the following properties:
  (i) B_q is ambient isotopic to a ball in $\mathbb{R}^{2n}$;
  (ii) $q \in B_q$ and $\overline{B}_q \subset U_q$;
  (iii) f is transversal to the boundary ∂B_q;
  (iv) if q is of type $I^\pm$ or I and if $r_q(\cdot, \theta)$ is a negative local resolution of f at q with sufficiently small θ, then the pair $(\overline{B}_q, im(r_q(\cdot, \theta)) \cap \overline{B}_q)$ is ambient isotopic in $\mathbb{R}^{2n}$ to the canonical pair (D^{2n}, D^n) of discs;
  (v) if q is of type II and if $r_q(\cdot, \theta)$ is a negative local resolution of f at q with sufficiently small θ, then the pair $(\overline{B}_q, im(r_q(\cdot, \theta)) \cap \overline{B}_q)$ is ambient isotopic in $\mathbb{R}^{2n}$ to the sum of two canonically embedded disjoint and unlinked discs D^n in D^{2n}.

It is clear from the definition that for each typical singularity q, normal neighbourhoods B_q of q do exist. The following lemma is probably well known, although we have not found an appropriate reference in the literature.

LEMMA 2.2. *Local resolution $r_q(\cdot, \theta)$ of a given sign, after restriction to a normal neighbourhood B_q, for sufficiently small θ depends up to $\mathcal{A}_0$-equivalence only on the type of a singularity q. More precisely, if maps f and f' have singularities of the same type at q and q', respectively, and if B_q and $B_{q'}$ are normal neighbourhoods, then for sufficiently small θ and θ' we have*

$$r_q(\cdot, \theta)|_{f^{-1}(B_q)} = \psi \circ r_{q'}(\cdot, \theta') \circ \varphi|_{f^{-1}(B_q)}$$

for some $\psi \in \mathrm{Diff}_0\mathbb{R}^{2n}$ and $\varphi \in \mathrm{Diff}_0 M^n$, where $\varphi(f^{-1}(\overline{B}_q)) = f^{-1}(\overline{B}_{q'})$.

Note that the lemma is trivial for negative resolutions. It is less clear for positive ones, but we do not want to deal with proving it in this paper.

We will call $\mathcal{A}_0$-equivalence classes of resolutions as in the lemma *normal local resolutions*.

It is possible to draw pictures of stable immersions $M^n \to \mathbb{R}^{2n}$, at least if $n = 2$. Let P be a generic projection $P : \mathbb{R}^4 \to \mathbb{R}^3$. Composing a stable immersion $f : M^2 \to \mathbb{R}^4$ with such a P we get a generic stable mapping $P \circ f : M^2 \to \mathbb{R}^3$, i.e., the one with transversal double intersection lines, transversal triple intersection points, and cross-caps (also known as the Whitney umbrellas). Up to $\mathcal{A}_0$-equivalence f is completely determined by *markings* (u/d) on self-intersection lines of the projection $P \circ f$. They indicate which of the two sheets meeting along the line passes higher/lower with respect to the coordinate of the projection. Self-intersection points of f in $\mathbb{R}^4$ are indicated as special points on the self-intersection lines of $P \circ f$ in $\mathbb{R}^3$, at which the marking along the corresponding line changes. A generic projection onto $\mathbb{R}^3$ of the image of a stable immersion $f : M^2 \to \mathbb{R}^4$, equipped with markings, will be called a *diagram* of f. For immersions $M^n \to \mathbb{R}^{2n}$ with $n > 2$ one can "draw" diagrams in $\mathbb{R}^{2n-1}$ in the same way.

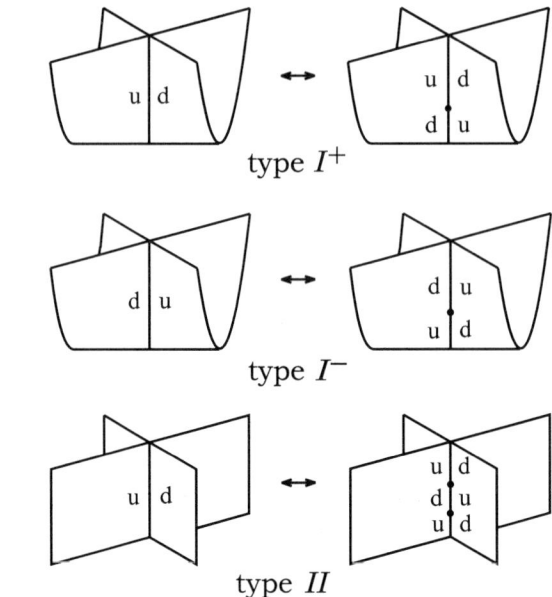

FIGURE 1. Diagrams of normal local resolutions

Figure 1 presents diagrams of normal local resolutions of typical singularities of the three types $I^{\pm}$ and II. Note that similar diagrams of normal local resolutions of typical singularities in dimension $n > 2$ will have a form of a transversal crossing of two n-dimensional sheets along a line or a higher dimensional cross-cap in $\mathbb{R}^{2n-1}$ with appropriate special points and markings.

For the rest of this section we deal simultaneously with even/odd dimensional manifolds, even though they behave in slightly different ways.

DEFINITIONS 2.3. If n is even, we call $f : M^n \to \mathbb{R}^{2n}$ a (p,q,r)-map if $f \in X^{(p+q+r)}$ (i.e., f has exactly $p + q + r$ singularities all of which are typical), and if there are exactly p, q and r singularities of types I^+, I^-, and II, respectively, among the singularities of f.

If n is odd, we define a (p,r)-map in an analogous way as above, as having p and r singularities of types I and II, respectively.

Two (p,q,r)-maps or two (p,r)-maps $f, f : M^n \to \mathbb{R}^{2n}$ are called *weakly isotopic* if there are $\varphi \in \mathrm{Diff}_0 M^n$ and $\psi \in \mathrm{Diff}_0 \mathbb{R}^{2n}$ such that:
  (i) there is a collection U_i of disjoint normal neighbourhoods of singularities of f such that each U_i is also a normal neighbourhood of a singularity of the map $\psi \circ f' \circ \varphi$;
  (ii) the corresponding singularities of f and $\psi \circ f' \circ \varphi$ inside each of the sets U_i have the same type;
  (iii) the maps f and $\psi \circ f' \circ \varphi$ coincide outside the set $\cup_i f^{-1}(U_i)$.
Resolutions of two weakly isotopic maps are *consistent* if they have the same signs at the pairs of singularities which correspond to each other by the weak isotopy.

The following is a direct consequence of the definitions above and Lemma 2.2.

COROLLARY 2.4. *Consistent resolutions of weakly isotopic (p,q,r)- or (p,r)-maps are $\mathcal{A}_0$-equivalent.*

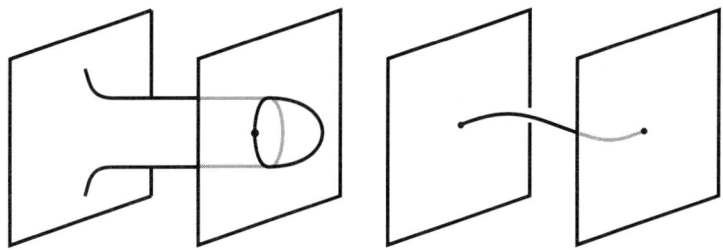

FIGURE 2. A representation of a (p, q, r)-map near a singularity of type II

REMARK. Corollary 2.4 is a useful local ingredient in the classification of maps with finitely many typical singularities. It says that from the point of view of finite type invariant maps, which are weakly ambient isotopic, cannot be distinguished. It also shows that a more detailed study of typical singularities is not necessary for our purposes.

The main result of this section is the following.

PROPOSITION 2.5. *Let p, q and r be fixed nonnegative integers. Then any two (p, q, r)-maps (or (p, r)-maps for odd n, respectively) of a connected closed manifold M^n into $\mathbb{R}^{2n}$ can be made weakly isotopic by a finite sequence of moves of the following form:*
  (a) *deformation by an $\mathcal{A}_0$-equivalence;*
  (b) *crossing a typical singularity: a local change in a small ball in $\mathbb{R}^{2n}$ disjoint from all images of singularities, corresponding to passing from a positive to a negative (or vice versa) normal local resolution of a typical singularity.*

PROOF. We start with a convenient *representation* of a singular map. We replace small normal neighbourhoods of all singularities of type I by *dots*. The part of the image of the manifold M^n in $\mathbb{R}^{2n}$ which is close to such dots is regularized, so that the dots are placed on regular sheets. One can identify these regularized sheets with sheets appearing in negative normal local resolutions of the corresponding singularities. The dots are signed: they contain information about the subtypes of singularities in the case where n is even.

A representation of a singularity of type II is slightly more complicated. It consists locally of two disjoint regular sheets of the manifold, a dot on each sheet, and a curve segment between those two dots; the segment is perpendicular to the sheets at the endpoints, it is injective and does not meet the sheets at its interior points. The two sheets in the representation correspond to sheets appearing in a negative local normal resolution of the singularity. The curve segment represents a thin pipe of form $S^{n-1} \times I$; the dot at one of its endpoints represents a gluing of this pipe along $S^{n-1} \times \{0\}$ to the sheet on which the dot lies (after removing a small n-disc from the sheet). The dot at the second endpoint of the segment represents the singularity or a very small normal neighbourhood of it. One should view the sheet on which this dot lies as extending one of the sheets of the manifold inside this normal neighbourhood. The other sheet inside this neighbourhood is viewed to meet the second end $S^{n-1} \times \{1\}$ of the pipe along its boundary.

For $n = 2$ a local part of the representation close to a singularity of type II is presented by means of a diagram in Figure 2.

Moreover, we assume that dots of any representation in $\mathbb{R}^{2n}$ are all distinct and distinct from stable double points of a represented map. Finally, a representation has to coincide with the represented map outside the preimage of the sum of small normal neighbourhoods of the singularities of this map.

It is clear that a representation allows us to recover all the resolutions of a represented map uniquely up to $\mathcal{A}_0$-equivalence.

We will construct a sequence of moves as in Proposition 2.5 from a homotopy between representations of the considered (p, q, r)- or (p, r)-maps. By a homotopy of a representation we mean a homotopy of the manifold part of this representation, with compatible homotopies of all curve segments together with dots at their endpoints. By contractibility of $\mathbb{R}^{2n}$, such a homotopy between representations of any two (p, q, r)- or (p, r)-maps, respecting the types of dots, always exists.

Putting a homotopy as above in general position, we can assume that during the homotopy:
  (i) the manifold part of the representation crosses only typical singularities;
  (ii) all the dots remain disjoint from each other, from stable self-intersentions, and from singularities of the manifold part;
  (iii) the curve segments are smoothly embedded and disjoint with each other.

Since the set of directions perpendicular to a subspace of codimension ≥ 2 is connected, we can modify a homotopy as above on the curve segments, close to their endpoints, so that additionally
  (iv) the curve segments are perpendicular to the manifold part at their endpoints during the whole homotopy.

Finally, if $n > 2$ we can assume also that
  (v) the curve segments remain disjoint from the manifold part, except at the endpoints.

If $n = 2$ then (v) has to be replaced by the following condition:
  (v′) the surface traced during the homotopy by the interior of any curve segment is transversal to the manifold part.

It is clear that a homotopy of representations which satisfies conditions (i) through (v) above (with (v) replaced by (v′) if $n = 2$) can be followed by a homotopy of represented mappings, so that the initial singularities are preserved together with their types. Such a homotopy of mappings is an $\mathcal{A}_0$-equivalence, except when it crosses events mentioned in (i) and (v′) above. Crossing a typical singularity as in (i) clearly corresponds to move (b) in the statement of Proposition 2.5. Crossing a manifold part by a curve segment as in (v′) corresponds to crossing a sheet of the manifold by a thin pipe. This deformation can be realized by two moves of form (b), with type II singularities involved, namely a creation and an annihilation of two intersections between the pipe and the sheet. These two moves are presented in Figure 3.

It is clear that generically local moves in a homotopy as above can be separated in time. Thus we get a sequence of moves (performed on actual maps rather than their representations) connecting the initial and terminal maps up to weak isotopy.

This finishes the proof of the proposition.

REMARK. Following Vassiliev, we can think of the sets $X^{(k)}$ of maps with k typical singularities as of the strata of the discriminant D of codimension k in the whole space $C^\infty(M^n, \mathbb{R}^{2n})$. Proposition 2.5 may be viewed then as a description of irreducible components of these strata. These irreducible components turn out to

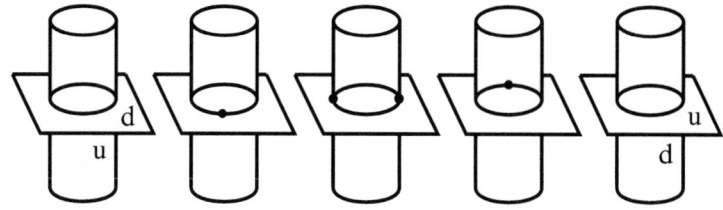

FIGURE 3. Crossing a manifold by a thin pipe

be indexed by the triples (p, q, r) of nonnegative integers with $p+q+r = k$. An easy computation shows that there are as many as $\frac{(k+2)(k+1)}{2}$ irreducible components in the stratum $X^{(k)}$.

Lemmas 3.1 and 3.2 in the next section show that we can interprete finite type invariants of order k (up to lower order invariants) as functions on the set of irreducible components of the stratum $X^{(k)}$. This shows the significance of the results obtained in this section for the classification of finite type invariants.

In the classical case of knots in $\mathbb{R}^3$ the important feature of a singular knot is its inner geometry represented by a chord diagram (see [**BN**]). This phenomenon is responsible for the fact that there are many irreducible components of the strata of the discriminant, and consequently plenty of finite type invariants. There is nothing like inner geometry in the case of stably immersed n-manifolds in $\mathbb{R}^{2n}$, for $n \geq 2$. Proposition 2.5 justifies this observation and tells us also that no other phenomena appear—irreducible components are determined simply by the types of singularities. We think that this is the essential reason why the finite type invariant approach does not work well here.

3. Calculation of finite type invariants

Throughout this section we assume that M^n is a closed connected manifold and all invariants of finite type are the invariants of $\mathcal{A}_0$-equivalence for stable immersions of the manifold M^n into $\mathbb{R}^{2n}$. We follow the notation used in the first two sections.

We start this section with identifying finite type invariants of order k (viewed up to invariants of lower order) with functions on the triples (p, q, r) (pairs (p, r) if n is odd) of nonnegative integers with $p + q + r = k$ ($p + r = k$, respectively). We then show that functions which represent the invariants satisfy some relations, which are the analogs of the four-term relations for Vassiliev weight systems for knots. Using these relations we estimate the dimension of the space of finite type invariants of given order. Finally, we show that the obvious invariants saturate the estimated dimension, thus proving the main theorems of the paper.

The following lemma exploits the results of the previous section.

LEMMA 3.1. *If n is even, then for arbitrary nonnegative integers p, q and r satisfying $p+q+r = k$, any finite type invariant $V \in V^{(k)}$ of order k is constant on the set $X^{(p,q,r)}$ of all (p, q, r)-maps. Similarly, if n is odd, then for any nonnegative integers p and r with $p + r = k$, any finite type invariant $V \in V^{(k)}$ of order k is constant on the set $X^{(p,r)}$ of all (p, r)-maps.*

PROOF. It follows from Corollary 2.4 that any invariant V, after extension, has equal values at weakly isotopic (p, q, r)- or (p, r)-maps. In view of Proposition

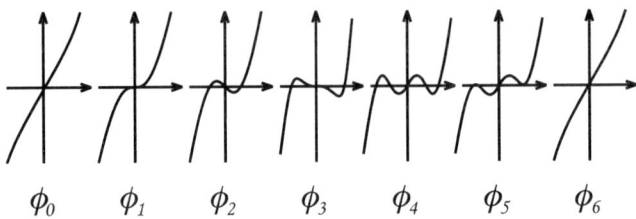

FIGURE 4. A 1-parameter family of odd functions ϕ_t

2.5 it is enough then to prove that V does not change under a single move. But this follows directly from the fact that $V|_{X^{(k+1)}} \equiv 0$.

If n is even, put $B_k = \{(p,q,r) : p,q,r \in \mathbb{Z},\ p,q,r \geq 0,\ p+q+r = k\}$, and denote by $\mathbb{R}^{B_k}$ the vector space of real functions on B_k. For $V \in V^{(k)}$ let $\phi(V) \in \mathbb{R}^{B_k}$ be the function defined by $\phi(V)(p,q,r) = V(f_{p,q,r})$, for arbitrary map $f_{p,q,r} \in X^{(p,q,r)}$. Similarly, if n is odd, put $B_k = \{(p,r) : p,r \in \mathbb{Z},\ p,r \geq 0,\ p+r = k\}$, and for $V \in V^{(k)}$ define the function $\phi(V)$ on B_k in an analogous way.

LEMMA 3.2. *The kernel of the map* $\phi : V^{(k)} \to \mathbb{R}^{B_k}$ *is equal to* $V^{(k-1)}$.

PROOF. Note that $\phi(V) \equiv 0$ iff $V|_{X^{(k)}} \equiv 0$, that is iff $V \in V^{(k-1)}$.

As a consequence of Lemma 3.2 we see that the function $\phi(V)$ represents an invariant $V \in V^{(k)}$ faithfully up to invariants of lower order. The next lemma shows that the functions appearing as such representatives have to satisfy certain relations.

LEMMA 3.3. *Let* $V \in V^{(k)}$ *be a finite type invariant of order* k. *If* n *is even, then for any nonnegative integers* p, q *and* r *with* $p+q+r = k-1$ *we have*

$$\phi(V)(p,q,r+1) = \phi(V)(p+1,q,r) + \phi(V)(p,q+1,r).$$

Similarly, if n *is odd, then for any nonnegative integers* p *and* r *with* $p+r = k-1$ *we have*

$$\phi(V)(p,r+1) = 2\phi(V)(p+1,r).$$

PROOF. Let ϕ_t with $t \in [0,6]$ be a smooth 1-parameter family of odd functions in which the pattern of zeroes and extrema changes as in Figure 4. Assume that the family is *cyclic* in the sense that the terminal function ϕ_6 coincides with the initial one ϕ_0. Consider then a 1-parameter family of mappings $f_t : \mathbb{R}^n \to \mathbb{R}^{2n}$ given by the following formula

$$f_t(x_1,\ldots,x_n) = (\phi_t(x_1), x_1^2, x_1 x_2, \ldots, x_1 x_n, x_2, \ldots, x_n).$$

Note that the mappings f_1 and f_3 have singularities of types $I^{\pm}$, respectively (type I if n is odd), while f_5 has a singularity of type II. For other values of t the maps f_t are stable immersions. It follows that f_t defines a sequence of three moves under which the initial map f_0 is transformed back to itself. For $n = 2$ these three moves are illustrated by the diagrams in Figure 5.

We view the family f_t as a local deformation in some local coordinates. Then clearly it can be extended to a deformation $\tilde{f}_t$ for which $\tilde{f}_0$ is a (p,q,r)-map (or a

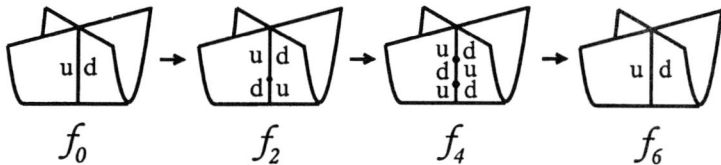

FIGURE 5. The cycle of moves which implies the relations of Lemma 3.3

(p,r)-map if n is odd). We assume here that the maps $\tilde{f}_t$ for different t coincide with each other outside the region in which the local deformation f_t takes place.

Recalling that $\tilde{f}_6 = \tilde{f}_0$, we have for even n

$$0 = [V(\tilde{f}_0) - V(\tilde{f}_2)] + [V(\tilde{f}_2) - V(\tilde{f}_4)] + [V(\tilde{f}_4) - V(\tilde{f}_0)]$$
$$= -\phi(V)(p+1,q,r) - \phi(V)(p,q+1,r) + \phi(V)(p,q,r+1),$$

and similarly for n odd

$$0 = -2\phi(V)(p+1,r) + \phi(V)(p,r+1),$$

and the lemma follows.

The next lemma gives an estimate for the dimension of the space of finite type invariants of given order.

LEMMA 3.4. *If n is even then $\dim V^{(k)}/V^{(k-1)} \leq k+1$. If n is odd then $\dim V^{(k)}/V^{(k-1)} \leq 1$.*

PROOF. By Lemma 3.2, the quotient space $V^{(k)}/V^{(k-1)}$ maps injectively into the subspace $F_k \subset \mathbb{R}^{B_k}$ of functions satisfying the equations in Lemma 3.3.

For any $V \in V^{(k)}$, the function $\phi(V)$ is determined by its values on the triples of form $(p,0,r)$, since we can calculate the values of $\phi(V)$ at triples (p,q,r) with $q \neq 0$ inductively, using the equations in Lemma 3.3. Since there are $k+1$ triples of the form $(p,0,r)$ in B_k, the lemma follows.

We finish this section by proving the main results of the paper.

3.5. Proof of Theorems 1 and 2. Consider the case when n is even. Denote by $K[\sigma,\epsilon]^{\leq k}$ the space of invariants obtained from all two variable polynomials of degree not greater than k, with invariants σ and ϵ as variables. Then $K[\sigma,\epsilon]^{\leq k}$ is a linear subspace in $V^{(k)}$ isomorphic to the corresponding polynomial space $K[x,y]^{\leq k}$. A large family of immersions necessary to establish this isomorphism is constructed as follows. First close up the first two pictures in Figure 1 to immersions of S^2 with $+1$ and -1 self-intersection points, respectively. Then by connected sums we get immersions of S^2 with p positive and q negative self-intersections, for arbitrary nonnegative integers p,q. This family suffices. An analogous construction can also be performed in higher dimensions.

The arguments above imply the following inequality

$$\dim V^{(k)} \geq \dim K[\sigma,\epsilon]^{\leq k} = \dim K[x,y]^{\leq k} = \frac{(k+2)(k+1)}{2}.$$

On the other hand, by Lemma 3.4 we have an estimate

$$\dim V^{(k)} = \sum_{i=0}^{k} \dim V^{(i)}/V^{(i-1)} \le \sum_{i=0}^{k}(i+1) = \frac{(k+2)(k+1)}{2}.$$

It follows that the first inequality is an equality. It implies then that $V^{(k)} = K[\sigma,\epsilon]^{\le k}$, and the theorem follows in this case.

The argument for the case when n is odd runs similarly. It uses the space $K[\sigma]^{\le k}$ of one variable polynomials of degree not bigger than k with the invariant σ as a variable. This finishes the proof.

4. The nonconnected case

Let $M_1 \cup M_2 \cup \cdots \cup M_m$ be the decomposition of a closed manifold M^n into connected components. Consider the invariants σ_{ij} for $1 \le i \le j \le m$, equal to the total numbers of intersections between images of components M_i nad M_j (or to the total numbers of self-intersections of components M_i if $i = j$). Moreover, if n is even, consider the invariants ϵ_i for $1 \le i \le m$ equal to the Euler numbers of normal bundles of components M_i (perhaps twisted, depending on orientability of M_i).

In this section we prove the following result.

THEOREM 4.1. *If n is even, then the only finite type invariants for stable immersions of a manifold M^n into $\mathbb{R}^{2n}$ are the polynomials in $\frac{m^2+3m}{2}$ variables with the invariants σ_{ij} and ϵ_i as variables. Similarly, if n is odd, then the only finite type invariants are the polynomials in $\frac{m^2+m}{2}$ variables with the invariants σ_{ij} as variables.*

To prove Theorem 4.1 we follow the arguments of Sections 1–3. We concentrate on the case when n is even, and omit the obvious modifications of the argument for the case when n is odd.

We start with discerning more subtypes of typical singularities:
(i) subtypes I_i^+ and I_i^-, for $1 \le i \le m$, according to a component M_i at which a singularity of type I appears (the meaning of signs is the same as before, see Definition 1.3);
(ii) subtypes II_{ij}, for $1 \le i \le j \le m$, according to components M_i and M_j to which the two sheets involved in a singularity of type II belong.

Let $p = (p_i)_{1 \le i \le m}$, $q = (q_i)_{1 \le i \le m}$ and $r = (r_{ij})_{1 \le i \le j \le m}$ be strings of nonnegative integers. Then a map $f : M^n \to \mathbb{R}^{2n}$ is a (p,q,r)-*map*, if it has exactly p_i, q_i and r_{ij} typical singularities of types I_i^+, I_i^- and II_{ij}, respectively, and if it is stable otherwise. In this notation, Proposition 2.5 holds for nonconnected closed manifolds, with the same proof. Similarly, the obvious analogues of Lemmas 3.1 and 3.2 hold, with the same proofs.

To state the proper analogue of Lemma 3.3, for given l with $1 \le l \le m$, denote by r^{ll} the string $(r'_{ij})_{1 \le i \le j \le m}$ defined by $r'_{ij} = r_{ij}$ for $ij \ne ll$, and $r'_{ll} = r_{ll} + 1$. Similarly, define $p^l = (p'_i)_{1 \le i \le m}$ by $p'_i = p_i$ for $i \ne l$, and $p'_l = p_l + 1$, and the same for q^l.

LEMMA 4.2. *For each $V \in V^{(k)}$, for any strings p, q and r of nonpositive integers with $\sum p_i + \sum q_i + \sum r_{ij} = k - 1$, and for any $1 \le l \le m$, we have*

$$\phi(V)(p, q, r^{ll}) = \phi(V)(p^l, q, r) + \phi(V)(p, q^l, r).$$

The proof is similar to that of Lemma 3.3.

An analogue of Lemma 3.4 is the following result.

LEMMA 4.3. *Let $d_{m,k}$ denote the number of all pairs of strings (p,r) of non-positive integers, such that $\sum p_i + \sum r_{ij} = k$. Then $\dim V^{(k)}/V^{(k-1)} \leq d_{m,k}$.*

PROOF. We modify the proof of Lemma 3.4. Denote by 0 the string $(0,\ldots,0)$ with m zeros. For any $V \in V^{(k)}$, the function $\phi(V)$ is determined by its values on the triples of strings of form $(p,0,r)$. We calculate the values of $\phi(V)$ at triples (p,q,r) with $q \neq 0$ inductively, using the equations in Lemma 4.2.

Since there are $d_{m,k}$ triples $(p,0,r)$ (recall that we assume that $\sum p_i + \sum q_i + \sum r_{ij} = k$), the lemma follows.

4.4. Proof of Theorem 4.1. Consider the linear subspace $K[\sigma_{ij}, \omega_i]^{\leq k} \subset V^{(k)}$ of all polynomials of $\frac{m^2+3m}{2}$ variables and degree not bigger than k, with invariants σ_{ij} and ϵ_i as variables. The dimension of $K[\sigma_{ij}, \omega_i]^{\leq k}$ is equal to that of $K[x_{ij}, y_i]^{\leq k}$ since those linear spaces are isomorphic (the argument for this is an obvious modification of the one used in the proof of Theorem 1). This dimension is equal to $\sum_{l=0}^{k} d_{m,l}$, since there are $d_{m,l}$ distinct monomials of degree l.

On the other hand, by Lemma 4.3 we have an estimate

$$\dim V^{(k)} = \sum_{l=0}^{k} \dim V^{(l)}/V^{(l-1)} \leq \sum_{l=0}^{k} d_{m,l},$$

and thus there are no other invariants in $V^{(k)}$ than the polynomials mentioned above.

The analogous argument works in the case when n is odd.

5. Finite type invariants of the regular homotopy

Let D' denote the set of all mappings $g \in C^{\infty}(M^n, \mathbb{R}^{2n})$ of a closed manifold M^n which are not immersions. Then the complement $C^{\infty}(M^n, \mathbb{R}^{2n}) \setminus D'$ is open and dense, and its components correspond to the regular homotopy classes of immersions. We then view the singularities of type II no longer as singularities, so that we are left only with the singularities of type I as typical ones. In particular, we have the following.

5.1. Fact. Any two immersions of a closed manifold M^n into $\mathbb{R}^{2n}$ can be deformed to one another by a finite sequence of moves of the following forms:

(a) a regular homotopy;

(b) a move of type I.

Let M^n be a closed manifold with m connected components. If n is even, we consider subtypes I_i^+ and I_i^- of typical singularities, as in Section 4. Given strings $p = (p_i)$ and $q = (q_i)$ of nonnegative integers, we say that $f : M^n \to \mathbb{R}^{2n}$ is a (p,q)-*map*, if it has exactly p_i and q_i singularities of types I_i^+ and I_i^-, respectively, and it is an immersion otherwise. Similarly, if n is odd, we consider subtypes I_i only, and given a string $p = (p_i)$ define a p-*map* in an obvious way.

We define finite type invariants of the regular homotopy similarly to Definition 1.6, using extensions of invariants to (p,q)- and p-maps. We have also the (obvious) appropriate version of Proposition 2.5, referring only to singularities of type I, and proved in the same way (or even easier, since the painful part regarding singularities of type II is no longer necessary). Equipped with this, we repeat the arguments of Sections 3 and 4 and get the folowing.

THEOREM 5.2. *Let M^n be a closed manifold with m connected components, and let ϵ_i for $1 \leq i \leq m$ be the Euler numbers from the beginning of Section 4. If n is even, then the only finite type invariants of the regular homotopy of immersions $M^n \to \mathbb{R}^{2n}$ are the polynomials in m variables with invariants ϵ_i substituted for the variables. If n is odd, the only finite type invariants are the constants.*

We omit the details.

REMARK. One should compare the above result with the Hirsh–Smale theorem. Among the many invariants of immersions (often depending on the source manifold) only the Euler number turns out to be of finite type.

References

[BN] D. Bar-Natan, *On the Vassiliev knot invariants*, Topology **34** (1995), 423–472.
[G–G] M. Golubitsky and V. Guillemin, *Stable mappings and their singularities*, Springer-Verlag, New York, 1973.
[M–S] J. W. Milnor and J. D. Stasheff, *Characteristic classes*, Princeton Univ. Press, Princeton, NJ, 1974.
[Sp] E. H. Spanier, *Algebraic topology*, Mc-Graw Hill, New York, 1966.
[Wh] H. Whitney, *The self-intersections of a smooth n-manifold in $2n$-space*, Ann. of Math. **45** (1944), 220–246.

On Enumeration of Unicursal Curves

Sergei K. Lando

1. Introduction

Consider an immersion $u : S^1 \to \mathbb{R}^2$ of the circle into the plane, $u'(x) \neq 0$ for all $x \in S^1$. Suppose also that the image $u(S^1)$ has only double points of intersection and the intersection in each double point is transversal (i.e., the tangent vectors to branches of the image are not collinear); see Figure 1.

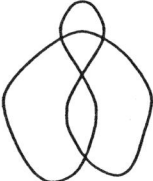

FIGURE 1. A unicursal curve

The image of such an immersion is called a *unicursal curve*.

Considered up to isotopies of the plane, there are 2 curves with one double point, 5 curves with two double points and 21 curves with three double points (see Figure 2).

Note that on the sphere there is 1 curve with one double point, 2 curves with two double points and 6 curves with three double points. Similarly one can investigate unicursal curves on (closed) surfaces of higher genera.

Let n be the number of double points of a unicursal curve.

Table 1 contains the numbers of plane and spherical unicursal curves with given number n of double points, as well as the numbers of marked spherical unicursal curves (see below) and the value of the upper estimate for the last number proved in the paper. The numbers in the first two rows are borrowed from [1], where they are obtained by direct drawing of all the curves. The third row is the result of computations from [3].

A complicated problem follows.

How many unicursal curves (considered up to isotopies) are there with a given number of double points?

1991 *Mathematics Subject Classification*. Primary 57M25.

The research is partly supported by the Russian Fund of Basic Investigation (project 95-01-008 46a) and the INTAS Grant (project # 4373).

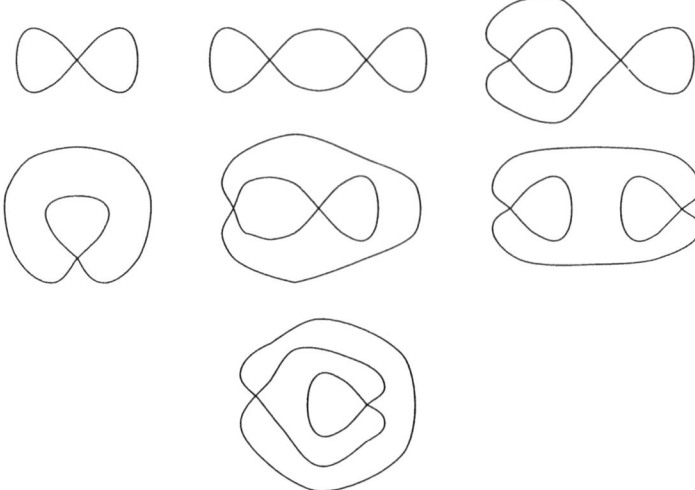

FIGURE 2. Plane unicursal curves with one and two points of self-intersection

TABLE 1. The number of plane and spherical unicursal curves

n	0	1	2	3	4	5
plane curves	1	2	5	21	102	639
spherical curves	1	1	2	6	21	99
marked spherical curves	1	2	8	42	260	1796
upper estimate	1	2	9	54	378	3888

This problem is yet not solved. In the present paper we give an upper estimate for the number of these curves and write out a matrix integral enumerating them.

First, let us describe more precisely the class of objects we are enumerating. We deal with *marked spherical unicursal curves*, i.e., with oriented spherical unicursal curves with a marked point (which must be different from all double points of the curve). Denote the number of marked spherical unicursal curves with n double points by U_n. Note that U_n is not greater than $4n$ times the number of spherical unicursal curves since there are $2n$ ways to choose the marked point, and 2 ways to choose the orientation of the curve (of course, some of these ways can provide the same marked curve).

THEOREM 1.1. *The number U_n of marked spherical unicursal curves satisfies the estimate*

$$U_n < 4 \cdot 3^n \frac{(2n-1)!}{(n+2)!(n-1)!} \sim C \cdot 12^n \cdot n^{-5/2}$$

for some constant C.

This estimate is proved in Section 2. It is a consequence of the enumeration theorem for spherical curves obtained in [**2**] by means of integration over the space of Hermitian matrices.

As a development of this method we prove the following result.

THEOREM 1.2. *The number U_n of unicursal curves with n double points on a surface of genus g is equal to the coefficient at the monomial $cN^{n+1-2g}s^n$ in the expansion of*

$$\int_{\mathcal{H}_N}\cdots\int_{\mathcal{H}_N} e^{\frac{s}{4}\operatorname{tr}\sum_{i,j=1}^{c} H_i H_j H_i H_j}\,d\mu(H_1)\ldots d\mu(H_c).$$

Here $\mathcal{H}_N$ denotes the space of $N\times N$ Hermitian matrices, $H_i\in\mathcal{H}_N$, and μ is a (standard) Gaussian measure on $\mathcal{H}_N$. The proof follows the coloring idea of V. Kazakov developed in [**5**].

2. Graph embeddings enumeration via matrix integration

Matrix integration as a way of separating graph embeddings in surfaces of different genera was invented by physicists [**6, 2**] and reinvented independently by mathematicians [**4**]. We start with the clearer example studied in [**4**].

Let us fix a $2n$-star on the plane. Gluing the needles of the star pairwise we obtain a graph with one vertex and n edges. This graph can be made into a *ribbon graph* by replacing each edge by a strip so that in the neighborhood of the center of the star the $2n$ strips are glued naturally respecting the topology of the plane. We also suppose that each of the strips is flat. This means that each of them can be put on the plane without self-intersection (but, generally, not all the strips together).

The ribbon graph is a two-dimensional surface with boundary. The number of connected components of the boundary determines the genus of the surface by the Euler formula.

The number of ways to glue $2n$ needles pairwise is $(2n-1)!! = 1\cdot 3\cdot 5\cdots(2n-1)$. Note that this number is precisely

$$p_n(1) = \frac{1}{\sqrt{2\pi}}\int_{-\infty}^{\infty} x^{2n} e^{-x^2/2}\,dx.$$

The last integral can be generalized to the integral

(2.1) $$p_n(N) = c_N \int_{\mathcal{H}_N} \operatorname{tr} H^{2n} e^{-\frac{1}{2}\operatorname{tr} H^2}\,dH,$$

where $\mathcal{H}_N$ is the space of $N\times N$ Hermitian matrices, H runs over $\mathcal{H}_N$, dH is the standard matrix measure on $\mathcal{H}_N$, and c_N is the normalizing factor determined by the condition, $c_N\int_{\mathcal{H}_N} e^{-\frac{1}{2}\operatorname{tr} H^2}\,dH = 1$.

It turns out that integration over the space of Hermitian matrices splits ribbon graphs with respect to their genera. To be more precise, the following theorem holds.

THEOREM 2.1 (see [**2, 4**]). *For each n the function $p_n(N)$ is a polynomial in N. The coefficient at N^{n-2g+1} of this polynomial is the number of ribbon graphs of genus g.*

A similar result holds in a more general situation, namely, for enumeration of gluings of arbitrary set of stars with respect to the genera of the resulting ribbon graph. We shall make use of one of the results of this kind, which serves as the main example in [**2**].

THEOREM 2.2 (see [2]). *The coefficient at s^n in the power series expansion of*

$$\log\left(c_N \int_{\mathcal{H}_N} e^{s\,\mathrm{tr}\,H^4} e^{-\frac{1}{2}\,\mathrm{tr}\,H^2} dH\right)$$

is a polynomial in N. The coefficient at $n! N^{n+2-2g}$ of this polynomial is the number of connected gluings of fixed n four-stars on the plane providing ribbon graphs of genus g.

The leading coefficient (i.e., the coefficient at N^{n+2}) enumerates connected spherical gluings and it is equal to $12^n \frac{(2n-1)!}{(n+2)!}$.

The function log in the generating function of the theorem selects all "connected" gluings from arbitrary ones. An arbitrary gluing of n disjoint 4-stars can provide a disconnected graph.

The estimate of Theorem 1.1 now follows immediately. Indeed, Theorem 2.2 enumerates connected plane gluings of 4-stars independently of the number of "components" in the resulting curve. Unicursal curves are precisely the curves with one "component". However each unicursal curve with n self-intersections can be realized by different gluings of the set of n 4-stars.

Namely, consider a set of n fixed 4-stars marked by numbers $1, \ldots, n$ on the plane and a unicursal curve with n double points. A small neighborhood of each double point of the curve can be considered as a 4-star. We are going to associate to the curve $4^{n-1}(n-1)!$ different gluings of the stars. Among the double points of the curve there is one such that its leaving vector contains the starting point. Let us position this double point in the center of the star number 1 and fix the direction of the leaving vector. After that there are $4^{n-1}(n-1)!$ variants to position other $n-1$ double points at the stars $2, \ldots, n$. All these ways determine different gluings of the set of n marked 4-stars. The estimate follows.

3. Integral enumerating unicursal curves

Consider n 4-stars on the plane. We know already the way of selecting spherical gluings of the needles (by taking the coefficient at the highest degree of N in a matrix integral). What we need is to select unicursal gluings, i.e., gluings with "one component". In order to do this, let us color each pair of opposite needles in one of c colors. We consider further only *admissible* gluings, i.e., those that glue the needles of the same color.

Each gluing provides a curve with some number k of "components". Unicursal curves are precisely those that consist of one "component" only. A k-"component" curve admits c^k admissible colorings. Therefore, a unicursal curve admits precisely c colorings.

To incorporate colorings in the integral, consider an integrand of the form

$$\sum_{i,j=1}^{c} \mathrm{tr}(H_i H_j H_i H_j)^n.$$

Here we substitute the product $H_i H_j H_i H_j$ for H^4, and interchanging of indices matches admissible gluings. Integrating this expression over all spaces $\mathcal{H}_{N,i}$, $i = 1, \ldots, c$ we obtain a function, which will be a polynomial both in N and in c. The coefficient at the monomial c is precisely the generating function for unicursal gluings. Note that a unicursal curve is automatically connected.

Some of these gluings provide the same unicursal spherical curve. Each unicursal curve is represented by precisely $n!4^n$ gluings. Indeed, in each gluing we are free to permute 4-stars and to rotate each star.

Theorem 1.2 is proved.

The coefficient at the highest degree in N provides the number of spherical unicursal gluings.

References

[1] V. I. Arnold, *Plane Curves: Their Invariants, Perestroikas and Classifications*, Advances in Soviet Mathematics, **21**, (1994), 33–91.

[2] D. Bessis, C. Itzykson, J. B. Zuber, *Quantum Field Theory Techniques in Graphical Enumerations*, Adv. Appl. Math., **1** (1980), 109–157.

[3] S. M. Gusein-Zade, *On the enumeration of curves from infinity to infinity*, Advances in Soviet Mathematics, **21**, (1994), 189–198.

[4] J. Harer, D. Zagier, *The Euler Characteristic of the Moduli Space of Curves*, Inv. Math., **85** (1986), 457–485.

[5] S. K. Lando, A. K. Zvonkin, *Meanders*, Selecta Mathematica Sovietica, **11**, (1992), no. 2, 117–144.

[6] G. 't Hooft, *A Planar Diagram Theory for Strong Interactions*, Nuclear Physics B, **72**, (1974), 461–473.

INDEPENDENT UNIVERSITY OF MOSCOW AND MOSCOW INSTITUTE FOR SYSTEM RESEARCH
E-mail address: lando@mccme.ru

Vassiliev Invariants Classify Flat Braids

Alexander B. Merkov

ABSTRACT. A new series of Vassiliev invariants of ornaments, generalizing most previously known invariants, is introduced. These invariants classify decompositions of permutations into products of transpositions, formulated by S. Fomin, which can obviously be reduced to homotopy classification of certain collections of plain curves.

1. Decompositions of permutations into products of transpositions

The following problem was formulated by S. Fomin in [**F95**]. Let us consider the action of the symmetric group S_k on the k-element set $\{1,\ldots,k\}$. It is obvious that any permutation can be decomposed into a sequence of transpositions of neighboring elements. Denote by $\langle i \rangle$ the transposition of i and $i+1$, where $1 \leq i < k$. Two sequences $\langle i_1 \rangle \langle i_2 \rangle \ldots \langle i_n \rangle$ and $\langle i'_1 \rangle \langle i'_2 \rangle \langle i'_3 \rangle \ldots \langle i'_{n'} \rangle$ are said to be equivalent if one can be transformed into the other by several steps of the following types:

1. Insertion or deletion of a pair of identical transpositions at any place; that is, $\langle i \rangle \langle i \rangle \leftrightarrow \emptyset$.
2. Transposing a pair of neighboring "independent" transpositions; that is $\langle i \rangle \langle j \rangle \leftrightarrow \langle j \rangle \langle i \rangle$ whenever $|i - j| > 1$.

Such steps obviously preserve the product of the transpositions. The problem is if there exists an efficient way to verify whether two given sequences of transpositions are equivalent or not.

The answer is positive. We will show the existence of a simple algorithm that verifies the equivalence of transpositions and construct a system of integer-valued functions defined on such sequences of transpositions which are constant on the equivalence classes and distinguish elements of different classes. But first we give some other formulations of this combinatorial problem.

Algebraic: To solve the identity problem in the group defined by $k-1$ generators $g_1, \ldots, g_{k-1}$ and $(k-1)+(k-2)(k-3)/2$ relations $(g_i)^{-1} = g_i$ and $g_i g_j = g_j g_i$ when $|i-j| > 1$.

1991 *Mathematics Subject Classification.* Primary 57M25.

Supported by INTAS grant, project #4373, Netherland Organization for Scientific Research (NWO), project 47.03.005, and by RFBR grant, project 95-01-00846.

©1999 American Mathematical Society

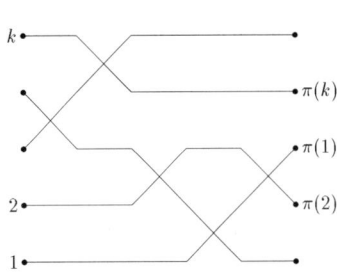

FIGURE 1. Decomposition of a permutation π into transpositions

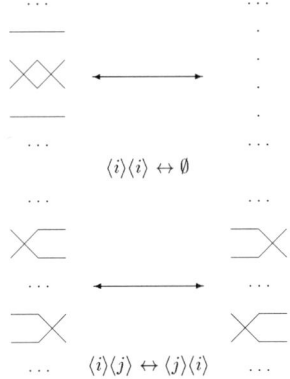

FIGURE 2. Equivalent fragments

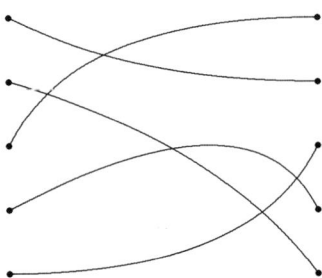

FIGURE 3. Flat braid

FIGURE 4. Admissible local move

Combinatorial-geometric: To classify the collections of k piecewise linear curves shown in Figure 1 up to equivalence generated by moves shown in Figure 2. This formulation is just a picture of the original one: the collection is composed of a number of copies of (some of) $k-1$ vertical strips of unit width $s_1, \ldots, s_{k-1}$; within strip s_i points $(x-1, 1), (x-1, 2), \ldots, (x-1, k)$ are connected to points $(x, 1), (x, 2), \ldots, (x, k)$ in such a way that $(x-1, i)$ is connected to $(x, i+1)$, $(x-1, i+1)$ to (x, i), and the y-coordinates of other points are preserved.

Geometric: To classify up to homotopy the collections of k smooth functions $\phi_1, \ldots, \phi_k : [0,1] \to \mathbf{R}$, whose graphs are similar to the ones shown in Figure 3, with fixed endpoints ($\phi_j(0) = j$ for $1 \leq j \leq k$, and $\{\phi_1(1), \ldots, \phi_k(1)\} = \{1, \ldots, k\}$), finitely many simple intersections of the graphs and forbidden multiple intersections. (The generic admissible homotopy does not change the topology of the collection of the graphs except for finitely many instants, when the move shown in Figure 4 occurs.) We will call such a collection a *flat braid*[1] of k threads.

It is obvious that for all the formulations above the equivalence classes to be classified form naturally isomorphic groups. The isomorphism is generated by the maps

$$g_{i_1} \ldots g_{i_n} \quad \mapsto \quad \langle i_1 \rangle \ldots \langle i_n \rangle$$
$$\mapsto \quad s_{i_1} \ldots s_{i_n} \text{concatenated from right to left}$$
$$\mapsto \quad \text{its smooth approximation scaled horizontally by } 1/n.$$

---

[1]The same object was called *twin* by Khovanov in [**Kh94**] and [**Kh96**].

This isomorphism commutes with the natural homomorphisms of each of the groups to S_k.

Although the classification of the collections of smooth functions looks the most terrible (the classified subject is neither discrete nor even finite-dimensional), it is a rather simple case of homotopy classification of collections of plane curves developed in [**V93**] and [**M95**]. A brief review of this classification is given in Section 2. In Section 3 some new Vassiliev invariants are introduced. Since Vassiliev invariants classify dyed braids up to homotopy (see [**B-N95**]), one can expect that they classify flat braids too. This is true and will be proved in Section 4 using a special case of invariants of Section 3.

2. Vassiliev invariants of ornaments and doodles

First, two ambiguities in the title of this section should be resolved. According to Fenn and Taylor ([**FT77**]), a *doodle* is a finite collection of closed curves in the 2-sphere or plane (or in another two-dimensional manifold) with no self-intersections and no triple intersections.[2] According to Vassiliev ([**V93**]) an *ornament* is a more general object. Namely, it is a finite collection of closed curves in the plane with no three different curves meeting at one point; self-intersections and triple intersection points of not more than two curves are allowed.

The notions "finite order invariant", "finite type invariant" and "Vassiliev invariant" coincide identically. The first one was introduced by V.Vassiliev himself, and the other two are used by the majority of the other authors.

2.1. Homological theory.
What follows is an informal sketch of the theory of finite order invariants of ornaments. See [**V93**] and [**M95**] for details and [**V94**] for more general theory.

All invariants of ornaments can be expressed via the Alexander duality in terms of closed homology of the *discriminant*, i.e., of the set of collections of curves with forbidden triple points. The *finite order invariants* are exactly those which can be expressed in terms of a finite number of strata of the natural stratification of the discriminant induced by the classification of singular points. Since the discriminant is a very singular space, it is more convenient to study its *resolution*, which is a topological space homotopy (and hence, homology) equivalent to the discriminant. Although these spaces are infinite dimensional, the approximation technique developed in [**V93**] allows one to consider them as finite dimensional spaces of some big dimension ∞. There exists a natural filtration of the resolution space, and the corresponding homological spectral sequence and the dual cohomological spectral sequence $E_r^{p,q}$, $r \geq 1$, $p \leq 0$, $p + q \geq 0$, can be written. There exists an algorithm calculating this spectral sequence. For each n the group $\oplus_{p=0}^{n} E_\infty^{-p,p}$ is the graded group adjoined to the group of *invariants of order* n. There is an interesting question of whether or not this spectral sequence converges to cohomology of the space of ornaments. The question is still open. The same question is also open for the parallel theory of invariants of knots. It was answered positively for invariants of string links by Bar-Natan in [**B-N95**].

Another problem is to describe all finite order invariants explicitly. By now even the groups $E_1^{-p,p}$ are not calculated in general, though their dimensions are estimated from above by some combinatorial formulas. Nor the conjecture that

---

[2]According to Khovanov ([**Kh94**]) self-intersections are allowed.

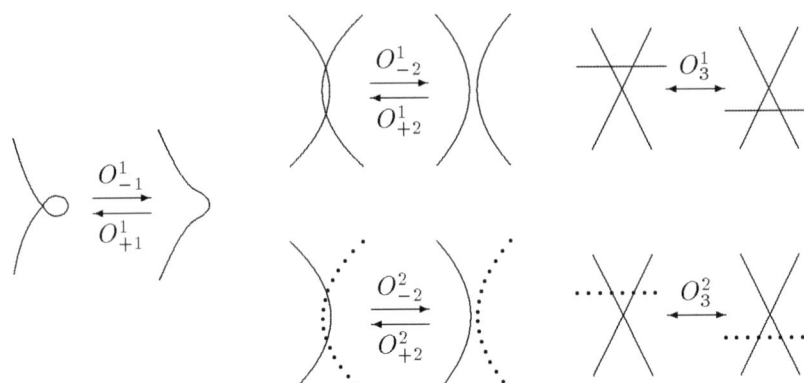

FIGURE 5. Admissible local moves for ornaments. The upper index indicates the number of participating components. The lower index indicates the number of participating intersection points or its jump.

$E_1^{-p,p} = E_\infty^{-p,p}$ is proved (a similar fact for knots was proved by Kontsevich in [**K93**]).

2.2. Elementary theory. The homological definition of the order of invariants is not convenient for calculation of the order of an explicitly constructed invariant. Fortunately, it can be reformulated in elementary terms, see [**V93**]. To do this we need several definitions and some notation.

2.2.1. *Ornaments and quasiornaments.* Denote by C_k the disjoint union $c_1 \sqcup \cdots \sqcup c_k$ of k circles.

DEFINITION 1 ([**V93**]). A *k-ornament* (or simply *ornament*) is a C^∞-smooth map $C_k \to \mathbb{R}^2$ such that the images of no three different circles intersect at the same point in $\mathbb{R}^2$. Two ornaments are *equivalent*, if the corresponding maps $C_k \to \mathbb{R}^2$ can be connected by a homotopy $C_k \times [0,1] \to \mathbb{R}^2$ such that for any $t \in [0,1]$ the corresponding map of $C_k \times t$ is an ornament.

DEFINITION 2 ([**V93**]). A *k-quasiornament* (or simply *quasiornament*) is a C^∞-smooth map $C_k \to \mathbb{R}^2$.

DEFINITION 3 ([**V93**]). A *k*-ornament is *regular* if it is an immersion of C_k and all the multiple points of the image of C_k in $\mathbb{R}^2$ are simple transversal intersection points only. A *k*-quasiornament is *regular* if it is an immersion of C_k, at any multiple point of the image all local components meet pairwise transversally, and the preimage of each multiple point either consists of two points, or contains points of at least three circles c_l.[3]

Regular ornaments form a dense open set in the space of quasiornaments. Given a generic path in the space of ornaments, almost all ornaments along this path remain regular. At a finite number of singular instants one of the local moves shown in Figure 5 occurs. Below we consider regular (quasi)ornaments only.

---
[3]The last condition was not stated explicitly in [**V93**].

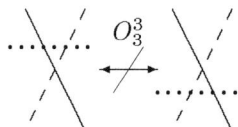

FIGURE 6. Forbidden local moves for ornaments

REMARK. Ornaments and quasiornaments can be defined similarly in any 2-dimensional manifold, e.g., in the sphere $\mathbf{S}^2$, instead of the plane $\mathbb{R}^2$. Though the homological theory is the simplest for plane ornaments, some elementary theorems are more natural for ornaments in $\mathbf{S}^2$.

2.2.2. *Examples of invariants.* Let us fix an orientation of $\mathbb{R}^2$. Let x be a point of $\mathbb{R}^2$. Recall that the index of a point $x \in \mathbb{R}^2$ with respect to a closed oriented curve $c: \mathbf{S}^1 \to \mathbb{R}^2$ not containing x is the rotation number of the vector $c(t) - x$ as t runs along $\mathbf{S}^1$. Denote by $\mathrm{ind}_l(x) = \mathrm{ind}_{\phi(c_l)}(x)$ the index of x with respect to the l-th component c_l of a regular k-quasiornament ϕ if $x \notin \phi(c_l)$, and the smallest value of the index in a small neighborhood of x is $x \in \phi(c_l)$.

If x is a simple intersection of the i-th component c_i and j-th component c_j, let $\sigma_{ij}(x)$ equal 1 if the orientation given by the tangent frame $(t_x(c_i), t_x(c_j))$ coincides with the orientation of $\mathbb{R}^2$, and -1 otherwise.

To any regular k-ornament ϕ and numbers $i, j, 1 \leq i < j \leq k$, there corresponds the following integer-valued function $I_{ij}(b_1, \ldots, b_k)$ of integer arguments:

$$I_{ij}(b_1, \ldots, b_k)(\phi) = \sum_{\substack{x \in \phi(c_i) \cap \phi(c_j) \\ \mathrm{ind}_l(x) = b_l, 1 \leq l \leq k}} \sigma_{ij}(x).$$

THEOREM 1 ([**V93**, Theorem 3]). *For any $1 \leq i < j \leq k$ and integer $b_1, \ldots, b_k$, the function $I_{ij}(b_1, \ldots, b_k)$ is an invariant of ornaments.*

For any k integer nonnegative exponents $\beta = (\beta_1, \ldots, \beta_k)$ we can define the functions $V_{ij}(\beta)$ of k-ornaments as

$$V_{ij}(\beta) = \sum_{b_1, \ldots, b_k = -\infty}^{\infty} \binom{b_1}{\beta_1} \cdot \ldots \cdot \binom{b_k}{\beta_k} \cdot I_{ij}(b_1, \ldots, b_k),$$

where $1 \leq i < j \leq k$ and

$$\binom{b_l}{\beta_l} = \frac{b_l \cdot (b_l - 1) \cdot \ldots \cdot (b_l - \beta_l + 1)}{\beta_l!},$$

regardless of whether b_l is positive or negative. Obviously, $V_{ij}(\beta)$ are integer-valued invariants. These *index momenta invariants* are the simplest finite order invariants (see Theorem 2 below).

2.2.3. *Singular points and degeneration modes.*

DEFINITION 4 ([**V93**]). A point $x \in \mathbb{R}^2$ is a *singular point of complexity j* of a regular quasiornament $\phi: C_k \to \mathbb{R}^2$, if $\phi^{-1}(x)$ consists of exactly $j + 1$ points, at least three of which belong to different components of C_k. The *complexity* $\mathrm{compl}(\phi)$ of a regular quasiornament ϕ is the sum of the complexities of all its singular points.

DEFINITION 5 ([**V93**]). Suppose that a regular quasiornament ϕ has m singular points $x^1,\ldots,x^m$. A *degeneration mode* of ϕ is an order of marking all the points of their preimages, satisfying the following conditions. At any step we mark either three points of $\phi^{-1}(x^l)$ for x^l belonging to three different components of C_k (if none of the points of $\phi^{-1}(x^l)$ is already marked), or one point (if three or more points of $\phi^{-1}(x^l)$ are already marked).

2.2.4. *Characteristic numbers.* The *characteristic numbers* assigned to regular quasiornaments and their degeneration modes by invariants are integer numbers defined inductively by the complexity of quasiornaments.

The characteristic number assigned to an ornament (whose only degeneration mode is empty) is the value of the invariant of the ornament. The last step of the degeneration mode of the k-quasiornament is either marking of a triple of points on different components c_i, c_j and c_l, $i < j < l$, or marking a single point on c_l. Consider a local transversal perturbation of the l-th component of the k-quasiornament ϕ in a small neighborhood of one of the marked points, i.e., a shift of a small piece of the image $\phi(c_l)$ through the intersection point x transversally to the tangent line to $\phi(c_l)$ in x. The perturbed quasiornaments are of smaller complexity than the initial one, and belong to two different equivalence classes. One of them can be called "positive" and the other "negative". Namely, if x is a triple intersection of $\phi(c_i)$, $\phi(c_j)$, and $\phi(c_l)$, then the "positive" k-quasiornaments are those with greater value of $\sigma_{ij}(x)\operatorname{ind}_l(x)$, and if x is a more complicated intersection, then the "positive" k-quasiornaments are those with greater value of $\operatorname{ind}_l(x)$. The characteristic number of the given k-quasiornament and the degeneration mode is defined as the difference of the characteristic numbers of the "positive" and "negative" k-quasiornaments for the degeneration modes, obtained from the given one by removing the last step.

2.2.5. *Finite order invariants.*

DEFINITION 6 ([**V93**]). An invariant of ornaments is an *invariant of order i* if all characteristic numbers assigned by it to any regular quasiornament of complexity $> i$ vanish.

THEOREM 2 ([**V93**, Theorem 4]). *$V_{ij}(\beta)$ is an invariant of order $|\beta|+1$, where $|\beta| = \beta_1 + \cdots + \beta_k$.*

3. New finite order invariants of ornaments

3.1. Chord diagrams. Different kinds of diagrams of ornaments were introduced in [**FT77**], [**M94**] and [**M95**]. Here and in subsection 3.2 we describe the construction of [**M95**] using the notation of [**PV94**] and [**P94**].

DEFINITION 7. A *configuration of length n* is a collection of $2n$ different points in C_k divided into n pairs of *adjoined points*. An ornament $\phi: C_k \to \mathbb{R}^2$ *respects* the configuration if it sends each of the pairs of points into one point in $\mathbb{R}^2$. Two configurations are *isomorphic* if one can be mapped onto the other by a diffeomorphism $C_k \to C_k$ preserving each circle of C_k and its orientation.

These configurations are a natural generalization of *chord diagrams* (or *unoriented Gauss diagrams*) to collections of curves: each pair of adjoined points can be substituted by a chord (i.e., a homeomorphic image of a segment) connecting

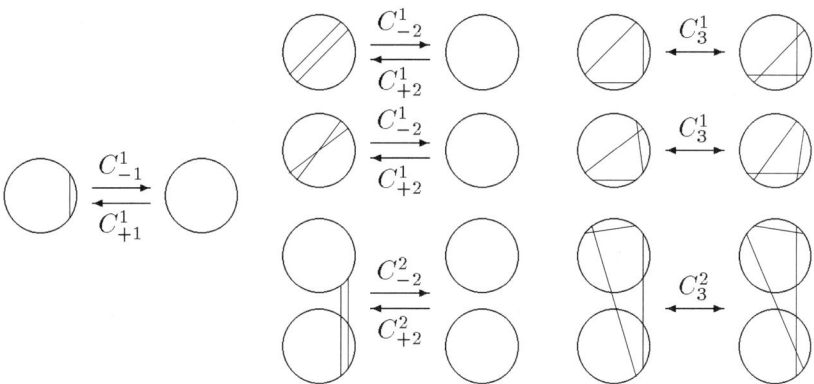

FIGURE 7. Admissible moves of chord diagrams. Unchanged chords are omitted. For moves of types $C^2_{\pm 2}$ and C^2_3 only one of several possible pairs of diagrams is shown.

them. From now on by "chord diargam" we mean a class of isomorphic configurations. They may be also thought of as classes of homeomorphic 1-dimensional cell complexes with k oriented circles marked and numbered.

The chord diagram corresponding to the full preimage of the union of all intersection points of a regular k-ornament is called its *full chord diagram*.

Let us define equivalence of chord diagrams in a way, guaranteeing that the full diagrams of equivalent ornaments are also equivalent. The equivalence is defined by the following *admissible moves*, corresponding to admissible local moves of ornaments, shown in Figures 7 and 5, respectively.

C^1_{-1} (See local move O^1_{-1} in Figure 5.) Deletion of a chord of some of the circles whose ends are neighboring (i.e., are not separated by the ends of other chords).

C^1_{+1} (See local move O^1_{+1} in Figure 5.) The move opposite to C^1_{-1}.

C^1_{-2} (See local move O^1_{-2} in Figure 5.) Annihilation of two chords of some of the circles whose ends form two pairs of neighboring points.

C^1_{+2} (See local move O^1_{+2} in Figure 5.) The move opposite to C^1_{-2}.

C^2_{-2} (See local move O^2_{-2} in Figure 5.) Annihilation of two chords connecting two circles whose ends form two pairs of neighboring points.

C^2_{+2} (See local move O^2_{+2} in Figure 5.) The move opposite to C^2_{-2}.

C^1_3 (See local move O^1_3 in Figure 5.) Permutation $(u, v, w, x, y, z) \mapsto (v, u, x, w, z, y)$ of the ends of the chords $[z, u]$, $[v, w]$ and $[x, y]$, where (y, z), (u, v) and (w, x) are pairs of neighboring points on the same circle (this is an involution).

C^2_3 (See local move O^2_3 in Figure 5.) Permutation $(u, v, w, x, y, z) \mapsto (v, u, x, w, z, y)$ of the ends of the chords $[z, u]$, $[v, w]$ and $[x, y]$, where (u, v) and (w, x) are pairs of neighboring points on one circle, and (y, z) is a pair of neighboring points on another circle (this also is an involution).

The upper index of a move denotes the number of participating circles and the lower one denotes either the number of participating chords (if unsigned), or its jump (if signed).

The equivalence class of the full chord diagram is obviously an invariant of the ornament. This invariant is rather rough: it describes the inner topology of the image of the ornament and completely ignores the topology of the ambient manifold: plane, sphere, etc. This invariant is nonnumerical. The description of the classes is simplified by the fact that each equivalence class contains only a finite number of chord diagrams of minimal length, but the problem of verifying whether or not two given chord diagrams are equivalent is hard.

Below the following operations over this chord-diargam-valued invariant will be undertaken:

- the chord diagrams will be simplified to provide easy verification of their equivalence (see the present subsection);
- the chord diagrams will be complexified to take into account the topology of small neighborhoods of the images of ornaments (see subsection 3.2);
- the invariant will be turned into numeric invariants. These invariants will be of finite order (see subsection 3.4).

For any set of pairs of circles in C_k one can delete from the full chord diagram the chords connecting these pairs of circles. The admissible moves of such *partial chord diagrams* can be obtained naturally from the moves of full chord diagrams. The corresponding equivalence classes are rougher invariants of ornaments.

The following type of partial chord diagrams deserves a separate name.

DEFINITION 8. A *non-self-intersecting chord diagram* is a partial chord diagram with no chords whose both ends belong to the same circle. The non-self-intersecting chord diagram $\mathrm{CDiag}(\phi)$ of an ornament ϕ is the partial chord diagram corresponding to the full preimage of intersection points of a regular ornament ϕ except the self-intersections of its components.

The non-self-intersecting chord diagram of an ornament with no self-intersections of the components (i.e., of a doodle in the sense of [**FT77**]) coincides with its full diagram. The set of the admissible moves of a non-self-intersecting chord diagram is simplified significantly. The moves C_*^1 vanish, the moves $C_{\pm 2}^2$ remain unchanged, and the moves C_3^2 are simplified as follows:

c_3^2 Permutation of two neighboring ends of two chords connecting the same pair of different circles.

Notation: non-self-intersecting chord diagram X is denoted by $(x^1, \ldots, x^n)$, where $x^l = [x_{i^l}^l, x_{j^l}^l]$, $l = 1, \ldots, n$, are all its chords, $x_{i^l}^l \in c_{i^l}$ and $x_{j^l}^l \in c_{j^l}$ are the ends of the l-th chord, $1 \leq i^l < j^l \leq k$. The length of the diagram X is denoted by $L(X) = n$.

PROPOSITION 1 (Maximum principle, [**M95**, Proposition 38]). *For each sequence of admissible moves of non-self-intersecting chord diagrams there exists a sequence of moves with the same ends such that the length of each intermediate diagram does not exceed the maximum of the lengths of the end diagrams.*

To prove this proposition we introduce a more general class of diagrams.

DEFINITION 9. A *multichord diagram* is a set of k oriented circles connected by finitely many chords with a positive *multiplicity* assigned to each chord, satisfying the following conditions:
- each chord connects two different circles;
- if two chords have a common endpoint on some circle, then the other two endpoints are different points on another circle.

Similar to the usual (non-self-intersecting) chord diagrams, which are a special case of multichord diagrams, the latter are defined up to orientation preserving diffeomorphisms of the circles.

If an arc on a circle is bounded by two different endpoints of two chords of a multichord diagram, whose other endpoints belong to another circle, the same for both chords, we can define a *contraction* of the diagram along this arc in the following way:
- the arc is removed and its ends are pasted together;
- if the endpoints of the two chords on the other circle coincide, then the chords are pasted together and the multiplicity of the new chord is the sum of the multiplicities of the two old chords.

Similarly, a multichord diagram can be contracted along several chords at once. If a multichord diagram cannot be contracted at all, it is called *uncontractible*.

LEMMA 1. *Among the diagrams to which the given diagram can be contracted, there is exactly one uncontractible diagram.*

PROOF. Let us mark all the arcs along which the given diagram can be contracted. Then for those circles whose arcs are all marked, we unmark an arbitrary arc. After that we contract the diagram along all arcs that are still marked. It is easy to see that the diagram obtained is the required uncontractible diagram, and does not depend on the choice of the unmarked chords. □

The uncontractible multichord diagram provided by Lemma 1 is called the *passport* of the given diagram.

LEMMA 2. *Two non-self-intersecting chord diagrams can be transformed one into another by a sequence of moves of type c_3^2 if and only if they have the same passport.*

PROOF. The "only if" part is obvious because moves of the type c_3^2 preserve the passport. Since the preimages of each endpoint of the passport can be permuted by moves of the type c_3^2 independently, the "if" part follows from the existence of decomposition of a permutation into transpositions of neighboring elements. □

LEMMA 3. *Proposition 1 holds for a sequence of admissible moves of the types C_{+2}^2, c_3^2, ... ,c_3^2, C_{-2}^2.*

PROOF. We have a sequence $X_0, X_1, \ldots, X_q$ of non-self-intersecting chord diagrams connected by a sequence $M_1, \ldots, M_q$ of admissible moves, the first move M_1 is of the type C_{+2}^2, the last move M_q is of the type C_{-2}^2, and all the intermediate moves are of the type c_3^2. The diagrams $X_1, \ldots, X_{q-1}$ consist of the same chords: only their mutual position is changed by the moves. To prove the lemma we consider two cases.

1. The images of the pair of chords born at the first move M_1, and of the pair of chords killed at the last move M_q, coincide in the common passport of diagrams $X_1, \ldots, X_{q-1}$.
2. They are different.

In the first case the passports of diagrams X_0 and X_q are obtained from the passport of X_1, which coincides by Lemma 2 with the passport of X_{q-1}, by decreasing by 2 the multiplicity of the same chord (or by deleting it and possibly further contraction of the passport if its multiplicity is 2). Hence, by Lemma 2, X_0 can be transformed to X_q by a sequence of moves of type c_3^2. Such moves preserve the length of the diagram.

In the second case we choose in the sequence $M_2, \ldots, M_{q-1}$ of moves of type c_3^2 three subsequences $M_1^i, \ldots, M_{n^i}^i$, $i = 0, 1, 2$, as follows:

- M^0 is the naturally renumbered subsequence, consisting of all moves, in which none of the two born or two killed chords participates;
- M^1 is the naturally renumbered subsequence, consisting of all moves, in which none of two killed chords participates;
- M^2 is the naturally renumbered subsequence, consisting of all moves, in which none of two born chords participates.

(Obviously, M^0 is a subsequence of each of M^1 and M^2, and $M = M^1 \cup M^2$.) Then the sequence of moves

$$M_1^2, \ldots, M_{n^2}^2, M_q, (M_{n^0}^0)^{-1}, \ldots, (M_1^0)^{-1}, M_1, M_1^1, \ldots, M_{n^1}^1$$

being applied to X_0, transforms it to X_q, and the length of each intermediate diagram is less than or equal to $L(X_0) = L(X_q)$. $\square$

PROOF OF PROPOSITION 1. Let $X_0, X_1, \ldots, X_z$ be a sequence of non-self-intersecting chord diagrams connected by a sequence $M_1, \ldots, M_z$ of admissible moves, L_j stands for the length $L(X_j)$, and $L_X = \max_{0 \leq j \leq z}(L_j)$. It follows from Lemma 3, that any segment $X_p, X_{p+1}, \ldots, X_{q-1}, X_q$ of the sequence, such that $L_p = L_q < L_X$ and $L_{p+1} = \cdots = L_{q-1} = L_X$ can be replaced by a sequence $X_p = Y_0, Y_1, \ldots, Y_r = X_q$ of diagrams, connected by admissible moves, with $L_Y < L_X$. Indeed, applying finitely many such substitutions, one can decrease the number of intermediate diagrams X_j of length $L_j = L_X$ to zero, and hence decrease L_X. This process can be repeated until $L_X = \max(L_0, L_z)$. $\square$

COROLLARY 1. *Among non-self-intersecting chord diagrams equivalent to the diagram of an ornament, there is a finite number of pairwise nonisomorphic diagrams of a minimal length. The sequence of moves turning the diagram to a minimal diagram does not contain moves of the type C_{+2}^2. There exists an algorithm producing all the minimal diagrams equivalent to a given one. In particular, there exists an algorithm checking the equivalence of two given diagrams.* $\square$

DEFINITION 10. A non-self-intersecting chord diagram is *prime* if no move of the type c_3^2 or C_{-2}^2 can be applied to it. A diagram is *minimal* if moves of the type c_3^2 are possibly applicable, but no move of the type C_{-2}^2 can be applied after any number of moves of the type c_3^2.

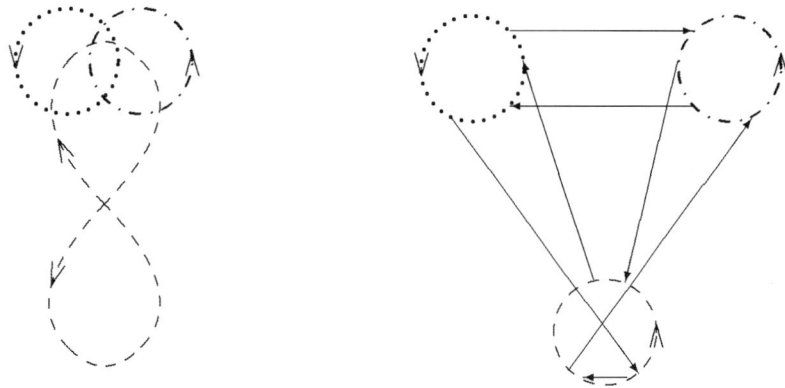

FIGURE 8. An ornament and its arrow diagram

3.2. Arrow diagrams. From now on we consider ornaments in an oriented surface (in the plane, unless otherwise stated). Since the orientations of the components of ornaments are fixed, the chords of the full diagram of an ornament ϕ can be oriented canonically. Namely, if $t(x)$ stands for the unit positive tangent vector at the point $x \in C_k$, then a chord $[x_i, x_j]$ is oriented from x_i to x_j (the arrow is directed from x_i to x_j) if the tangent frame $(\phi_*(t(x_i)), \phi_*(t(x_j)))$ is positive, and vice versa. Such diagrams are called (full) *arrow diagrams* or *oriented Gauss diagrams*.

EXAMPLE. An ornament and its full arrow diagram are shown in Figure 8.

The theory of arrow diagrams is parallel to the theory of chord diagrams. The admissible local moves of ornaments induce the set of *admissible moves* A_*^* of arrow diagrams. The only difference between the moves A_*^* and C_*^* is that the directions of the arrows in moves $A_{\pm 2}^*$ must be opposite. The equivalence class of the arrow diagram of an ornament is a stronger invariant than the class of its chord diagram.

The boundary of each component of the complement of the image of an ornament consists of several canonically oriented non-self-intersecting piecewise-smooth closed curves. The preimage of such a curve in the full arrow diagram of the ornament is a 1-cycle. The set of all such cycles is determined by the arrow diagram unambiguously. Indeed, the positive orientation of each such cycle is defined by its regular point and the direction in it in the following way. The running point moves along the components of the arrow diagram; every time it meets an endpoint of an arrow, it jumps to its other endpoint and goes on along the other (or, maybe, the same) component in the same (positive or negative) direction as before if it has jumped along the arrow, and in the opposite direction otherwise. And conversely, the directions of the arrows can be reconstructed by the preimages in the chord diagram of the canonically oriented boundary components.

LEMMA 4. *Let ϕ and ψ be regular k-ornaments in the sphere $\mathbf{S}^2$ with connected images $\phi(C_k)$ and $\psi(C_k)$ and the same full arrow diagram. Then the map $\psi \circ \phi^{-1} : \phi(C_k) \to \psi(C_k)$ can be extended to an orientation preserving diffeomorphism $\mathbf{S}^2 \to \mathbf{S}^2$.*

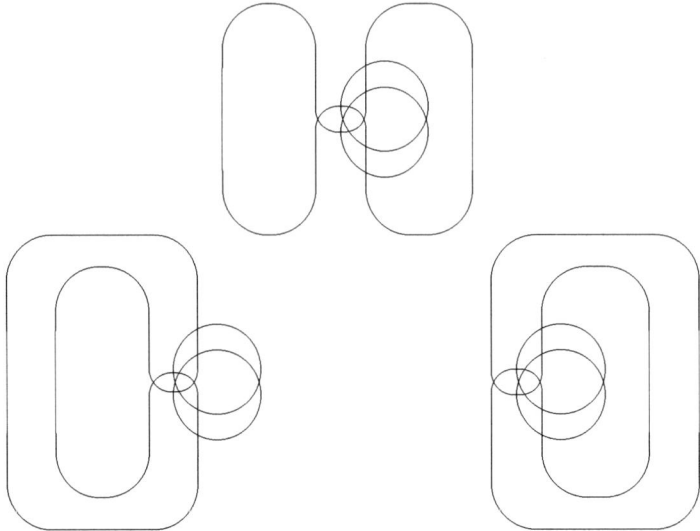

FIGURE 9. $\mathbf{S}^2$-equivalent pairwise nonequivalent ornaments with the same arrow diagram. The indexed arrow diagram of the upper ornament differs from the indexed arrow diagram of the lower ornaments

PROOF. Since $\psi \circ \phi^{-1}$ preserves the orientation of tangent frames at each intersection or self-intersection point, it can be extended to a diffeomorphism of a thin neighbourhood U of $\phi(C_k)$ with smooth boundary. The arrow diagram determines which component of $\mathbf{S}^2 \setminus U$ must go to which component of $\mathbf{S}^2 \setminus \psi \circ \phi^{-1}(U)$. Because the ornament is connected, each component has connected boundary and is diffeomorphic to a disk, and each diffeomorphism of the boundaries can be extended to a diffeomorphism of the components. Standard technical tricks provide smoothness on ∂U. □

The following theorem is probably known, but I have no exact reference.

THEOREM 3. *A connected ornament in sphere $\mathbf{S}^2$ is determined up to equivalence by its full arrow diagram.*

PROOF. The theorem follows from Lemma 4 and the well-known fact that the space of orientation preserving diffeomorphisms $\mathbf{S}^2 \to \mathbf{S}^2$ can be contracted onto $SO(3)$ and hence is connected (see, e.g., [**S59**]). □

REMARK. An ornament in $\mathbb{R}^2$ cannot be reconstructed from its full arrow diagram unless its exterior contour is marked on the diagram (see Figure 9). A disconnected ornament cannot be reconstructed from its arrow diagram even in $\mathbf{S}^2$ (see Figure 10).

As in the case of chord diagrams, certain *partial arrow diagrams*, namely the diagrams ignoring self-intersections, are much more convenient to study than full

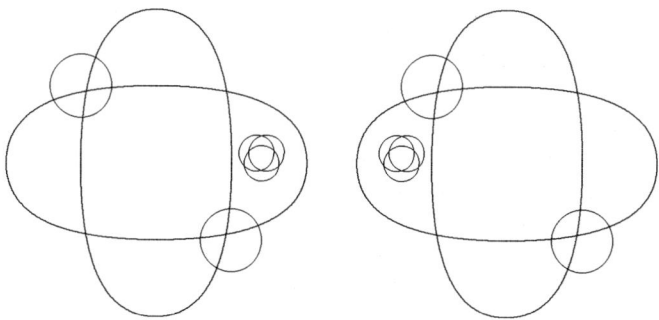

FIGURE 10. Nonequivalent ornaments with the same indexed arrow diagram. The three-component subdoodle is moved as a whole within the four-component one

arrow diagrams. Such a *non-self-intersecting arrow diagram* of a regular ornament ϕ is denoted by $\mathrm{ADiag}(\phi)$. These non-self-intersecting diagrams can be considered also as non-self-intersecting chord diagrams with the following signs assigned: the sign $\sigma(x)$ of a chord $x = [x_i, x_j]$, $i < j$, is equal to 1 if the arrow is directed from x_i to x_j, and to -1 otherwise. Then the notation for arrow diagrams is as follows: $X = ((x^1, \sigma(x^1)), \ldots, (x^n, \sigma(x^n)))$, where $x^l = [x_{i^l}^l, x_{j^l}^l]$, $l = 1, \ldots, n$ are all its chords. If the signs are induced by a regular ornament ϕ respecting the chord diagram $x^1, \ldots, x^n$, then $\sigma(x^l)$ should be substituted by $\sigma(x^l, \phi)$. Obviously, $\sigma(x^l, \phi) = \sigma_{i^l j^l}(\phi(x^l))$ (see 2.2.2).

Let us list the admissible moves of non-self-intersecting arrow diagrams explicitly:

A_{-2}^2 Annihilation of two arrows connecting two circles, if the beginning of each arrow is a neighbor of the end of the other one.

A_{+2}^2 The move opposite to A_{-2}^2.

A_3^2 (coincides with c_3^2) Permutation of two neighboring ends of two chords (= arrows with the orientation ignored), connecting the same pair of different circles.

Proposition 1, Corollary 1 and their proofs remain valid for arrow diagrams without any change. Moreover, Proposition 38 in [**M95**], which is the predecessor of Proposition 1, deals with arrow diagrams.

Another kind of partial arrow diagram is shown in the following example. It is not used in this paper.

EXAMPLE. Let us consider the following k partial arrow diagrams of a k-ornament: the l-th diagram consists of those arrows of the full diagram, which are directed to or from the l-th circle c_l. The moves A_*^* can be naturally redefined for such diagrams. Namely, in the move A_{-2}^2 the end of one arrow must be neigboring to the beginning of the other arrow on the l-th circle only. The k equivalence classes obtained are the Fenn–Taylor invariant defined in [**FT77**].

3.3. Indexed arrow diagrams. To each intersection point x of a regular ornament ϕ, or to each arrow of its diagram, a k-dimensional integer vector $\mathrm{Ind}(x, \phi) = (\mathrm{ind}_1(x), \ldots, \mathrm{ind}_k(x))$ can be assigned. Its values are neighboring for neighboring intersection points in a natural sense. In fact, the index-vectors of one intersection point per connected component of the image of the ornament determine all the other index-vectors unambiguously. The definitions of admissible moves and equivalence might be refined easily to reflect what must happen with the indices of intersection points. The analogue of Proposition 1 remains true. The indexed arrow diagram of an ornament (considered up to equivalence) is a stronger invariant than the simple arrow diagram. The latter can never distinguish $\mathbf{S}^2$-equivalent ornaments in $\mathbb{R}^2$, while the former sometimes can (see Figure 9).

Again, it is useful to consider *non-self-intersecting indexed arrow diagrams*, ignoring all the information about self-intersections of ornaments. Namely, to each arrow x^l of the non-self-intersecting diagram $\mathrm{ADiag}(\phi)$ of an ornament ϕ, the $(k-2)$-dimensional integer vector

$$\mathrm{Ind}'(x,\phi) = (\mathrm{ind}_1(x^l), \ldots, \widehat{\mathrm{ind}_{i^l}(x^l)}, \ldots, \widehat{\mathrm{ind}_{j^l}(x^l)}, \ldots, \mathrm{ind}_k(x^l))$$

is assigned.

3.4. Invariants generated by chord diagrams.
Chord or arrow diagrams are 1-dimensional cell complexes with certain additional structure. Let us consider embeddings of a chord diagram X into an arrow diagram Y, which map each circle c_l onto itself, preserving its orientation. Following Polyak, let us call a class of homotopy equivalence of such embeddings a *representation* of a chord diagram X in an arrow diagram Y.

Given a representation $\mu : X \to Y$, denote by $\sigma(\mu)$ the product

$$(1) \qquad \sigma(\mu) = \prod_{l=1}^{L(X)} \sigma(\mu(x^l)),$$

and by $\langle X, Y \rangle$ the sum

$$(2) \qquad \langle X, Y \rangle = \sum_{\mu : X \to Y} \sigma(\mu)$$

over all representations $\mu : X \to Y$.

Given a regular k-ornament ϕ, let us define $\langle X, \phi \rangle$ by

$$(3) \qquad \langle X, \phi \rangle = \langle X, \mathrm{ADiag}\,\phi \rangle = \sum_{\mu : X \to \mathrm{ADiag}\,\phi} \prod_{l=1}^{L(X)} \sigma(\mu(x^l), \phi).$$

Practically the same construction was used by Polyak and Viro in [**PV94**] for invariants of knots and links and by Polyak in [**P94**] for invariants of plane curves.

PROPOSITION 2. *If X is a prime chord diagram, then $\langle X, \cdot \rangle$ is an invariant of ornaments.*

PROOF. We need to verify that the admissible local moves of the ornaments do not change $\langle X, \cdot \rangle$. To do this, it is enough to verify that moves of the types A_3^2, A_{-2}^2 and A_{+2}^2 of the arrow diagrams of ornaments do not change $\langle X, \cdot \rangle$. This is elementary. □

This class of invariants can be easily generalized. Let us fix a numbering of chords of the chord diagram X of length n. Let F be an integer valued function of $(k-2) \times n$ integer variables. Given a k-ornament ϕ, let us define $\langle (X, F), \phi \rangle$ as the sum

(4) $$\langle (X, F), \phi \rangle = \sum_{\mu: X \to \mathrm{ADiag}(\phi)} \sigma(\mu) F(\mathrm{Ind}'(\mu(x^1), \phi), \ldots, \mathrm{Ind}'(\mu(x^n), \phi))$$

over all representations $\mu: X \to \mathrm{ADiag}(\phi)$. In fact, $\langle (X, F), \phi \rangle$ depends not on the ornament ϕ itself but on its indexed diagram.

For instance, if $F = \times_{l=1}^n f^l$, where each f^l is a function of $(k-2)$ integer variables, the formula (4) may be rewritten as

(5) $$\langle (X, F), \phi \rangle = \sum_{\mu: X \to \mathrm{ADiag}(\phi)} \sigma(\mu) \prod_{l=1}^n f^l(\mathrm{Ind}'(\mu(x^l), \phi))$$
$$= \sum_{\mu: X \to \mathrm{ADiag}(\phi)} \prod_{l=1}^n \sigma(\mu(x^l), \phi) f^l(\mathrm{Ind}'(\mu(x^l), \phi)).$$

THEOREM 4. *If X is a prime chord diagram of n chords and F is an integer valued function of $(k-2) \times n$ integer variables, then $\langle (X, F), \cdot \rangle$ is an invariant of ornaments.*

PROOF. Similarly to the proof of Proposition 2, one needs just to verify that moves of the types A_3^2, A_{-2}^2 and A_{+2}^2 of arrow diagrams do not change $\langle (X, F), \cdot \rangle$. □

EXAMPLE. All invariants $V_{ij}(\beta)$ with $\beta_i = \beta_j = 0$ are defined by formula (4) for chord diagrams of length 1.

Another way to generalize invariants (3) is to allow nonprime chord diagrams. Here is an evident example of such a generalization.

PROPOSITION 3. *Let $X_1, \ldots, X_m$ be the set of all minimal chord diagrams equivalent to a minimal diagram X_1. Then*

(6) $$\sum_{j=1}^m \langle X_j, \cdot \rangle$$

is an invariant of ornaments.

PROOF. The same as the proof of Proposition 2. □

Other generalizations may be obtained by applying the formula (4) to nonprime chord diagrams with F satisfying certain conditions.

EXAMPLE. Let us denote by X_{ij} the subdiagram of the chord diagram X consisting of all chords $x^l \in X$, connecting c_i and c_j. Let us call two chords x^l and x^m of a diagram X *permutable* if they belong to the same chord subdiagram X_{ij} and neither x_{il}^l and x_{im}^m in c_i, nor x_{jl}^l and x_{jm}^m in c_j are separated by the ends of the chords of $X \setminus X_{ij}$. Let X be a minimal chord diagram and $f^1, \ldots, f^{L(X)}$ be a set of functions of $(k-2)$ integer variables, such that $f^l = f^m$, when x^l and x^m are permutable. Then the formula

$$\sum_{j=1}^{m} \langle (X_j, \times_{l=1}^{n} f^l), \cdot \rangle,$$

where $X_1, \ldots, X_m$ is the set of all minimal chord diagrams, equivalent to X, defines an invariant of ornaments.

Finally we generalize the invariant (4) in such a way, that the generalization depends not only on the indexed diargam of the ornament, but contains more information. For instance, such generalizations can distingush the ornaments shown in Figure 10.

The function $\mathrm{ind}_*(\cdot)$ of a closed curve and a point in $\mathbb{R}^2$ (see 2.2.2) can be extended by linearity to a pairing defined on pairs of nonintersecting 1-cycles and 0-chains (similarly to a pairing of 1-cycles and 0-cycles in $\mathbf{S}^2$). Hence for each chord diagram X, regular ornament ϕ, and representation $\mu : X \to \mathrm{ADiag}(\phi)$ the pairing $\mathrm{ind} \circ (\phi_* \circ \mu_* \otimes \phi_* \circ \mu_*) : Z_1(X) \times_0 (X) \to \mathbb{Z}$ is defined. We have already used this pairing when the 1-cycle is a circle of the chord diagram.

PROPOSITION 4. *Let a non-self-intersecting chord diagram X be a disjoint union of prime chord diagrams $X_0 = \{x^1, \ldots, x^n\}$ and X_1, ξ a 1-cycle in X_1, and F an integer valued function of n integer variables. Then the sum*

$$(7) \qquad \langle ((X_0, X_1), \xi, F), \phi \rangle = \sum_{\mu} F(\mathrm{ind}_{\mu(\xi)}(\mu(x^1)), \ldots, \mathrm{ind}_{\mu(\xi)}(\mu(x^n)))$$

over all representations $\mu : X \to \mathrm{ADiag}(\phi)$ is an invariant of ornaments.

PROOF. It is easy to check that the sum (7) remains unchanged under admissible local moves of ϕ. □

3.5. Order of invariants.

THEOREM 5. *Let X be a prime chord diagram and F be a polynomial. Then $\langle (X, F), \cdot \rangle$ is an invariant of order $\deg(F) + L(X)$.*

COROLLARY 2. *If X is a prime chord diagram, then $\langle X, \cdot \rangle$ is an invariant of order $L(X)$.*

REMARK. The bound given by Corollary 2 seems to be not sharp.

PROPOSITION 5. *The invariant given by formula (6) for a minimal chord diagram of n chords is an invariant of order n.*

PROPOSITION 6. *If F is a polynomial, then formula (7) gives an invariant of order $\deg(F) + L(X)$.*

PROOF. of Theorem 5 The proof essentially repeats the proof of Lemma 31 and Theorem 7 in [M95], so only the outline will be given. The sum (4) is defined for regular quasiornaments as well. It is enough to consider only the case $F = \times_{l=1}^{n} f^l$, where each f^l is a monomial of $(k-2)$ variables and $n = L(X)$. The proof is done inductively in the complexity of quasiornaments. The inductive assumption is as follows.

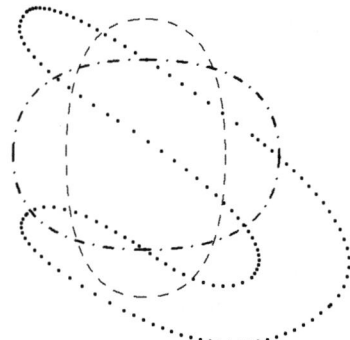

FIGURE 11. There exists an invariant of order 16 not vanishing on this ornament

Let ϕ be a quasiornament. For any representation $\mu : X \to \mathrm{ADiag}(\phi)$ denote by $n(\mu)$ the number of chords of X going to simple intersection points of ϕ. Then the characteristic number assigned to a degeneration mode of ϕ is equal to a sum

$$(8) \qquad \sum_j \prod_{l=1}^n \sigma(\mu_j(x^l), \phi_j) p_j^l(\mathrm{Ind}'(\mu_j(x^l), \phi_j)),$$

where the ornaments ϕ_j are perturbations of ϕ (see 2.2.4), p_j^l are polynomials, and $\mu_j : X \to \mathrm{ADiag}(\phi_j)$ are representations such that for each summand in (8)

$$(9) \qquad \sum_{l=1}^n \deg(p_j^l) + n(\mu_j) \le \sum_{l=1}^n \deg(f^l) + n - \mathrm{compl}(\phi).$$

Now the induction base ($\mathrm{compl}(\phi) = 0$, i.e., ϕ is an ornament) is trivial and the induction step is elementary. Then if $\mathrm{compl}(\phi) > \deg(f^l) + n$, then the left-hand side of (9) is zero, and hence the sum (8) vanishes. $\square$

Propositions 5 and 6 are not formal consequences of Theorem 5, but their proofs are exactly the same.

Corollary 1 and Proposition 5 allow one to go from the classification of ornaments by their chord diagrams to classification by finite order invariants.

PROPOSITION 7. *If ornaments ϕ and ψ have nonequivalent minimal chord diagrams X and Y, then ϕ and ψ can be distinguished by some invariant of order $\max(L(X), L(Y))$.*

PROOF. Assume $L(X) \le L(Y)$. Then the desired invariant is given by Proposition 3 for $X_1 = Y$: it vanishes on ϕ and does not vanish on ψ, because the sum (6), calculated for $\mathrm{ADiag}(\psi)$, consists of several equal nonzero summands. $\square$

COROLLARY 3. *If the non-self-intersecting chord diagram of an ornament with n intersection points is not equivalent to the trivial non-self-intersecting chord diagram (i.e., to the diagram with no chords), then there exists an invariant of order n, not vanishing on this ornament.*

EXAMPLE. The chord diagram of the ornament shown in Figure 11 is minimal, and its length is equal to 16. Hence there exists an invariant of order 16, not vanishing on this ornament. All previously known finite order invariants (see [**M95**]) vanish on it.

4. Finite order invariants of flat braids

The content of this section is as follows. We transfer the constructions of the previous section from ornaments to flat braids. Then it turns out that the flat braid can be reconstructed by its chord diagram up to equivalence, so the analogue of Proposition 7 means that finite order invariants classify flat braids.

It is convenient to consider the space of *generalized flat braids*, which is the space of collections ϕ of smooth maps (*threads*) $\phi_j : [0,1] \to \mathbb{R}^2$, $1 \leq j \leq k$, such that $\phi_j(0) = (0,j)$, $\{\phi_1(1),\ldots,\phi_k(1)\} = \{(1,1),\ldots,(1,k)\})$, the first coordinate functions ϕ_j^1 are strictly monotonic (hence the threads are non-self-intersecting), and triple intersections are forbidden. In other words, we allow each thread to reparametrize independently. The space of generalized flat braids can be obviously contracted onto the space of flat braids. We will omit the word "generalized" below.

4.1. Chord diagrams.
Similarly to 3.1, the chord diagram is a collection of oriented "horizontal" segments connected by finitely many "nonhorizontal" chords. All the endpoints of the chords and the segments are different, and the diagram is considered up to orientation preserving diffeomorphisms of the segments. The chord diagram CDiag(ϕ) of a flat braid ϕ of k threads with intersection points $x^1,\ldots,x^L \in \mathbb{R}^2$ is the chord diagram with k segments and L chords, connecting the preimages of each of the intersection points.

The equivalence of chord diagrams is defined by the following local moves:

FBC^2_{-2} Annihilation of two chords connecting the same two segments, whose ends form two pairs of neighboring points.

FBC^2_{+2} The move opposite to FBC^2_{-2}.

All other definitions and notation repeat those in subsection 3.1. Obviously, each minimal chord diagram is prime. Hence, for each flat braid ϕ its *minimal diagram* (i.e., the unique minimal chord diagram, equivalent to CDiag(ϕ)) is well defined. Proposition 1, Corollary 1 and their proofs remain true after changing the words "ornament" to "flat braid" and "circle" to "segment", though each can be simplified significantly.

Not any chord diagram is the chord diagram of a flat braid.

PROPOSITION 8. *In the chord diagram* CDiag(ϕ) *the chords cannot be crossed in the following sense. If* $x^1 = (x^1_{i^1}, x_{i^2}), \ldots, x^{n-1} = (x^{n-1}_{i^{n-1}}, x_{i^n}), x^n = (x^n_{i^n}, x_{i^1})$ *is a sequence of chords, connecting i^1-th, $\ldots$, i^n-th segments, and $x^l_{i^{l+1}} < x^{l+1}_{i^{l+1}}$ for $l = 1,\ldots, n-1$, then $x^n_{i^1} > x^1_{i^1}$.*

PROOF. This property is obvious for "nongeneralized" canonically parametrized flat braids. It remains true after a reparametrization. □

The composition of flat braids can be extended trivially to composition of their chord diagrams. The following proposition is obvious.

PROPOSITION 9. *If both ϕ and ϕ', and ψ and ψ' are pairs of equivalent flat braids, then the composition $\phi\psi$ is equivalent to $\phi'\psi'$, and $\mathrm{CDiag}(\phi)$ is equivalent to $\mathrm{CDiag}(\phi')$.* □

THEOREM 6. *Any two flat braids with the same chord diagram are equivalent.*

PROOF. The proof is an elementary induction in the length of a chord diagram. If the length of the chord diagram is equal to 0 or 1 (i.e., the flat braids have at most one intersection point), the claim is obvious. Now consider the general case. By Proposition 8, the linear order on the segments of the chord diagram of a flat braid induces a partial order of its chords. Let us choose one of the right-most chords of the diagram and reparametrize the diagram and the two flat braids in such a way that the endpoints of this chord become the right-most of all the endpoints. Then the braids and the diagram can be decomposed into the product of braids (respectively, a diagram) of smaller length and braids (a diagram) of length 1. The inductive assumption and Proposition 9 complete the proof. □

LEMMA 5. *If X is the minimal diagram of a flat braid ϕ, then there exists a flat braid ψ, such that $X = \mathrm{CDiag}(\psi)$.*

PROOF. By modified Corollary 1, the chord diagram X is obtained from $\mathrm{CDiag}(\phi)$ by a sequence of moves of type FBC^2_{-2}. Each such move can obviously be lifted to the corresponding move of the flat braids. □

Corollary 1, as modified for flat braids, Lemma 5, and Theorem 6 immediately imply the following theorem.

THEOREM 7. *Two flat braids are equivalent if and only if their minimal chord diagrams coincide. There exists an algorithm verifying the equivalence of flat braids.*
□

4.2. Arrow diagrams. As in subsection 3.2, the arrow diagram is a collection of oriented "horizontal" segments connected by finitely many "nonhorizontal" arrows. All the endpoints of the arrows and the segments are different, and the diagram is considered up to orientation preserving diffeomorphisms of the segments. The arrow diagram $\mathrm{ADiag}(\phi)$ of a flat braid ϕ is its chord diagram $\mathrm{CDiag}\,\phi$ with the following directions of chords: if in a small left half-neighborhood of the intersection x of i-th and j-th threads the i-th one goes over the j-th one (or, equivalently, $\sigma_{ij}(x) = 1$), the arrow is directed from the i-th segment to the j-th one.

By Theorem 6 the chord diagram of a flat braid uniquely determines its arrow diagram. For instance, for any pair of threads (i-th and j-th, $i < j$) and x running from left to right, the signs $\sigma_{ij}(x)$ alternate starting with -1.

4.3. Invariants. For any chord diagram X let us define the function $\langle X, \cdot \rangle$ of flat braids by (3). Proposition 2, Corollary 2 and their proofs remain true without any change. The following theorem, similar to Proposition 7, gives full classification of flat braids.

THEOREM 8. *Two flat braids ϕ and ψ with m and n intersection points are not equivalent if and only if they can be distinguished by some invariant of order $\max(m,n)$ given by (3).*

PROOF. If ϕ and ψ are not equivalent, their minimal chord diagrams X and Y are not isomorphic by Theorem 7. Assume $L(X) \geq L(Y)$. Then the invariant $\langle X, \cdot \rangle$ distinguishes the flat braids ϕ and ψ. □

The results and methods of Section 3 can be applied to the classification of doodles. It turns out that Vassiliev invariants, namely certain generalizations of the invariants (7), classify doodles, and that an analogue of Theorem 8 is true. This is the subject of another paper.

References

[B-N95] D. Bar-Natan, *Vassiliev homotopy string link invariants*, J. Knot Theory and Ramifications, **4** (1995), 13–32.

[F95] S. Fomin, *Piecewise-linear maps, total positivity, and pseudoline arrangements*, Conf. Algebraic and Geometric Combinatorics, Oberwolfach, 1995.

[FT77] R. Fenn and P. Taylor, *Introducing doodles*, Topology of low-dimensional manifolds, R. Fenn (ed.), Lecture Notes in Math., vol. 722, Springer–Verlag, Berlin, Heidelberg, New York, 1977, pp. 37–43.

[Kh94] M. Khovanov, *Doodle groups*, preprint, Yale Univ., 1994; Trans. Amer. Math. Soc., **349** (1997), 2297–2315.

[Kh96] _____, *Real $K(\pi, 1)$ Arrangements from finite root systems*, Math. Res. Letters, **3** (1996), 261–274.

[K93] M. Kontsevich, *Vassiliev's knot invariants*, Adv. in Soviet Math., vol. 16, Part 2, Amer. Math. Soc., Providence, RI, 1993, pp. 137–150.

[M94] A. B. Merkov, *On classification of ornaments*, Singularities and Bifurcations, V. I. Arnold (ed.), Adv. in Soviet Math., vol. 21, Amer. Math. Soc., Providence, RI, 1994, pp. 199–211.

[M95] _____, *Finite order invariants of ornaments*, 1995, J. Math. Sciences (to appear).

[P94] M. Polyak, *Invariants of plane curves via Gauss diagrams*, Preprint, Max-Plank-Institut MPI/116-94, Bonn, 1994.

[PV94] M. Polyak and O. Viro, *Gauss diagram formulas for Vassiliev invariants*, Internat. Math. Res. Notices, **1994**, no. 11, 445–454.

[S59] S. Smale, *Diffeomorphisms of the 2-sphere*, Proc. Amer. Math. Soc., **10** (1959), 621–626.

[V93] V. A. Vassiliev, *Invariants of ornaments*, Singularities and Bifurcations, V. I. Arnold (ed.), Adv. in Soviet Math., vol. 21, Amer. Math. Soc., Providence, RI, 1994, pp. 225–261.

[V94] _____, *Complements of discriminants of smooth maps: topology and applications* (revised edition), Transl Math. Monographs, vol. 98, Amr. Math. Soc., Providence, RI, 1994.

INSTITUTE FOR SYSTEM ANALYSIS OF RUSSIAN ACADEMY OF SCIENCES, 9, PROSP. 60-LETIYA OKTYABRYA, MOSCOW 117312, RUSSIA

E-mail address: `merx@ium.ips.ras.ru`

New Whitney-type Formulas for Plane Curves

Michael Polyak

ABSTRACT. The classical Whitney formula relates the algebraic number of self-intersections of a generic plane curve to its winding number. We generalize it to an infinite family of identities expressing the winding number in terms of the internal geometry of a plane curve. This enables us to split the Whitney formula by some characteristic of double points. It turns out, that only crossings of a very specific type contribute to the computation of the winding number. We also provide a "difference integration" of these formulas, establishing a new family of simple formulas with the base point pushed off the curve. Similar new identities are obtained for Arnold's invariant St of plane curves.

1. Whitney formula and its generalizations

1.1. Introduction. The classical Whitney formula [4] relates the algebraic number of times that a generic immersed plane curve intersects the Whitney index, or winding number, of this curve. Since it was discovered in 1937, this formula remained more of an isolated curious fact rather than a part of a more general picture. In particular, its generalizations remained unknown. The first step in this direction was made only recently [2], with an introduction of higher-dimensional versions of the Whitney formula. Here we take a different direction toward its generalization, showing that this is just the simplest one in an infinite family of identities. These identities express the Whitney index of a plane curve in terms of some functions defined at double points of the curve. A particular choice of these functions as elementary bump functions allow us to split the sum in the Whitney formula over different types of double points (see subsection 1.4). It turns out that only a very specific type of double points contributes to the computation of the index.

All these formulas, just as the original one, involve a choice of a base point on the curve. We introduce another type of formula, this time with the base point pushed off the curve. Taking the difference of two such formulas with the base point placed in a pair of adjacent regions, one gets a formula of the first type, with the base point on the arc of the curve between these regions. Thus the second type of formulas can be considered as a difference integration of formulas of the first type with respect to position of the base point (see Section 2).

1991 *Mathematics Subject Classification.* Primary 57M25.

©1999 American Mathematical Society

Finally, we turn to Arnold's invariant St of plane curves [1]. Noticing that Shumakovitch's formula [3] for St looks similar to the Whitney formula for index, we generalize it in a similar manner.

The main part of this work was done when the author was visiting the Fields Institute in Toronto and the Max-Planck-Institut für Matematik in Bonn, which he wishes to thank for their hospitality.

1.2. The classical Whitney formula. Let C be a generic oriented plane curve, i.e., an oriented circle S^1 immersed in the (oriented) plane $\mathbb{R}^2$, with the only singularities being transversal double points. The direction of the tangent vector defines a Gauss map $S^1 \to \mathbb{R}^2 - \{0\} \cong S^1$. The *Whitney index* index(C) of C is the degree of this map, i.e., the number of turns made by the tangent vector as we pass along C. It is the only homotopy invariant of C in the class of immersions; see [4].

For $x \in \mathbb{R}^2 - C$ the direction of the vector connecting x with a point y moving along C defines another map $S^1 \to \mathbb{R}^2 - \{0\} \cong S^1$. The *index* $\text{ind}_C(x)$ *of x with respect to C* is the degree of this map, i.e., the number of turns made by the vector $\vec{xy}$ as we pass along C. Extend ind_C to $\text{ind}_C : \mathbb{R}^2 \to \frac{1}{2}\mathbb{Z}$ by averaging the values of two adjacent components of $\mathbb{R}^2 - C$ for a generic point on C and four adjacent components for a double point.

Fix a generic base point $x \in C$. For a double point $d \in C$, this determines an ordering of two outgoing branches of C in d, thus the sign $\varepsilon_d(x)$ of d is determined by the orientation of the corresponding frame of tangents to C in d. Following Whitney [4], we choose the sign convention shown in Figure 1a.

The Whitney formula is formulated as follows.

THEOREM 1 (Whitney [4]).

(1) $$\text{index}(C) = \sum_d \varepsilon_d(x) + 2\,\text{ind}_C(x).$$

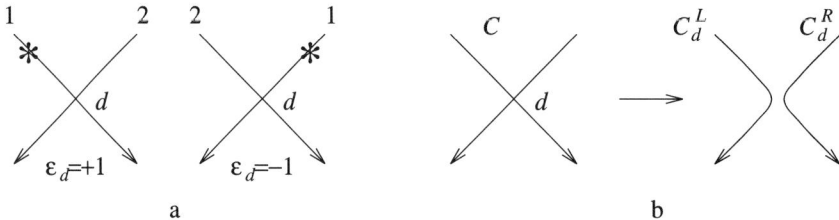

FIGURE 1. Signs and splitting of double points

1.3. Generalized Whitney formulas. Smoothing C at a double point d so that the orientation is respected, we get two curves C_d^L and C_d^R, where the tangent vector of C_d^L turns clockwise in a neighborhood of d (and counter-clockwise for C_d^R); see Figure 1b.

The *partial indices* $(L_d(x), R_d(x))$ *of x in d* are the indices $L_d(x) = \text{ind}_{C_d^L}(x)$ and $R_d(x) = \text{ind}_{C_d^R}(x)$ of the base point x with respect to the curves C_d^L and C_d^R. Note that one of them is an integer, while another is a (noninteger) half-integer, depending on whether x lies on C_d^L or C_d^R, i.e., according to the sign $\varepsilon_d(x)$ of d.

Conversely, we can consider $\varepsilon_d(x)$ as a function $\varepsilon(L_d(x), R_d(x))$ of these indices. It clearly extends to a (rather trivial) function $\varepsilon : \mathbb{Z} \times (\mathbb{Z} + \frac{1}{2}) \cup (\mathbb{Z} + \frac{1}{2}) \times \mathbb{Z} \to \mathbb{R}$, defined by $\varepsilon(i, j) = +1$ if $i \in \mathbb{Z} + \frac{1}{2}$ and $\varepsilon(i, j) = -1$ if $i \in \mathbb{Z}$. This function is obviously skew-symmetric $\varepsilon(i, j) = -\varepsilon(j, i)$; also, $\varepsilon(\text{ind}_C(x), 0) = +1$. It allows us to rewrite the Whitney formula (1) in the following way.

THEOREM 1'. *Let $(L_d(x), R_d(x))$ be the partial indices with respect to x in d and let $\varepsilon : \mathbb{Z} \times (\mathbb{Z} + \frac{1}{2}) \cup (\mathbb{Z} + \frac{1}{2}) \times \mathbb{Z} \to \mathbb{R}$ be defined by $\varepsilon(i, j) = +1$ if $i \in \mathbb{Z} + \frac{1}{2}$ and $\varepsilon(i, j) = -1$ if $i \in \mathbb{Z}$. Then*

$$(2) \qquad \varepsilon(\text{ind}_C(x), 0)\,\text{index}(C) = \sum_d \varepsilon(L_d(x), R_d(x)) + 2\,\text{ind}_C(x).$$

As it turns out, any other skew-symmetric function leads to a similar new formula. Denote by $I(m)$ the set of integers on the open interval between 0 and m, i.e., $I(m) = \{i \in \mathbb{Z} | m < i < 0\}$ for $m < 0$ and $I(m) = \{i \in \mathbb{Z} | 0 < i < m\}$ for $m \geq 0$.

THEOREM 2. *Let $(L_d(x), R_d(x))$ be the partial indices with respect to x in a double point d of C and let $f : \mathbb{Z} \times (\mathbb{Z} + \frac{1}{2}) \cup (\mathbb{Z} + \frac{1}{2}) \times \mathbb{Z} \to \mathbb{R}$ be skew-symmetric. Then*

$$(3) \qquad f(\text{ind}_C(x), 0)\,\text{index}(C) = \sum_d f(L_d(x), R_d(x)) + g_f(\text{ind}_C(x)),$$

where $g_f : \mathbb{Z} + \frac{1}{2} \to \mathbb{R}$ is defined by

$$g_f(m) = (m + \frac{1}{2}\,\text{sgn}(m))f(m, 0) + \text{sgn}(m)\sum_{i \in I(m)} f(m - i, i).$$

Note that for $f(i, j) = \varepsilon(i, j)$ we have $f(m - i, i) = 1$ for any $i \in \mathbb{Z}$, so $\text{sgn}(m)\sum_{i \in I(m)} f(m - i, i) = m - \frac{1}{2}\text{sgn}(m)$ and the function g_f becomes $g_\varepsilon(m) = 2m$. Hence we obtain Theorem 1' as a corollary.

The proof of this theorem is similar to the one of Theorem 4 and is given in subsection 2.3.

1.4. Splitting the Whitney formula by partial indices. We show below that Theorem 2 is equivalent to an interesting splitting of the Whitney formula by partial indices of double points. This is done in the following way. Various choices of f lead to new identities for the winding number. Obviously, skew-symmetric functions on $\mathbb{Z} \times (\mathbb{Z} + \frac{1}{2}) \cup (\mathbb{Z} + \frac{1}{2}) \times \mathbb{Z}$ are in one-to-one correspondence with functions on $\mathbb{Z} \times (\mathbb{Z} + \frac{1}{2})$. Any such function can be written as a linear combination of elementary bump functions, taking value 1 in one point of $\mathbb{Z} \times (\mathbb{Z} + \frac{1}{2})$ and 0 in all the others. Thus we can reformulate Theorem 2 in terms of the corresponding elementary skew-symmetric functions.

Fix $x \in C$ and $j \in \mathbb{Z} + \frac{1}{2}$. Let $\text{ind}_C(x) = i$ and consider equation (3) for a function

$$f(k, l) = \delta_{k,j}\delta_{l,i-j} - \delta_{k,i-j}\delta_{l,j}.$$

The left-hand side of (3) vanishes unless $j = i$. Let D_k, $k \in \mathbb{Z}$, be the union of double points of C with either $L_d(x) = k$ or $R_d(x) = k$. Note that only double points in D_{i-j} give nonzero contribution to the sum in the right-hand side of (3). Let us substitute f into (3) first for $j = i$, and then for $j \neq i$. We learn that the Whitney formula can be rewritten by counting only the double points in D_0 and the sum over double points in $\cup_{k \neq 0} D_k$ also splits by k.

THEOREM 2′. *Let $x \in C$ be a generic base point with $\mathrm{ind}_C(x) = i \in \mathbb{Z} + \frac{1}{2}$. Then*

$$\mathrm{index}(C) = \sum_{d \in D_0} \varepsilon_d(x) + i + \frac{1}{2}\,\mathrm{sgn}(i).$$

Also, for $k \neq 0$ we have

$$\sum_{d \in D_k} \varepsilon_d(x) = \begin{cases} -\mathrm{sgn}(i) & \text{if } k \in I(i), \\ 0 & \text{otherwise.} \end{cases}$$

2. Integrating Whitney-type formulas

2.1. Difference integration. Our initial motivation for studying the Whitney formula was the following. Note that two terms in (1), namely $\mathrm{index}(C)$ and $2\,\mathrm{ind}_C(x)$, are still well defined when the base point x is chosen in $\mathbb{R}^2 - C$. So it is tempting to look for some formula similar to (1) in this case. At first, it remains unclear how to adapt $\sum_d \varepsilon_d(x)$ to such a situation. To narrow the search, we can require that a new formula should be a "difference integration" of (1) with respect to the position of the base point. In other words, when we consider formulas corresponding to x being placed in two adjacent components of $\mathbb{R}^2 - C$, their difference should give the Whitney formula (1) with the base point on the arc of C between these regions. The functions $\mathrm{index}(C)$ and $2\,\mathrm{ind}_C(x)$, being constant and linear functions of $\mathrm{ind}_C(x)$ respectively, are then integrated in an obvious way. It remains to make an important observation that the function $L_d(x) - R_d(x)$ is an integral of the sign $\varepsilon_d(x) = \varepsilon(L_d(x), R_d(x))$. Thus we arrive naturally at the following theorem.

THEOREM 3. *Fix a base point $x \in \mathbb{R}^2 - C$ and let $(L_d(x), R_d(x))$ be the partial indices in a double point d of C. Then*

(4) $$\mathrm{ind}_C(x)\,\mathrm{index}(C) = \sum_d (L_d(x) - R_d(x)) + \mathrm{ind}_C(x)^2.$$

PROOF. Move the sum over the double points to the left-hand side. Lemma 1 below shows that the resulting function $\mathrm{i}^2(C, x)$ is invariant when we deform C by homotopy in $\mathbb{R}^2 - x$ (and the same, obviously, holds for $\mathrm{ind}_C(x)^2$). Any curve in $\mathbb{R}^2 - x$ is homotopic to one of the standard curves C_n with $\mathrm{ind}_{C_n}(x) = n$, $n \in \mathbb{Z}$, shown in Figure 2a. Thus it suffices to compare the values of $\mathrm{i}^2(C, x)$ and $\mathrm{ind}_C(x)^2$ on the curves C_n. A simple verification shows that $\sum_d (L_d(x) - R_d(x))$ vanishes for each C_n, so $\mathrm{i}^2(C_n, x) = n^2 = \mathrm{ind}_{C_n}(x)^2$. □

Further on by a homotopy of C we mean a generic homotopy that may be nonregular only at a finite number of moments when a kink is deleted or added.

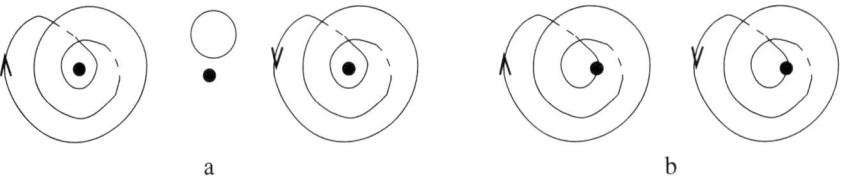

FIGURE 2. Standard curves C_n for an integer and half-integer n

FIGURE 3. Elementary deformations of curves

LEMMA 1. *Fix a point x in $\mathbb{R}^2 - C$. The function*
$$i^2(C, x) = \text{ind}_C(x)\,\text{index}(C) - \sum_d (L_d(x) - R_d(x))$$
is an invariant of homotopy type of C in $\mathbb{R}^2 - x$.

PROOF. It suffices to verify the invariance under the Reidemeister-type moves Ω_1–Ω_3 of curves in $\mathbb{R}^2 - x$ (see Figure 3). Under Ω_1, the contribution of a new double point to the sum equals $\text{ind}_C(x) - 0$ if $\text{index}(C)$ increases by 1, and $0 - \text{ind}_C(x)$ if $\text{index}(C)$ decreases by 1. This is obviously compensated by the corresponding change of $\text{ind}_C(x)\,\text{index}(C)$. Under Ω_2 two new double points d_1, d_2 have the opposite pair of partial indices: $(L_{d_1}(x), R_{d_1}(x)) = (R_{d_2}(x), L_{d_2}(x))$. Thus the corresponding terms cancel out. The move Ω_3 does not change any of the relevant indices. □

2.2. Generalizing the integration formula. Returning to the proof of Lemma 1, we see that homotopy invariance followed from just a few properties of the function $\text{diff}(L_d(x), R_d(x))$, where $\text{diff}(i,j) = i - j$. Namely, its skew-symmetry provides invariance under Ω_2, and the formula
$$\text{ind}_C(x) = \text{diff}(\text{ind}_C(x), 0) = -\text{diff}(0, \text{ind}_C(x))$$
provides invariance under Ω_1. So, any other skew-symmetric function $F : \mathbb{Z} \times \mathbb{Z} \to \mathbb{R}$ could be used in its place.

LEMMA 2. *Fix $x \in \mathbb{R}^2 - C$. The function*
$$G_F(C, x) = F(\text{ind}_C(x), 0)\,\text{index}(C) - \sum_d F(L_d(x), R_d(x))$$
is an invariant of the homotopy type of C in $\mathbb{R}^2 - x$.

Now, computing the values of $G_F(C, x)$ on the standard curves C_n, $n \in \mathbb{Z}$, we obtain the following generalization of Theorem 3.

THEOREM 4. *Fix a base point $x \in \mathbb{R}^2 - C$. Let $(L_d(x), R_d(x))$ be the partial indices in a double point d of C and let $F : \mathbb{Z} \times \mathbb{Z} \to \mathbb{R}$ be a skew-symmetric function. Then*

(5)
$$F(\text{ind}_C(x), 0)\,\text{index}(C) = \sum_d F(L_d(x), R_d(x)) + F(\text{ind}_C(x), 0)\,\text{ind}_C(x).$$

Repeating the idea of subsection 1.4, we want to split (5) over partial indices of double points. We must adjust the construction used in Theorem 2′ to the case $x \in \mathbb{R}^2 - C$. We keep the notation D_k for the set of double points with either $L_d(x) = k$ or $R_d(x) = k$. This time, however, the partial indices $L_d(x)$, $R_d(x)$ are both integers, so to get rid of the symmetry $D_k = D_{\text{ind}_C(x)-k}$ we restrict ourselves only to $k \in \mathbb{Z}$ such that $\text{sgn}(i) \cdot k \in [0, \frac{|i|}{2}] \cup (|i|, \infty)$. This allows us to assign a sign ε_d to any double point $d \in D_k$. Namely (as in the case $x \in C$), for $k \ne \frac{i}{2}$ we define $\varepsilon_d(x) = +1$ if $R_d(x) = k$, and $\varepsilon_d(x) = -1$ if $L_d(x) = k$. For even k we extend it to $k = \frac{i}{2}$ by $\varepsilon_d(x) = 0$ for any $d \in D_{\frac{i}{2}}$.

THEOREM 4'. *Fix $x \in \mathbb{R}^2 - C$. Then*
$$\sum_{d \in D_k} \varepsilon_d(x) = \begin{cases} \text{index}(C) - \text{ind}_C(x) & \text{if } k = 0, \\ 0 & \text{otherwise.} \end{cases}$$

2.3. Toward the proof of Theorem 2. Consider the partial derivative
$$f(i + \frac{1}{2}, j) = F(i+1, j) - F(i, j)$$
of the function F in Theorem 4. Using the initial condition $F(j,j) = 0$ we can express F via f by means of an integral sum. For $n \ge j$ we get
$$F(n, j) = \sum_{i=0}^{n-1} f(j + \frac{1}{2} + i, j),$$
thus the partial derivative of $nF(n, j)$ is given by
$$(n+1)F(n+1, j) - nF(n, j) = (n+1)f(n + \frac{1}{2}, j) + \sum_{i=1}^{n-j} f(n + \frac{1}{2} - i, j + i).$$

Taking the difference of formulas (5) with base points in two adjacent regions, we obtain a formula with a base point on the curve C; it is easy to identify it as the formula in Theorem 2. Thus, if the function f in Theorem 2 integrates to a skew-symmetric function F, Theorem 2 follows from Theorem 4. However, this is not always the case: an integral F of the function f of Theorem 2 is skew-symmetric only if the following condition is satisfied for any integers j and n:
$$\sum_{i=0}^{n-1}(f(j + \frac{1}{2} + i, j) - f(j + \frac{1}{2} + i, n+j)) = 0.$$
Therefore, in the general case Theorem 2 does not follow from Theorem 4 and should be proven separately.

PROOF OF THEOREM 2. The proof is based on the same idea as in Theorems 3 and 4.

Fix $x \in \mathbb{R}^2$ and denote by $\mathcal{C}$ the space of *generic based immersions* $C : S^1 \to \mathbb{R}^2$, i.e., immersions with the base point x being a generic point of C. Two immersions C and C' in $\mathcal{C}$ may be connected by isotopy and a sequence of Reidemeister-type moves Ω_1–Ω_3 (not involving the base point x) and their inverses if and only if $\text{ind}_C(x) = \text{ind}_{C'}(x)$. Thus it suffices to verify the invariance of
$$f(\text{ind}_C(x), 0) \text{index}(C) - \sum_d f(L_d(x), R_d(x)) - g_f(\text{ind}_C(x))$$
under Ω_1–Ω_3 and to show that its value is 0 for each curve C_n with $\text{ind}_{C_n}(x) = n$, $n \in \mathbb{Z} + \frac{1}{2}$ as depicted in Figure 2b.

The proof of invariance under Ω_1–Ω_3 repeats the proof of Lemmas 1 and 2. Namely, under Ω_1, the contribution of a new double point to the sum equals

$f(\mathrm{ind}_C(x), 0)$ if index(C) increases by 1 and $f(0, \mathrm{ind}_C(x))$ if index(C) decreases by 1. Due to the skew-symmetry of f, this cancels out with the corresponding change of $f(\mathrm{ind}_C(x), 0)\,\mathrm{index}(C)$. Under Ω_2 two new double points d_1, d_2 have the opposite pair of partial indices: $(L_{d_1}(x), R_{d_1}(x)) = (R_{d_2}(x), L_{d_2}(x))$. Again, due to the skew-symmetry of f, the corresponding terms cancel out. The move Ω_3 does not change any of the relevant indices, so each summand is preserved.

Finally, the function g_f is chosen so that its values on C_n, $n \in \mathbb{Z} + \frac{1}{2}$ coincide with those of $f(\mathrm{ind}_C(x), 0)\,\mathrm{index}(C) - \sum_d f(L_d(x), R_d(x))$. $\square$

3. New formulas for strangeness of plane curves

3.1. Strangeness of plane curves.
Immersions $S^1 \to \mathbb{R}^2$ with more than two pints of S^1 mapped into the same point of $\mathbb{R}^2$ form a *discriminant hypersurface* D in the space of all plane curves. The main (open) stratum D^0 of D consists of plane curves with exactly one triple point and several double points, all of them transversal. Arnold [1] observed that this discriminant is coorientable, i.e., there exists a consistent choice of one—called *positive*—of the two parts separated by D^0 in a neighborhood of any of its points C_{sing}. Indeed, the orientation of the singular curve C_{sing} determines the cyclic order (t_1, t_2, t_3) of tangents t_i to its three branches in the triple point t. Pushing the third branch off t in a (generic) direction n gives a positive deformation of C_{sing} if and only if the orientation of the frames (t_1, t_2) and (t_3, n) coincide.

This observation allowed Arnold in [1] to define axiomatically a new basic invariant St of generic plane curves. *Strangeness St* is an invariant of regular homotopy in the class of curves without triple points, which increases by 1 under a positive triple-point deformation Ω_3 (positive crossing of D^0). This, together with the additivity of St under the operation of band-summation of curves, defines strangeness uniquely.

3.2. Formulas for St.
A simple explicit formula for St was obtained by Shumakovitch [3].

THEOREM 5 (Shumakovitch [3]).

(6) $$St(C) = \sum_d \varepsilon_d(x)\,\mathrm{ind}_C(d) + \mathrm{ind}_C(x)^2 - \frac{1}{4}.$$

Since expression (6) is quite similar to Whitney's expression (1) for the winding number, one can expect for St results similar to Theorems 2–4. Indeed, the proofs of the following statements basically repeat the ones of Lemma 2 and Theorems 2–4.

LEMMA 3. *Fix $x \in \mathbb{R}^2 - C$ and let $F : \mathbb{Z} \times \mathbb{Z} \to \mathbb{R}$ be skew-symmetric. The function*
$$H_F(C, x) = F(\mathrm{ind}_C(x), 0) St(C) - \sum_d F(L_d(x), R_d(x))\,\mathrm{ind}_C(d)$$
is an invariant of homotopy of C in the class of curves in $\mathbb{R}^2 - x$ without triple points.

Under a positive triple-point deformation of C, $H_F(C, x)$ increases by
$$F(i_1 + i_2 + i_3, 0) + F(i_1, i_2 + i_3) + F(i_2, i_3 + i_1) + F(i_3, i_1 + i_2),$$

where i_1, i_2, i_3 are indices of x with respect to three subcurves determined by the appearing triple point.

PROOF. Under Ω_1, the contribution of a new double point to the sum is equal to $\pm F(\text{ind}_C(x), 0)\text{ind}_C(d)$ if index(C) changes by ± 1. This is compensated by the corresponding change of $F(\text{ind}_C(x), 0)St(C)$ (since (6) implies that St changes by $\pm\text{ind}_C(d)$). Under Ω_2, both St and the sum over the double points (due to the skew-symmetry of g) are preserved. Calculating the change under the positive Ω_3 move, we get the second statement of the lemma. $\square$

PROPOSITION 1. *Fix a base point* $x \in \mathbb{R}^2 - C$. *Let* $F : \mathbb{Z} \times \mathbb{Z} \to \mathbb{R}$ *be a skew-symmetric function, satisfying*

$$(7) \qquad F(i+j+k, 0) + F(i, j+k) + F(j, k+i) + F(k, i+j) = 0$$

for any $i, j, k \in \mathbb{Z}$. *Then*

$$(8) \qquad F(\text{ind}_C(x), 0)St(C) = \sum_d F(L_d(x), R_d(x))\text{ind}_C(d) + H_F(\text{ind}_C(x)),$$

where $H_F(n) = \frac{n^2 - |n|}{2}F(n, 0) + \sum_{i \in I(n)} |i| \cdot F(n-i, i)$.

PROOF. In view of Lemma 3, it suffices to compute the values of $G_F(C_n, x)$ on the standard curves C_n, $n \in \mathbb{Z}$, and to compare it with the expression for $H_F(n)$ above. The equality $St(C_n) = \frac{n^2 - |n|}{2}$ can be easily deduced from (6). $\square$

In the class of functions $F(i, j) = h(i) - h(j)$ the only solutions of (7) are given by polynomials in i of degree at most 2. Similarly to Theorem 3, equation (8) for a linear function $h(i)$ gives a difference integration of Shumakovitch's formula (6). Differentiating (8) for a quadratic function $h(i)$, we get a new formula for St (compare with (6)).

COROLLARY 1. *Let* $x \in C$ *be a generic base point. Split* C *in a double point* d *and denote by* C_d^* *the part* C_d^L *or* C_d^R *containing* x. *Then*

$$(9) \qquad \text{ind}_C(x)St(C) = \sum_d \varepsilon_d(x)\text{ind}_{C_d^*}(x)\text{ind}_C(d) + h(\text{ind}_C(x)),$$

where $h : \mathbb{Z} + \frac{1}{2} \to \mathbb{R}$ *is defined by* $h(m) = \frac{2}{3}m(m^2 - \frac{1}{4})$.

Reformulating Lemma 3 for $x \in C$ and carrying out the computations for standard curves C_n (this time for $n \in \mathbb{Z} + \frac{1}{2}$), we easily adapt Proposition 1 to this case.

PROPOSITION 2. *Fix a base point* $x \in C$. *Let* $f : \mathbb{Z} \times (\mathbb{Z} + \frac{1}{2}) \cup (\mathbb{Z} + \frac{1}{2}) \times \mathbb{Z} \to \mathbb{R}$ *be a skew-symmetric function, satisfying*

$$f(i+j+k, 0) + f(i, j+k) + f(j, k+i) + f(k, i+j) = 0$$

for any $i, j \in \mathbb{Z}$ *and* $k \in \mathbb{Z} + \frac{1}{2}$. *Then*

$$(10) \qquad f(\text{ind}_C(x), 0)St(C) = \sum_d f(L_d(x), R_d(x))\text{ind}_C(d) + h_f(\text{ind}_C(x)),$$

where $h_f(m) = \frac{m^2 - \frac{1}{4}}{2}f(m, 0) + \sum_{i \in I(m)} |i| \cdot f(m-i, i)$.

References

[1] V. I. Arnold, *Topological invariants of plane curves and caustics*, University Lecture Series, vol. 5, Amer. Math. Soc., Providence, RI 1994.
[2] G. Mikhalkin and M. Polyak, *Whitney formula in higher dimensions*, J. Diff. Geom. **44** (1996), 583–594.
[3] A. Shumakovitch *Explicit formulas for strangeness of plane curves*, Algebra i Analiz **7** (1995), 165–199, English transl. in St. Petersburg Math. J. **7** (1996).
[4] H. Whitney, *On regular closed curves in the plane*, Compositio Math. **4** (1937), 276–284.

SCHOOL OF MATHEMATICAL SCIENCES, TEL-AVIV UNIVERSITY, TEL-AVIV 69978, ISRAEL
E-mail address: polyak@math.tau.ac.il

Tree-Like Curves and Their Number of Inflection Points

Boris Shapiro

To Vladimir Igorevich Arnold with love and admiration

ABSTRACT. In this short note we give a criterion when a planar tree-like curve, i.e., a generic curve in $\mathbb{R}^2$ each double point of which cuts it into two disjoint parts, can be sent by a diffeomorphism of $\mathbb{R}^2$ onto a curve with no inflection points. We also present some upper and lower bounds for the minimum number of inflection points on curves which are unremovable by diffeomorphisms of $\mathbb{R}^2$.

1. Introduction

This paper provides a partial answer to the following question posed by Arnold to the author in June 1995. Given a generic immersion $c : S^1 \to \mathbb{R}^2$ (i.e., with double points only) let $\sharp_{inf}(c)$ denote the number of inflection points on c (assumed finite) and let $[c]$ denote the class of c, i.e., the connected component in the space of generic immersions of S^1 to $\mathbb{R}^2$ containing c. Finally, let $\sharp_{inf}[c] = \min_{c' \in [c]} \sharp_{inf}(c')$.

PROBLEM. Estimate $\sharp_{inf}[c]$ in terms of the combinatorics of c.

The problem itself is apparently motivated by the following classical result due to Möbius (see, e.g., [**Ar3**]).

THEOREM. *Any embedded noncontractible curve on $\mathbb{RP}^2$ has at least 3 inflection points.*

The present paper contains some answers for the case when c is a tree-like curve, i.e., when it satisfies the condition that if p is any double point of c, then $c \setminus p$ has 2 connected components. We plan to drop the restrictive assumption of tree-likeness in our next paper (see [**Sh**]). Classes of tree-like curves are naturally enumerated by partially directed trees with a simple additional restriction on directed edges; see Section 2. It was a pleasant surprise that for the classes of tree-like curves there exists a (relatively) simple combinatorial criterion characterizing when $[c]$ contains a nonflattening curve, i.e., $\sharp_{inf}[c] = 0$ in terms of its tree. (Following Arnold, we use the word "nonflattening" in this text as a synonym for the absence of inflection

---

1991 *Mathematics Subject Classification.* Primary 53A04.
Key words and phrases. Tree-like curves, Gauss diagram, Inflection points.

©1999 American Mathematical Society

points.) On the other hand, all attempts to find a closed formula for $\natural_{inf}[c]$ in terms of partially directed trees failed. Apparently such a formula does not exist; see Section 5, Concluding Remarks.

The paper is organized as follows. Section 2 contains some general information on tree-like curves. Section 3 contains a criterion of nonflattening. Section 4 presents some upper and lower bounds for $\natural_{inf}[c]$. Finally, in Section 6, the Appendix, we calculate the number of tree-like curves having the same Gauss diagram.

For general background on generic plane curves and their invariants the author would recommend [**Ar1**] and especially [**Ar2**], which is excellent reading for a newcomer to the subject.

ACKNOWLEDGMENTS. It is a great pleasure to thank V. Arnold for the formulation of the problem and for his support and encouragement during the many years of our contact. His papers [**Ar1**] and [**Ar2**], in which he develops Vassiliev-type theory for plane curves, revived the whole area. It is difficult to overestimate the role of F. Aicardi for this project. Her suggestions have essentially improved the original text and eliminated some mistakes in the preliminary version. Finally, the author wants to acknowledge the hospitality and support of the Max-Planck-Institut during the preparation of this paper.

2. Some generalities on planar tree-like curves

Recall that a generic immersion $c : S^1 \to \mathbb{R}^2$ is called a *tree-like curve* if by removing any of its double points p we get that $c \setminus p$ has 2 connected components, where c also denotes the image set of c; see Figure 1. Some of the results below were first proved in [**Ai**] and were later found by the author independently. Recall that *the Gauss diagram* of a generic immersion $c : S^1 \to \mathbb{R}^2$ is the original circle S^1 with the set of all preimages of its double points, i.e., with the set of all pairs $(\phi_1, \phi_2), \ldots, (\phi_{2k-1}, \phi_{2k})$ where ϕ_{2j-1} and ϕ_{2j} are mapped to the same point on $\mathbb{R}^2$ and k is the total number of double points of $c(S^1)$. One might think that every pair of points (ϕ_{2j-1}, ϕ_{2j}) is connected by an edge.

STATEMENT 2.1 (see Proposition 2.1 in [**Ai**]). *A generic immersion $c : S^1 \to \mathbb{R}^2$ is a tree-like curve if and only if its Gauss diagram is planar, i.e., can be drawn (including edges connecting preimages of double points) on $\mathbb{R}^2$ without selfintersections; see Figure 2.*

a) tree-like curve b) nontree-like curve

FIGURE 1. Illustration of the notion of a tree-like curve

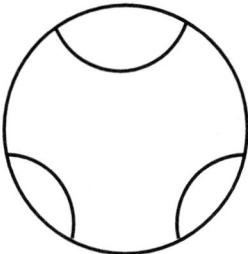

FIGURE 2. Planar Gauss diagram for the case in Figure 1a

REMARK 2.2. There is an obvious isomorphism between the set of all planar Gauss diagrams and the set of all planar connected trees. Namely, each planar Gauss diagram GD corresponds to the following planar tree. Let us place a vertex in each connected component of $D^2 \setminus GD$ where D^2 is the disc bounded by the basic circle of GD and connect by edges all vertices lying in the neighboring connected components. The resulting planar tree is denoted by $\text{Tr}(GD)$. Leaves of $\text{Tr}(GD)$ correspond to the connected components with one neighbor. On GD these connected components inherit the natural cyclic order according to their position along the basic circle of GD. This cyclic order induces a natural cyclic order on the set Lv of leaves of its planar tree $\text{Tr}(GD)$.

Decomposition of a tree-like curve. Given a tree-like curve $c : S^1 \to \mathbb{R}^2$ we decompose its image into a union of curvilinear polygons bounding contractable domains as follows. Take the planar Gauss diagram $GD(c)$ of c and consider the connected components of $D^2 \setminus GD(c)$. Each such component has a part of its boundary lying on S^1.

DEFINITION 2.3. The image of the part of the boundary of a connected component in $D^2 \setminus GD(c)$ lying on S^1 forms a closed non-self-intersecting piecewise smooth curve (a curvilinear polygon) called *a building block* of c; see Figure 3. (We call vertices and edges of building blocks *corners* and *sides* to distinguish them from vertices and edges of planar trees used throughout the paper.)

The union of all building blocks constitutes the whole tree-like curve. Two building blocks have at most one common corner. If they have a common corner then they are called *neighboring*.

DEFINITION 2.4. A tree-like curve c is called cooriented if every its side is endowed with a coorientation, i.e., with a choice of local connected component of the complement $\mathbb{R}^2 \setminus c$ along the side. (Coorientations of different sides are, in general, unrelated.) There are two *continuous* coorientations of c obtained by choosing one of two possible coorientations of some side and extending it by continuity; see Figure 3.

LEMMA 2.5. *Given a continuous coorientation of a tree-like curve c one gets that all sides of any building block are either inward or outward cooriented with respect to the interior of the block. (Since every building block bounds a contractible domain, the outward and inward coorientation have a clear meaning.)*

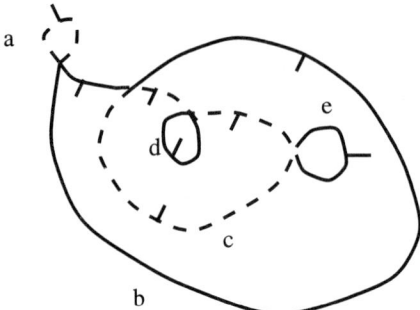

FIGURE 3. Splitting a tree-like curve into building blocks and their coorientation

PROOF. Simple induction on the number of building blocks. □

DEFINITION 2.6. Given a tree-like curve c we associate to it the following planar partially directed tree $\text{Tr}(c)$. At first we take the undirected tree $\text{Tr}(GD(c))$ where $GD(c)$ is the Gauss diagram of c; see Remark 2.2. (Vertices of $\text{Tr}(GD(c))$ are in one-to-one correspondence with building blocks of c; neighboring blocks correspond to adjacent vertices of $\text{Tr}(GD(c))$.) For each pair of neighboring building blocks b_1 and b_2 we do the following. If a building block b_1 contains a neighboring building block b_2 then we direct the corresponding edge (b_1, b_2) of $\text{Tr}(GD(c))$ from b_1 to b_2. The resulting partially directed planar tree is denoted by $\text{Tr}(c)$. Since $\text{Tr}(c)$ depends only on the class $[c]$ we will also use the notation $\text{Tr}[c]$.

We associate with each of two possible continuous coorientations of c the following labeling of $\text{Tr}[c]$. For an outward (resp. inward) cooriented building block we label by "+" (resp. "−") the corresponding vertex of $\text{Tr}[c]$; see Figure 4.

DEFINITION 2.7. Consider a partially directed tree Tr (i.e., some of its edges are directed). Tr is called *a noncolliding partially directed tree or NCPD tree* if no path of Tr contains edges pointing at each other. The usual tree Tr' obtained by forgetting directions of all edges of Tr is called *underlying*.

LEMMA 2.8. a) *For any tree-like curve c its $\text{Tr}(c)$ is noncolliding*;
b) *The set of classes of nonoriented tree-like curves is in one-to-one correspondence with the set of all NCPD trees on the nonoriented $\mathbb{R}^2$.*

PROOF. A connected component of tree-like curves with a given Gauss diagram is uniquely determined by the enclosure of neighboring building blocks. The obvious restriction that if two building blocks contain the third one then one of them is contained in the other is equivalent to the noncolliding property (see the example in Figure 4). □

REMARK 2.9. In terms of the NCPD tree above one can easily describe the Whitney index (or the total rotation) of a given tree-like curve c as well as the coorientation of its building blocks. Namely, fixing the inward or outward coorientation of some building block we determine the coorientation of any other building block as follows. Take the (only) path connecting the vertex corresponding to the

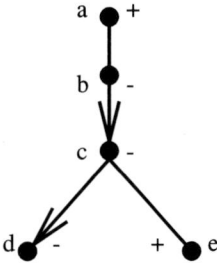

FIGURE 4. NCPD tree Tr[c] for the example in Figure 3 with coorientation of its vertices

fixed block with the vertex corresponding to the other block. If the number of undirected edges in this path is odd, then the coorientation changes, and if this number is even, then it is preserved. (In other words, $\mathrm{Coor}(b_1) = (-1)^{q(b_1,b_2)} \mathrm{Coor}(b_2)$, where $q(b_1, b_2)$ is the number of undirected edges on the above path.)

LEMMA 2.10 (see Theorem 3.1 in [**Ai**]).
$$\mathrm{ind}(c) = \sum_{b_i \in \mathrm{Tr}(c)} \mathrm{Coor}(b_i).$$

PROOF. Obvious. □

3. Nonflattening of tree-like curves

In this section we give a criterion for nonflattening of a tree-like curve in terms of its NCPD tree, i.e., we characterize all cases when $\sharp_{inf}[c] = 0$. (The author is aware of the fact that some of the proofs below are rather sloppy since they are based on very simple explicit geometric constructions on $\mathbb{R}^2$ which are not so easy to describe with complete rigor.)

DEFINITION 3.1. A tree-like curve c (or its class $[c]$) is called *nonflattening* if $[c]$ contains a generic immersion without inflection points.

DEFINITION 3.2. *The convex coorientation* of an arc A, an image of a smooth embedding $(0, 1) \to \mathbb{R}^2$ without inflection points, is defined as follows. The tangent line at any point $p \in A$ divides $\mathbb{R}^2$ into two parts. We choose at p a vector transverse to A and belonging to the halfplane not containing A. *The convex coorientation* of a nonflattening tree-like curve $c : S^1 \to \mathbb{R}^2$ is the convex coorientation of its arbitrary side extended by continuity to the whole curve; see Figure 5.

DEFINITION 3.3. Given a building block b of a tree-like curve c we say that a corner v of b is *of $\vee$-type* (resp., *of $\wedge$-type*) if the interior angle between the tangents to its sides at v is larger (resp., smaller) than $180°$; see Figure 6. (The interior angle is the one contained in the interior of b.)

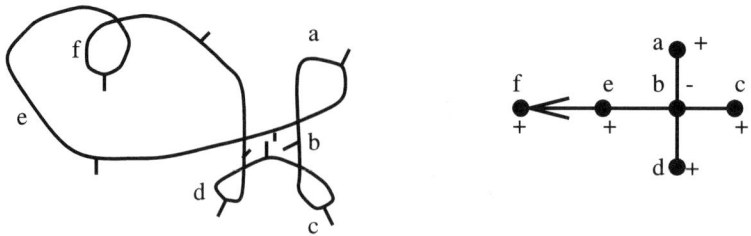

FIGURE 5. Nonflattening curve with the convex coorientation and its NCPD tree.

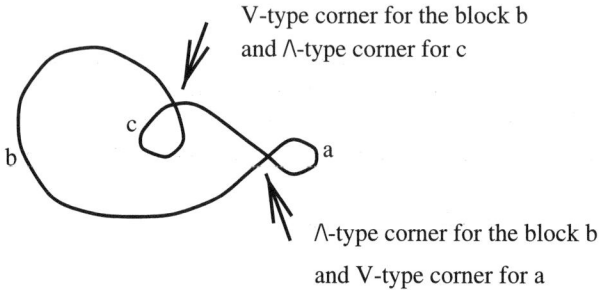

FIGURE 6. ∨- and ∧-type corners of building blocks

REMARK 3.4. If v is a ∨-type corner then the neighboring block b' sharing the corner v with b lies inside b, i.e., ∨-type corners are in one-to-one correspondence with edges of the NCPD tree of c directed from the vertex corresponding to b.

CRITERION 3.5 (For nonflattening). *A tree-like curve c is nonflattening if and only if the following three conditions hold for one of two possible coorientations of its NCPD tree (see Lemma 2.5).*

a) all building 1-gons (i.e., building blocks with one side) are outward cooriented or, in terms of its tree, all vertices of degree 1 of $\mathrm{Tr}[c]$ *are labeled with* "+";

b) all building 2-gons are outward cooriented or, in terms of its tree, all vertices of degree 2 are labeled with "+";

c) the interior of any concave building block (= all sides are concave) with $k \geq 3$ sides contains at most $k-3$ other neighboring blocks or, in terms of its tree, any vertex labeled by "−" *of degree $k \geq 3$ has at most $k-3$ leaving edges (i.e., edges directed from this vertex).*

PROOF. The necessity of a)–c) is rather obvious. Indeed, in the cases a) and b) a vertex of degree ≤ 2 corresponds to a building block with at most two corners. If such a building block belongs to a nonflattening tree-like curve, then it must be globally convex and therefore outward cooriented with respect to the above convex coorientation. For c) consider an inward cooriented (with respect to convex coorientation) building block b of a nonflattening curve. Such b is a curvilinear polygon with locally concave edges. Assuming that b has k corners one gets that the sum of its interior angles is less than $\pi(k-3)$. Therefore, the number of ∨-type

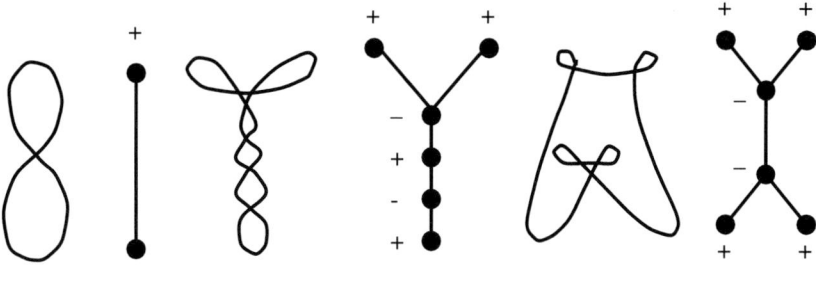

FIGURE 7. Curves violating each of three conditions of criterion 3.5 separately and their NCPD trees

corners (or leaving edges at the corresponding vertex) is less than $k-3$ (see Figure 7 for violations of conditions a)–c)).

Sufficiency of a)–c) is proved by a relatively explicit construction. Given an NCPD tree satisfying a)–c) let us construct a nonflattening curve with this tree using induction on the number of vertices. (This will imply the sufficiency according to Lemma 2.8b.) While constructing this curve inductively, we provide additionally that every building block is star-shaped with respect to some interior point, i.e., the segment connecting this point with a point on the boundary of the block always lies in the domain bounded by the block.

Case 1. Assume that the NCPD tree contains an outward cooriented leaf connected to an outward cooriented vertex (and therefore the connecting edge is directed). Obviously, the tree obtained by removal of this leaf is also an NCPD tree. Using our induction we can construct a nonflattening curve corresponding to the reduced tree and then, depending on the orientation of the removed edge, either glue inside the appropriate locally convex building block a small convex loop (which is obviously possible) or glue a big locally convex loop containing the whole curve. The possibility of gluing a big locally convex loop containing the whole curve is proved independently in Lemma 3.9 and Corollary 3.10.

Case 2. Assume that all leaves are connected to inward cooriented vertices. (By conditions a) and b) these vertices are of degree ≥ 3.) Using the NCPD tree we can find at least 1 inward cooriented vertex b which is not smaller than any other vertex, i.e., the corresponding building block contains at least 1 exterior side. Let k be the degree of b and $e_1, \ldots, e_k$ be its edges in the cyclic order. (Each e_i is either undirected or leaving.) By assumption c) the number of leaving edges is at most $k-3$. If we remove b with all its edges, then the remaining forest consists of k trees. Each of the trees connected to b by an undirected edge is an NCPD tree. We make every tree connected to b by a leaving edge into an NCPD tree by gluing the undirected edge instead of the removed directed one and we mark the extra vertex we get. By induction, we can construct k nonflattening curves corresponding to each of k obtained NCPD trees. Finally, we have to glue them to the corners of a locally concave k-gon with the sequence of ∨- and ∧-type corners prescribed by $e_1, \ldots, e_k$. The possibility of such a gluing is proved independently in Lemmas 3.11 and 3.12. □

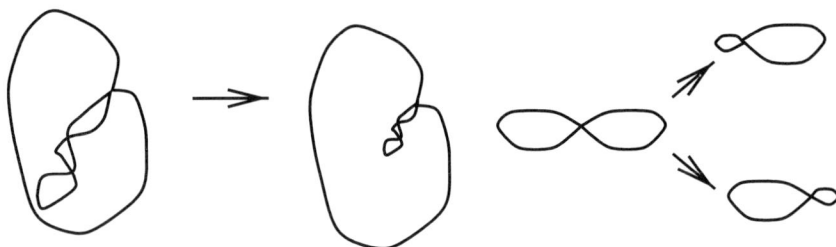

FIGURE 8. Contracting homothety for enclosed and not enclosed building blocks

3.6. AN IMPORTANT CONSTRUCTION. The following operation called *contracting homothety* will be used below extensively. It does not change the class of a tree-like curve and the number of inflection points.

Taking a tree-like curve c and some of its double points p we split $c \setminus p$ into two parts c^+ and c^- intersecting only at p. Let Ω^+ and Ω^- denote the domain bounded by the union of building blocks contained in c^+ and c^-, respectively. There are two options:

a) one of these unions contains the other, say, $\Omega^- \subset \Omega^+$; or

b) $\overline{\Omega}^- \cap \overline{\Omega}^+ = \{p\}$.

Let us choose some small neighborhood ϵ_p of the double point p such that c cuts ϵ_p into exactly four parts.

A contracting homothety with the center p in case a) consists of the usual homothety H applied to c^- which places $H(c^-)$ into ϵ_p followed by smoothing of the second and higher derivatives of the union $H(c^-) \bigcup c^+$ at p. (This is always possible by changing $H(c^-) \bigcup c^+$ in some even smaller neighborhood of p.) The resulting curve c_1 has the same NCPD tree as c and therefore belongs to $[c]$. (Note that we do not construct an isotopy of c and c_1 by applying a family of homotheties with the scaling coefficient varying from 1 to some small number. It suffices that the resulting curve c_1 has the same NCPD tree and therefore is isotopic to c.)

In case b) we can apply a contracting homothety to either of two parts and get two nonflattening tree-like curves c_1 and c_2 isotopic to c and such that either $c^+ = c_1^+$ while c_1^- lies in an arbitrary small neighborhood of p, or $c^- = c_2^-$ while c_2^+ lies in an arbitrary small neighborhood of p. See Figure 8 for the illustration of contracting homotheties.

DEFINITION 3.7. Consider a bounded domain Ω in $\mathbb{R}^2$ with a locally strictly convex piecewise C^2-smooth boundary $\partial \Omega$. Ω is called *rosette-shaped* if for any side e of $\partial \Omega$ there exists a point $p(e) \in e$ such that Ω lies in one of the closed half-spaces $\mathbb{R}^2 \setminus l_p(e)$ with respect to the tangent line $l_p(e)$ to $\partial \Omega$ at $p(e)$.

REMARK 3.8. For a rosette-shaped Ω there exists a smooth convex curve $\gamma(e)$ containing Ω in its convex hull and tangent to $\partial \Omega$ at exactly one point lying on a given side e of $\partial \Omega$.

LEMMA 3.9. *Consider a nonflattening tree-like curve c with convex coorientation and a locally convex building block b of c containing at least one exterior side, i.e., a side bounding the noncompact exterior domain on $\mathbb{R}^2$. There exists a* nonflattening curve c' isotopic to c such that its building block b' corresponding to b bounds a rosette-shaped domain.

PROOF. *Step* 1. Let k denote the number of corners of b. Consider the connected components $c_1, \ldots, c_k$ of $c \setminus b$. By the assumption that b contains an exterior side, one has that every c_i lies either inside or outside b (it cannot contain b). Therefore, using a suitable contracting homothety, we can make every c_i small and lying in a prescribed small neighborhood of its corner preserving the nonflattening property.

Step 2. Take the standard unit circle $S^1 \subset \mathbb{R}^2$ and choose k points on S^1. Then deform S^1 slightly into a piecewise smooth locally convex curve $\tilde{S}^1$ with the same sequence of $\vee$- and $\wedge$-type corners as on b. Now glue the small components $c_1, \ldots, c_k$ (after an appropriate affine transformation applied to each c_i) to $\tilde{S}^1$ in the same order as they sit on b. The resulting curve c' is a nonflattening tree-like curve with the same NCPD tree as c. □

COROLLARY 3.10. *Using Remark* 3.8, *one can glue a big locally convex loop containing the whole c' and tangent to c' at one point on any exterior edge and then deform this point of tangency into a double point and therefore get the nondegenerate tree-like curve required in case* 1.

Now we prove the supporting lemmas for case 2 of Criterion 3.5.

Take any polygon Pol with k vertices and with the same sequence of $\vee$- and $\wedge$-type vertices as given by $e_1, \ldots, e_k$; see the notations in the proof of case 2. The existence of such a polygon is exactly guaranteed by condition c), i.e., $k \geq 3$ and the number of interior angles $> \pi$ is less than or equal to $k - 3$. Deform it slightly to make it into a locally concave curvilinear polygon which we denote by $\widetilde{\text{Pol}}$.

LEMMA 3.11. *It is possible to glue a nonflattening curve $\tilde{c}$ (after an appropriate diffeomorphism) through its convex exterior edge to any $\wedge$-type vertex of $\widetilde{\text{Pol}}$ placing it outside $\widetilde{\text{Pol}}$ and preserving the nonflattening of the union.*

PROOF. We assume that the building block containing the side e of the curve $\tilde{c}$ to which we have to glue v is rosette-shaped. We choose a point p on e and substitute e by two convex sides meeting transversally at p. Then we apply to $\tilde{c}$ a linear transformation having its origin at v in order to a) make $\tilde{c}$ small, and b) to make the angle between the new sides equal to the angle at the $\wedge$-type vertex v to which we have to glue $\tilde{c}$. After that we glue $\tilde{c}$ by matching v and p, making the tangent lines of $\tilde{c}$ at p coinciding with the corresponding tangent lines of $\widetilde{\text{Pol}}$ at p and smoothing the higher derivatives. □

LEMMA 3.12. *It is possible to glue a nonflattening curve $\tilde{c}$ after cutting off its exterior building block with 1 corner to a $\vee$-type vertex of $\widetilde{\text{Pol}}$ and preserving the nonflattening of the union. The curve is placed inside $\widetilde{\text{Pol}}$.*

PROOF. The argument is essentially the same as above. We cut off a convex exterior loop from $\tilde{c}$ and apply to the remaining curve a linear transformation making it small and making the angle between two sides at the corner where we have cut off a loop equal to the angle at the $\vee$-type vertex. Then we glue the result to the $\vee$-type vertex and smooth the higher derivatives. □

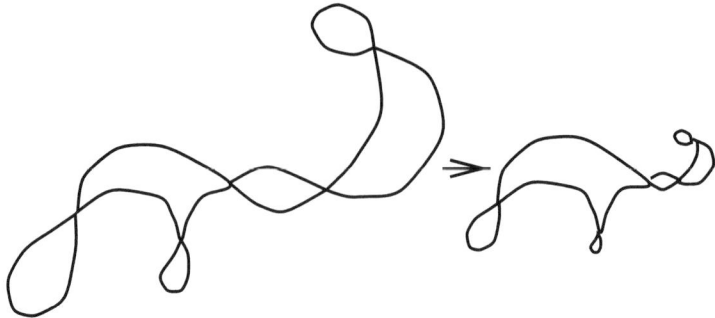

FIGURE 9. Separation of building blocks of different levels by contracting homothety

4. Upper and lower bounds of $\sharp_{inf}[c]$ for tree-like curves

Violation of any of the three conditions above of nonflattening leads to the appearance of inflection points on a tree-like curve which are unremovable by diffeomorphisms of $\mathbb{R}^2$. At first we reduce the question about the minimal number $\sharp_{inf}[c]$ of inflection points of a class of tree-like curves to a purely combinatorial problem and then we shall give some upper and lower bounds for this number. Some of the geometric proofs are only sketched for the same reasons as in the previous section. Since we are interested in inflections which survive under the action of diffeomorphisms of $\mathbb{R}^2$ we will assume from now on that all considered curves have only locally unremovable inflection points, i.e., those which do not disappear under arbitrarily small deformations of curves. (For example, the germ (t, t^4) is not interesting since its inflection disappears after an arbitrarily small deformation of the germ. One can assume that the tangent line at every inflection point of any curve we consider intersects the curve with the multiplicity 3.)

DEFINITION 4.1. A generic immersion $c : S^1 \to \mathbb{R}^2$ whose inflection coincides with some of its double points is called *normalized*.

PROPOSITION 4.2. *Every tree-like curve is isotopic to a normalized tree-like curve with at most the same number of inflection points.*

PROOF. *Step* 1. The idea of the proof is to separate building blocks as much as possible and then substitute every block by a curvilinear polygon with nonflattening sides. Namely, given a tree-like c let us partially order the vertices of its NCPD tree $\text{Tr}[c]$ by choosing one vertex as the root (vertex of level 1). Then we assign to all its adjacent vertices level 2, etc. The only requirement for the choice of the root is that all the directed edges point from the lower level to the higher. One can immediately see that the noncolliding property guarantees the existence of at least one root. Given such a partial order we apply consecutively a series of contracting homotheties to all double points as follows. We start with double points which are the corners of the building block b corresponding to the root. Then we apply contracting homotheties to all connected components of $c \setminus b$ placing them into the prescribed small neighborhoods of the corners of b. Then we apply contracting homothety to all connected components of $c \setminus (\cup b_i)$ where b_i has level less than or equal to 2, etc.; see the example in Figure 9. (Again we do not need to construct

FIGURE 10. Normal and abmnormal positions of tangent elements to a side

an explicit isotopy of the initial and final curves as soon as we know that they have the same NCPD tree and, therefore, are isotopic.) Note that every building block except for the root has its father to which it is attached through a ∧-type corner since the root contains an exterior edge. The resulting curve $\tilde{c}$ has the same type and number of inflection points as c since every building lies in a prescribed small neighborhood of the corresponding corner of its father and therefore does not intersect with other building blocks.

Step 2. Now we substitute every side of every building block by a nonflattening arc not increasing the number of inflection points. Fixing some orientation of $\tilde{c}$, we assign at every double point two oriented tangent elements to two branches of $\tilde{c}$ in the obvious way. Note that we can assume (after applying an arbitrarily small deformation of $\tilde{c}$) that any two of these tangent elements not sharing the same vertex are in general position, i.e., the line connecting the foot-points of the tangent elements is different from both tangent lines.

Initial change. At first we will substitute every building block of the highest level (which is necessarily a 1-gon) by a convex loop. According to our smallness assumptions, there exists a smooth (except for the corner) convex loop gluing which instead of the building block will make the whole new curve C^1-smooth and isotopic to $\tilde{c}$ in the class of C^1-smooth curves. This convex loop lies on the definite side with respect to both tangent lines at the double point. Note that if the original removed building block lies wrongly with respect to one (both resp.) tangent lines then it has at least one (two resp.) inflection points. After constructing a C^1-smooth curve we change it slightly in a small neighborhood of the double point in order to provide for each branch a) if the branch of $\tilde{c}$ changes convexity at the double point, then we produce a smooth inflection at the double point; b) if the branch does not change the convexity, then we make it smooth. The above remark guarantees that the total number of inflection points does not increase.

Typical change. Assume that all blocks of level $> i$ already have nonflattening sides. Take any block b of level i. By the choice of the root it has a unique ∧-type corner with its father. The block b has a definite sequence of its ∨- and ∧-type corners starting with the attachment corner and going around b clockwise. We want to cut off all connected components of $c \setminus b$ which have level $> i$ then substitute b by a curvilinear polygon with nonflattening sides and then glue back the blocks we cut off. Let us draw the usual polygon Pol with the same sequence of ∨- and ∧-type vertices as for b. Now we will deform its sides into convex and concave arcs depending on the sides of the initial b. The tangent elements to the ends of some side of b can be in one of 2 typical normal or 2 typical abnormal positions (up to orientation-preserving affine transformations of $\mathbb{R}^2$); see Figure 10.

If the position of the tangent elements is normal then we deform the corresponding side of Pol to get a nonflattening arc with the same position of the tangent elements as for the initial side of b. If the position is abnormal, then we deform the side of Pol to get a nonflattening arc which has the same position with respect to the tangent element at the beginning as the original side of b. Analogous considerations as before show that after gluing everything back and smoothing, the total number of inflections will not increase. $\square$

DEFINITION 4.3. A *local coorientation* of a generic immersion $c: S^1 \to \mathbb{R}^2$ is a free coorientation of each side of c (i.e., every arc between double points) which is, in general, discontinuous at the double points. *The convex coorientation* of a normalized curve is its local coorientation which coincides inside each (nonflattening) side with the convex coorientation of this side; see Definition 3.2.

Given a tree-like curve c with some local coorientation we want to understand when there exists a normalized tree-like curve c' isotopic to c whose convex coorientation coincides with a given local coorientation of c. The following proposition, which is closely connected with the criterion of nonflattening from Section 2, answers this question.

PROPOSITION 4.4 (Realizability criterion for a locally cooriented tree-like curve). *There exists a tree-like normalized curve with a given convex coorientation if and only if the following three conditions hold:*
  a) *Every building 1-gon is outward cooriented;*
  b) *At least one side of every building 2-gon is outward cooriented;*
  c) *If some building block on k vertices has only inward cooriented sides, then the domain it bounds contains at most $k-3$ neighboring building blocks.*

(Note that only condition b) is somewhat different from that of Criterion 3.5.)

SKETCH OF THE PROOF. The necessity of a)–c) is obvious. These conditions guarantee that all building blocks can be constructed. It is easy to see that they are, in fact, sufficient. Realizing each building block by some curvilinear polygon with nonflattening sides, we can glue them together in a global normalized tree-like curve. Namely, we start from some building block which contains an exterior edge. Then we glue all its neighbors to its corners. (In order to be able to glue them, we make them small and adjust the gluing angles by appropriate linear transformations.) Finally, we smooth higher derivatives at all corners and then proceed in the same way for all new corners. $\square$

COMBINATORIAL SETUP. Proposition 4.4 allows us to reformulate the question about the minimal number of inflection points $\sharp_{inf}[c]$ for tree-like curves combinatorially.

DEFINITION 4.5. A local coorientation of a given tree-like curve c is called *admissible* if it satisfies the conditions a)–c) of Proposition 4.4.

DEFINITION 4.6. Two sides of c are called *neighboring* if they share the same vertex and their tangent lines at this vertex coincide. We say that two neighboring sides in a locally cooriented curve c *create an inflection point* if their coorientations are opposite. For a given local coorientation Cc of a curve c let $\sharp_{inf}(Cc)$ denote the total number of created inflection points.

PROPOSITION 4.7 (Combinatorial reformulation). *For a given tree-like curve c one has*
$$\sharp_{inf}[c] = \min \sharp_{inf}(Cc),$$
where the minimum is taken over the set of all admissible local coorientations Cc of a tree-like curve c.

PROOF. This is the direct corollary of Propositions 4.2 and 4.4. Namely, for every tree-like curve $\tilde{c}$ isotopic to c there exists a normalized curve $\tilde{c}'$ with at most the same number of inflection points. The number of inflection points of $\tilde{c}'$ coincides with that of its convex coorientation Cc'. On the other side, for every admissible local coorientation Cc there exists a normalized curve c' whose convex coorientation coincides with Cc. □

A lower bound. A natural lower bound for $\sharp_{inf}[c]$ can be obtained in terms of the NCPD tree $\mathrm{Tr}[c]$. Choose any of two continuous coorientations of c and the corresponding coorientation of $\mathrm{Tr}[c]$; see Section 2. All 1-sided building blocks of c (corresponding to the leaves of $\mathrm{Tr}[c]$) have the natural cyclic order. (This order coincides with the natural cyclic order on all leaves of $\mathrm{Tr}[c]$ according to their position on the plane.)

DEFINITION 4.8. A neighboring pair of 1-sided building blocks (or of leaves on $\mathrm{Tr}[c]$) is called *reversing* if the continuous coorientations of these blocks are different. Let $\sharp_{rev}[c]$ denote the total number of reversing neighboring pairs of building blocks.

Note that $\sharp_{rev}[c]$ is even and independent on the choice of the continuous coorientation of c. Moreover, $\sharp_{rev}[c]$ depends only on the class $[c]$ and therefore we can use the notation above instead of $\sharp_{rev}(c)$.

PROPOSITION 4.9. $\sharp_{rev}[c] \leq \sharp_{inf}[c]$.

PROOF. Select a point p_i in each of the 1-sided building blocks b_i such that the side is locally convex near p_i with respect to the interior of b_i. (Such a choice is obviously possible since b_i has just one side.) The proof is accomplished by the following simple observation.

Take an immersed segment $\gamma : [0, 1] \to \mathbb{R}^2$ such that $\gamma(0)$ and $\gamma(1)$ are not inflection points and the total number of inflection points on γ is finite. At each nonflattening point p of γ we can choose the convex coorientation (see Section 3), i.e., since the tangent line to γ at p belongs locally to one connected component of $\mathbb{R}^2 \setminus \gamma$ we can choose a transversal vector pointing at that half-space. Let us denote the convex coorientation at p by $n(p)$.

LEMMA 4.10. *Assume that we have fixed a global continuous coorientation Coor of γ. If $\mathrm{Coor}(0) = n(0)$ and $\mathrm{Coor}(1) = n(1)$, then γ contains an even number of locally unremovable inflections. If $\mathrm{Coor}(0) = n(0)$ and $\mathrm{Coor}(1)$ is opposite $n(1)$, then γ contains an odd number of locally unremovable inflections.*

PROOF. Recall that we have assumed that all our inflection points are unremovable by local deformations of the curve. Therefore, when passing through such an inflection point, the convex coorientation changes to the opposite. □

An upper bound.

DEFINITION 4.11. If the number of 1-sided building blocks is larger than two, then each pair of neighboring 1-sided blocks of c (leaves of Tr$[c]$ resp.) is joined by the unique segment of c not containing other 1-sided building blocks. We call this segment *a connecting path*. If a connecting path joins a pair of neighboring 1-sided blocks with opposite coorientations (i.e., one block is inward cooriented and the other is outward cooriented with respect to the continuous coorientation of c), then it is called *a reversing connecting path*.

DEFINITION 4.12. Let us call a nonextendable sequence of 2-sided building blocks not contained in each other *a joint* of a tree-like curve. (On the level of its NCPD tree one gets a sequence of degree two vertices connected by undirected edges.)

Every joint consists of two smooth intersecting segments of c called *threads* belonging to two different connecting paths.

DEFINITION 4.13. For every nonreversing connecting path ρ in c we determine *its standard local coorientation* as follows. First we coorient its first side (which is the side of a 1-gon) outward and then extend this coorientation by continuity. (Since ρ is nonreversing its final side will be outward cooriented as well.)

DEFINITION 4.14. A joint is called *suspicious* if
a) both its threads lie on nonreversing paths and both sides of some 2-sided block from this joint are inward cooriented with respect to the above standard local coorientation of nonreversing paths; or
b) one thread lies on a nonreversing path and there exists a block belonging to this joint such that its side lying on the nonreversing path is inward cooriented (with respect to the standard local coorientation of nonreversing paths); or
c) both threads lie on reversing paths.

Let $\sharp_{jt}$ denote the total number of suspicious joints.

DEFINITION 4.15. A building block with k sides is called *suspicious* if it satisfies the following two conditions:
a) it contains at least $k-3$ other blocks, i.e., at least $4.15k-3$ edges are leaving the corresponding vertex of the tree;
b) all sides lying on nonreversing paths are inward cooriented with respect to the standard local coorientations of these nonreversing paths.

Let $\sharp_{bl}$ denote the total number of suspicious blocks.

PROPOSITION 4.16.
$$\sharp_{inf}[c] \leq \sharp_{rev}[c] + 2(\sharp_{jt} + \sharp_{bl}).$$

PROOF. According to Proposition 4.7, for any tree-like curve c one has $\sharp_{inf}[c] \leq \sharp_{inf}(Cc)$ where Cc is some admissible local coorientation of c. Let us show that there exists an admissible local coorientation with at most $\sharp_{rev}[c] + 2(\sharp_{jt} + \sharp_{bl})$ inflection points; see Definition 4.6. First we fix the standard local coorientations of all nonreversing paths. Then for each reversing path we choose any local coorientation with exactly one inflection point (i.e., one discontinuity of local coorientation on the reversing path) to get the necessary outward coorientations of all 1-sided blocks.

Now the local coorientation of the whole fat NCPD tree is fixed but it is not admissible, in general. In order to make it admissible we have to provide that conditions b) and c) of Proposition 4.4 are satisfied for at most $\sharp_{jt}$ suspicious joints and at most $\sharp_{bl}$ suspicious blocks. To make the local coorientation of each such suspicious joint or block admissible we need to introduce at most two additional inflection points for every suspicious joint or block. The proposition follows. □

5. Concluding remarks

In spite of the fact that there exists a reasonable criterion for nonflattening in the class of tree-like curves in terms of their NCPD trees, the author is convinced that there is no closed formula for $\sharp_{inf}[c]$. Combinatorial reformulation of Definition 4.6 reduces the calculation of $\sharp_{inf}[c]$ to a rather complicated discrete optimization problem which is hardly expected to have an answer in a simple closed form. (One can even make speculations about the computational complexity of the above optimization problem.)

The lower and upper bounds presented in Section 4 can be improved by using the much more complicated characteristics of an NCPD tree. On the other side, both of them are sharp on some subclasses of NCPD trees. Since in these cases a closed formula has not been obtained, the author did not try to get the best possible estimations here.

At the moment the author is trying to extend the results of this note to the case of all generic curves in $\mathbb{R}^2$; see [**Sh**].

6. Appendix: Counting tree-like curves with a given Gauss diagram

The combinatorial material of this section is not directly related to the main content of the paper. It is a side product of the author's interest in tree-like curves. Here we calculate the number of different classes of tree-like curves which have the same Gauss diagram.

PROPOSITION 6.1. *There exists a one-to-one correspondence between classes of oriented tree-like curves with $n-1$ double points on nonoriented $\mathbb{R}^2$ and the set of all planar NCPD trees with n vertices on oriented $\mathbb{R}^2$.*

PROOF. See [**Ai**].

DEFINITION 6.2. For a given planar tree Tr consider the subgroup Diff(Tr) of all orientation-preserving diffeomorphisms of $\mathbb{R}^2$ sending Tr homeomorphically onto itself as an embedded 1-complex. The subgroup PAut(Tr) of the group Aut(Tr) of automorphisms of Tr as an abstract tree induced by Diff(Tr) is called *the group of planar automorphisms of* Tr.

The following simple statement gives a complete description of different possible groups PAut(Tr). (Unfortunately, the author was unable to find a suitable reference for this but the proof is not too hard.)

STATEMENT 6.3.
1. *The group* PAut(Tr) *of planar automorphisms of a given planar tree* Tr *is isomorphic to* $\mathbb{Z}/\mathbb{Z}_p$ *and coincides (after an appropriate diffeomorphism of*

the plane) with the group of rotations about some center by multiples of $2\pi/p$.
2. If $\text{PAut}(\text{Tr}) = \mathbb{Z}/\mathbb{Z}_p$ for $p > 2$, then the center of rotation above is a vertex of Tr.
3. For $p = 2$ the center of rotation is either a vertex of Tr or the middle of its edge.
4. If the center of rotation is a vertex of Tr, then the action $\text{PAut}(\text{Tr})$ on Tr is free except for the center and the quotient can be identified with a connected subtree $\text{STr} \subset \text{Tr}$ containing the center.
5. For $p = 2$, if the center is the middle of an edge, then the action of $\text{PAut}(\text{Tr})$ on Tr is free except for this edge.

SKETCH OF THE PROOF. The action of $\text{PAut}(\text{Tr})$ on the set $Lv(\text{Tr})$ of leaves of Tr preserves the natural cyclic order on $Lv(\text{Tr})$ and thus reduces to the $\mathbb{Z}/\mathbb{Z}_p$-action for some p. Now each element $g \in \text{PAut}(\text{Tr})$ is determined by its action on $Lv(\text{Tr})$ and thus the whole $\text{PAut}(\text{Tr})$ is isomorphic to $\mathbb{Z}/\mathbb{Z}_p$. Indeed consider some $\mathbb{Z}/\mathbb{Z}_p$-orbit O on $Lv(\text{Tr})$ and all vertices of Tr adjacent to O. They are all pairwise different or all coincide since otherwise they cannot form an orbit of the action of diffeomorphisms on Tr. □

PROPOSITION 6.4. *The number $\sharp(GD)$ of all classes of oriented tree-like curves on nonoriented $\mathbb{R}^2$ with a given Gauss diagram GD on n vertices is equal to*

$$\sharp(GD) = \begin{cases} a) \ 2^{n-1} + (n-1)2^{n-2}, \text{ if } \text{PAut}(GD) \text{ is trivial}; \\ b) \ 2^{2k-2} + (2k-1)2^{2k-3} + 2^{k-2} \text{ where } n = 2k, \text{ if } \text{PAut}(GD) = \mathbb{Z}/2\mathbb{Z} \\ \quad \text{and the center of rotation is the middle of the side}; \\ c) \ 2^k + (2^{n-1} + (n-1)2^{n-2} - 2^k)/p, \text{ where } n = kp+1 \text{ and} \\ \quad \text{PAut}(GD) = \mathbb{Z}/p\mathbb{Z} \text{ for some prime } p \text{ (including } \text{PAut}(GD) = \mathbb{Z}/2\mathbb{Z} \\ \quad \text{with a central vertex}); \\ d) \ \text{see Proposition 6.5 for the general case.} \end{cases}$$

PROOF. By Proposition 6.1 we enumerate NCPD trees with a given underlying planar tree $\text{Tr}(GD)$.

Case a). Let us first calculate only those NCPD trees whose edges are all directed. The number of such NCPD trees equals the number n of vertices of $\text{Tr}(GD)$ since for any such tree there exists such a source-vertex (all edges are directed from this vertex). Now let us calculate the number of NCPD trees with l undirected edges. Since $\text{Aut}(GD)$ is trivial, we can assume that all vertices of $\text{Tr}(GD)$ are enumerated. There exist $\binom{n-1}{l}$ subgraphs in $\text{Tr}(GD)$ containing l edges, and for each of these subgraphs there exist $(n-l)$ NCPD trees with such a subgraph of undirected edges. Thus the total number

$$\sharp(GD) = \sum_{l=0}^{n-1} \binom{n-1}{l}(n-l) = 2^{n-1} + (n-1)2^{n-2}.$$

Case b). The $\mathbb{Z}/2\mathbb{Z}$-action on the set of all NCPD trees splits them into two classes according to the cardinality of orbits. The number of $\mathbb{Z}/2\mathbb{Z}$-invariant NCPD trees equals the number of all subtrees in a tree on k vertices where $n = 2k$ (since the source-vertex of such a tree necessarily lies in the center). The last number equals 2^{k-1}. This gives

$$\sharp(GD) = (2^{n-1} + (n-1)2^{n-2} - 2^{k-1})/2 + 2^{k-1} = 2^{2k-2} + (2k-1)2^{2k-3} + 2^{k-2}.$$

Case c). The $\mathbb{Z}/p\mathbb{Z}$-action on the set of all NCPD trees splits them into 2 groups according to the cardinality of orbits. The number of $\mathbb{Z}/p\mathbb{Z}$-invariant NCPD trees equals the number of all subtrees in a tree with $k+1$ vertices where $n = pk+1$ (since the source-vertex of such a tree lies in the center). The last number equals 2^k. This gives

$$\sharp(GD) = 2^k + (2^{n-1} + (n-1)2^{n-2} - 2^k)/p. \quad \square$$

PROPOSITION 6.5. *Consider a Gauss diagram GD with a tree $\mathrm{Tr}(GD)$ having n vertices which has $\mathrm{Aut}(\mathrm{Tr}) = \mathbb{Z}/p\mathbb{Z}$ where p is not a prime. Then for each nontrivial factor d of p the number of NCPD trees with $\mathbb{Z}/d\mathbb{Z}$ as their group of symmetry equals*

$$\sum_{d'|d} \mu(d') 2^{\frac{kp}{d'}},$$

where $\mu(d')$ is the Möbius function. Here, as above, $n = kp + 1$. (This gives a rather unpleasant expression for the number of all tree-like curves with a given GD if p is an arbitrary positive integer.)

PROOF. For each d such that $d|p$, consider a subtree STr_d with $m+1$ vertices $m = \frac{n-1}{d}$ "spanning" Tr with respect to the $\mathbb{Z}/d\mathbb{Z}$-action. The number of NCPD trees invariant at least with respect to $\mathbb{Z}/d\mathbb{Z}$ equals 2^m. Thus by the inclusion-exclusion formula one gets that the number of NCPD trees invariant exactly with respect to $\mathbb{Z}/d\mathbb{Z}$ equals $\sum_{d'|d} \mu(d') 2^{\frac{kp}{d'}}$. $\quad \square$

PROBLEM. Calculate the number of NCPD trees with a given underlying tree and of a given index.

References

[Ai] F. Aicardi, *Tree-like curves*, Adv. in Soviet Math., vol. 21, Amer. Math. Soc., Providence, RI, 1994, pp. 1–31.
[Ar1] V. Arnold, *Plane curves, their invariants, perestroikas and classification*, Adv. in Soviet Math., Amer. Math. Soc., Providence, RI, 1994, pp. 33–91.
[Ar2] _____, *Topological invariants of plane curves and caustics*, Univ. Lectures Series, vol. 5, Amer. Math. Soc., Providence, RI, 1994.
[Ar3] _____, *Topological problems in wave propagation*, Uspekhi Mat. Nauk **51** (1996), no. 1, 3–50; English transl, Russian Math. Surveys **51** (1996), no. 1, 1–50.
[Po] M. Polyak, *Invariants of generic curves via Gauss diagrams*, Preprint of MPI (1995), 1–10.
[Sh] B. Shapiro, *On the number of inflection points for generic points in $\mathbb{R}^2$*, in preparation.
[Shu] A. Shumakovich, *Explicit formulas for strangeness of plane curves*, Algebra i Analiz **7** (1995), no. 3, 165–199; English transl., St. Petersburg Math. J. **7** (1996), no. 3, 445–472.

DEPARTMENT OF MATHEMATICS, UNIVERSITY OF STOCKHOLM, S-10691, SWEDEN
E-mail address: shapiro@matematik.su.se

Geometry of Exact Transverse Line Fields and Projective Billiards

Serge Tabachnikov

1. Introduction and survey of results

1.1. We start with a description of the (usual) billiard transformation. A billiard table is a convex domain in $\mathbb{R}^n$ bounded by a smooth closed hypersurface M. The billiard transformation acts on the set of oriented lines (rays) in $\mathbb{R}^n$ that intersect M, and is determined by the usual law of geometric optics: the incoming and the outgoing rays lie in one 2-plane with the normal line to M at the impact point and make equal angles with this normal line. Equivalently, the reflection can be described as follows: at the impact point, the tangential component of a velocity vector along a ray is preserved while the normal component changes sign. An analogous law describes the billiard reflection in a Riemannian manifold with the boundary M. An interested reader may find a survey of the theory of mathematical billiards in [**T1**].

The billiard transformation is defined in metric terms and is equivariant with respect to the isometries of $\mathbb{R}^n$ that naturally act on the space of rays (and transform billiard tables to isometric ones). In [**T2**] we introduced a more general dynamical system, the projective billiard, equivariant under the greater group of projective transformations of $\mathbb{R}^n$. To define the projective billiard transformation, one needs an additional structure: a smooth transverse line field ξ along the billiard hypersurface $M \subset \mathbb{R}^n$ (respectively, a diffeomorphism f of $\mathbb{R}^n$ transforms a pair (M, ξ) to $(f(M), Df(\xi))$, where Df is the derivative of f).

Given such a "hairy" billiard table, the law of the projective billiard reflection reads

1) The incoming ray, the outgoing ray, and the transverse line $\xi(x)$ at the impact point x lie in one 2-plane π.
2) The three above lines and the line of intersection of π with the tangent hyperplane T_xM form a harmonic quadruple of lines.

(Four lines through a point is a harmonic quadruple if the cross-ratio of their intersection points with some, and hence with every, auxiliary line equals -1). Equivalently, a velocity vector along the incoming ray is decomposed at the impact point x into the tangential and transverse components, the latter being collinear

---

1991 *Mathematics Subject Classification.* Primary 58F05.

with $\xi(x)$; then the transverse component changes sign yielding a velocity vector along the outgoing ray.

If the transverse line field consists of the normals to M, then the projective billiard coincides with the usual one. More generally, consider a metric g in a domain containing M whose geodesics are straight lines. Then the (usual) billiard in this metric is a projective billiard with the transverse line field consisting of the g-normals to M. Examples of such metrics are the spherical and the hyperbolic ones; conversely, a metric in a domain in $\mathbb{R}^n$ whose geodesics are straight lines is a metric of constant curvature (see [**D**]). To further generalize, one may consider a Finsler metric whose geodesics are straight lines. Given a hypersurface M, the billiard transformation of the set of rays intersecting M is defined in this setting as well but, unless the Finsler metric is a Riemannian one, it is not a projective billiard transformation.

The space of rays in $\mathbb{R}^n$ has a canonical symplectic structure obtained from the one in the cotangent bundle $T^*\mathbb{R}^n$ by symplectic reduction (see, e.g., [**A-G**]). A fundamental property of the billiard transformation is that it preserves this symplectic form. Turning from the Hamiltonian to the Lagrangian point of view, the billiard trajectories are the extremals of the length functional. These properties make it possible to use methods of symplectic geometry and classical mechanics in the study of billiard dynamics.

It is therefore natural to ask for which pairs (M, ξ) the phase space of the respective projective billiard transformation T has a T-invariant symplectic structure. This difficult problem is far from being solved; some results in this direction from [**T2,3**] are surveyed in the next section. These results suggest that the following class of transverse fields plays a special role in the projective billiard problem, similar to that of the normal fields in the usual billiard setting.

The following notation will be used throughout the paper. Let $n(x)$, $x \in M$ denote the unit normal vector field along a cooriented hypersurface $M \subset \mathbb{R}^n$. Given a transverse line field ξ along M, let v be the vector field along ξ such that for the scalar product we have $v(x)\, n(x) = 1$ for all $x \in M$.

DEFINITION. A transverse field ξ is called *exact* if the 1-form $v\, dn$ on M is exact.

Here and elsewhere we adopt the following convention: v and n are thought of as vector-valued functions, and $v\, dn$ is the scalar product of a vector-function and a vector-valued 1-form.

Note that the form $v\, dn$ does not depend on the sign of the vector n. Thus the definition of exactness makes sense for noncooriented hypersurfaces as well. It also extends to the case when M is an immersed hypersurface and, more generally, a wave front. A wave front is the projection to $\mathbb{R}^n$ of a smooth immersed Legendrian submanifold L^{n-1} of the space of contact elements in $\mathbb{R}^n$. "Legendrian" means that L is tangent to the canonical contact structure in the space of contact elements (see [**A-G**]). A smooth hypersurface M canonically lifts to such a Legendrian manifold by assigning the tangent hyperplane T_xM to each point $x \in M$. Although a wave front may have singularities, it has a well-defined tangent hyperplane at every point, so, up to a sign, the unit normal vector is defined too. Then $v\, dn$ is considered to be a 1-form on L, and the field v along M is called exact if this 1-form on L is exact.

Geometry of exact transverse line fields is closely related to symplectic geometry of systems of rays, Finsler geometry, and projective geometry. We find it of interest on its own.

1.2. In this subsection we survey some results obtained in [**T2–5**]. Start with properties of exact transverse fields. Although they are defined in Euclidean terms, exactness is a projective property (see [**T3**]).

THEOREM 1.2.1. *Let ξ be an exact transverse line field along $M \subset \mathbb{R}^n$, and f be a projective transformation of $\mathbb{R}^n$ whose domain contains M. Then the line field $Df(\xi)$ along $f(M)$ is also exact.*

This result is sharp: we will show in Section 2 that if a local diffeomorphism f of $\mathbb{R}^n$ takes every exact transverse field along every hypersurface to an exact field, then f is projective.

Recall that a (not necessarily symmetric) Finsler metric in $\mathbb{R}^n$ is given by a smooth field of smooth star-shaped quadratically convex hypersurfaces $S_x \subset T_x\mathbb{R}^n$, $x \in \mathbb{R}^n$. These hypersurfaces, called the indicatrices, consist of the Finsler unit vectors and play the role of the unit spheres in Riemannian geometry. A Minkowski metric in $\mathbb{R}^n$ is a translation-invariant Finsler metric. Given a cooriented hypersurface $M \subset \mathbb{R}^n$, the Finsler normal to M at point $x \in M$ is defined as follows. There is a unique $u \in S_x$ such that the outward cooriented tangent hyperplane $T_u(S_x)$ is parallel to and has the same coorientation as T_xM in $T_x\mathbb{R}^n$. By definition, this u is the Finsler normal to M at x.

It is not true, in general, that Finsler normals to a hypersurface constitute an exact transverse field. However, the next result holds (see [**T3**]).

THEOREM 1.2.2. *Given a smooth (not necessarily symmetric) Minkowski metric in $\mathbb{R}^n$ and a cooriented hypersurface $M \subset \mathbb{R}^n$, the field of Minkowski normals to M is exact.*

The next property of exact transverse fields, implicit in [**T3**], generalizes the Huygens principle of wave propagation (and coincides with it when the exact field consists of the normals).

THEOREM 1.2.3. *Let v be the vector field along a hypersurface M generating an exact transverse line field such that $vn = 1$, and let f be a function on M such that $v\,dn|_M = df$. Fix a real t. For $x \in M$, let $x_t = x + te^{f(x)}v(x)$, and let M_t be the locus of points x_t. Then the tangent hyperplane to M_t at its every smooth point x_t is parallel to the tangent hyperplane to M at x.*

In the case of the usual wave propagation, that is, when $v = n$, the hypersurface M_t is t-equidistant from M. This M_t is a wave front, and the corresponding Legendrian manifold L_t is imbedded and consists of the tangent hyperplanes T_xM, parallel translated to the respective points x_t. A similar description applies to an exact transverse line field; however, the manifold L_t may be singular in the general case.

In the next two theorems $M \subset \mathbb{R}^2$ is a closed strictly convex smooth plane curve. The following result is contained in [**T2,4**].

THEOREM 1.2.4. *A transverse line field ξ along M is exact if and only if there exists a parameterization $M(t)$ such that $\xi(M(t))$ is generated by the acceleration vector $M''(t)$ for all t.*

Another specifically two-dimensional result from [**T4,5**] is a generalization of the classical 4-vertex theorem to exact transverse fields along curves (the 4-vertex theorem is the case of the normals).

THEOREM 1.2.5. *Let ξ be a generic exact transverse line field along a closed convex curve M. Then the envelope of the 1-parameter family of lines $\xi(x)$, $x \in M$, has at least 4 cusp singularities.*

Next we turn to the projective billiards. We already mentioned a class of projective billiard transformations with an invariant symplectic form, the usual billiards in a domain in $\mathbb{R}^n$ with a metric g of constant curvature whose geodesics are straight lines. The transverse field is the field of g-normals, and the invariant symplectic form is the canonical symplectic form on the space of oriented geodesics. Applying a projective transformation to such a billiard once again obtains a projective billiard with an invariant symplectic structure.

A different class of projective billiards with an invariant symplectic structure was introduced in [**T3**]. Let $M \subset \mathbb{R}^n$ be a sphere equipped with an exact transverse line field. Identify the interior of M with the hyperbolic space $\mathbb{H}^n$ so that the chords of M represent the straight lines in $\mathbb{H}^n$ (the Beltrami–Klein model of hyperbolic geometry). Then the respective projective billiard transformation T acts on oriented lines in $\mathbb{H}^n$, and the phase space of T is not compact, unlike that of the usual billiard transformation.

THEOREM 1.2.6. *The canonical symplectic structure on the space of rays in $\mathbb{H}^n$ is T-invariant. The map T has n-periodic orbits for every $n \geq 2$.*

Note that the supply of these projective billiards inside a sphere is, roughly, the same as that of the usual billiards: modulo a finite dimensional correction, both depend on a function on the sphere S^{n-1}. For the usual billiards, this is the support function of the billiard hypersurface; for the projective ones, it is the function f such that $v \, dn = df$.

This projective billiard inside a sphere can be described as follows. A transverse field along M can be thought of as a smooth family of rays L in $\mathbb{H}^n$ such that for every point x at infinity there is a unique ray $l_x \in L$ whose forward limit is x. Let σ_x be the involution of the space of rays in $\mathbb{H}^n$ that reflects the rays in l_x, and let τ be the involution of this space that reverses the orientation of the rays. Then, for a ray r with the forward limit point x, one has $T(r) = \tau(\sigma_x(r))$.

The next example is discussed in [**T2**]. Let $M \subset \mathbb{R}^2$ be a smooth closed convex curve and ξ be the field of lines along M through a fixed point x inside M. Then the corresponding projective billiard transformation has an invariant area form which blows up on the curve in the phase space that consists of the lines through x (see [**T2**] for details and, in particular, the relation with the dual, or outer, billiards). This construction generalizes to higher dimensions but the dynamics is rather dull: every orbit stays in a 2-plane.

Finally, the next result from [**T2**] is another manifestation of the relation between projective billiards and exact transverse line fields. Let M be a smooth closed convex plane curve with a transverse line field ξ. The phase space of the projective billiard map T is a cylinder whose two boundary circles consist of the oriented lines tangent to M.

THEOREM 1.2.7. *If there exists (the infinite jet of) a T-invariant area form along a boundary component of the phase cylinder, then ξ is exact.*

1.3. In this subsection we outline the contents of this paper.

Section 2 concerns properties of exact transverse line fields. First, we prove a result that is partially converse to Theorem 1.2.2. Let $v(x)$, $x \in M$, be an exact transverse field along a quadratically convex closed hypersurface M, and assume that the Gauss map

$$G: M \to S^{n-1}, \qquad G(x) = v(x)/|v(x)|,$$

is a diffeomorphism. Then there exists a Minkowski metric in $\mathbb{R}^n$ (in general, not symmetric) such that the transverse field is the field of Minkowski normals to M (the particular case of curves in the plane was considered in [**T3**]).

Next, we discuss the following local question: For which Finsler metrics in $\mathbb{R}^n$ is the field of Finsler normals to every hypersurface exact? We give a certain technical criterion for this to hold and deduce from it that the field of normals to a hypersurface in the hyperbolic or spherical metric in $\mathbb{R}^n$, whose geodesics are straight lines, is exact.

We also show that the field of the affine normals to a convex hypersurface is exact (the particular case of plane curves was considered in [**T3**]). Given a convex hypersurface $M \subset \mathbb{R}^n$ and a point $x \in M$, consider the $(n-1)$-dimensional sections of M by the hyperplanes, parallel to $T_x M$. The centroids of these sections lie on a curve starting at x, and the tangent line to this curve at x is, by definition, the affine normal to M at x.

We expect the exact transverse fields to enjoy many properties of the Euclidean normals. In particular, given an immersed closed hypersurface $M \subset \mathbb{R}^n$ and a point $p \in \mathbb{R}^n$, the number of normals to M from p is bounded below by the least number of critical points of a smooth function on M, and, generically, by the sum of the Betti numbers of M (the function in question is, of course, the distance squared from p to M). We extend this result to an exact transverse line field ξ along an immersed locally quadratically convex closed hypersurface M: for every point p the number of lines $\xi(x)$, $x \in M$, through p has the same lower bound as in the case of normals.

The nondegeneracy condition on the second quadratic form of M is used in the proof but may be redundant. We make a bolder conjecture. Let M be an immersed closed hypersurface in $\mathbb{R}^n$ and L the corresponding Legendrian submanifold in the space of cooriented contact elements in $\mathbb{R}^n$. Let L_1 be a Legendrian manifold that is Legendrian isotopic to L, and M_1 its wave front.

CONJECTURE. *For every exact transverse line field ξ along M_1 and a point $p \in \mathbb{R}^n$, the number of lines $\xi(x)$, $x \in M$, through p is bounded below by the least number of critical points of a smooth function on M.*

This estimate holds for the Euclidean normals and, more generally, the normals in a Finsler Hadamard metric; see [**Fe1**].

One may ask a similar question concerning the least number of "diameters" of a hypersurface M equipped with an exact transverse field ξ. A diameter of (M, ξ) is a line l, intersecting M at points x, y, such that $\xi(x) = \xi(y) = l$. We failed to extend the results from [**Fe2, Pu**] on the Euclidean diameters (double normals) to exact transverse fields.

Next, we turn to the projective invariance of exactness. The proof of Theorem 1.2.1 given in [**T3**] is purely computational. We give an intrinsic definition of exact transverse line fields along hypersurfaces in the projective space. This definiton agrees with the previous one for hypersurfaces that lie in an affine chart, and this implies Theorem 1.2.1. We extend the lower bound for the number of lines from an exact transverse field along a closed hypersurface through a fixed point to hypersurfaces in $\mathbb{RP}^n$ which are quadratically nondegenerate. An example of such a surface is the hyperboloid in $\mathbb{RP}^3$ or its small perturbation.

Finally, we consider a smooth transverse line field ξ along a sphere. Identifying the interior of the sphere with $\mathbb{H}^n$, one has an imbedding of the sphere to the space of oriented lines in $\mathbb{H}^n$: to a point $x \in S^{n-1}$ there corresponds the line $\xi(x)$, oriented outward. Let L be the image of this map. We show that if $n \geq 3$, then ξ is exact if and only if L is a Lagrangian submanifold of the space of oriented lines in $\mathbb{H}^n$ with its canonical symplectic structure, associated with the hyperbolic metric; if $n = 2$, then the condition is that L is exact Lagrangian for an appropriate choice of the 1-form whose differential is the symplectic form. It follows that if ξ is exact, then the lines from this field are the hyperbolic normals to a one-parameter family of equidistant closed wave fronts in $\mathbb{H}^n$. The case of an exact line field ξ along a circle in the plane is of interest. The envelope of the lines from ξ is then the caustic of a closed curve, and the algebraic length of this envelope in the hyperbolic metric equals zero (the sign of the length changes after each cusp).

The observation above offers a new look on the projective billiards inside a sphere, associated with exact transverse line fields. Let $M \subset \mathbb{H}^n$ be a closed convex hypersurface, ξ the family of its outward hyperbolic normals, and M_t its t-equidistant hypersurface. Then ξ determines an exact transverse field along the sphere at infinity, and the respective projective billiard inside this sphere may be considered, rather informally, as the limit $t \to \infty$ case of the usual billiard inside M_t.

Section 3 concerns projective billiards. Our first observation is as follows. Consider a closed quadratically convex hypersurface M with an exact transverse line field ξ, and let A and B be points inside M. Then there exist at least two rays through A that, after the projective billiard reflection in M, pass through B (the case of plane curves was considered in [**T2**]). If the points A and B coincide, one obtains two lines from ξ passing through a given point whose existence was already mentioned above.

The hyperbolic distance between points A and B inside a ball in the Beltrami-Klein model is $|\ln[A, B, X, Y]|$, where $[\cdot]$ denotes the cross-ratio, and X, Y are the points of intersection of the line AB with the boundary sphere. The same formula defines a Finsler metric inside any convex closed hypersurface M. This metric is called the Hilbert metric, and its geodesics are straight lines. The Hilbert metric determines a canonical symplectic structure ω on the space of oriented lines that intersect M. If M is a sphere with an exact transverse field ξ, then by Theorem 1.2.6 the respective projective billiard transformation preserves the form ω. It is an interesting problem to describe all pairs (M, ξ) such that the corresponding projective billiard map preserves this symplectic structure. We prove this to be the case for a closed convex plane curve M if and only if M is an ellipse and ξ is an arbitrary transverse field.

The phase space of the projective billiard inside a circle is a cylinder. This cylinder, being the space of rays in $\mathbb{H}^2$, has a canonical area form, preserved by the projective billiard transformation for every transverse line field along the circle. The flux of an area-preserving transformation T of a cylinder C is defined as follows. Choose a noncontractible simple closed curve $\gamma \subset C$. Then the flux of T is the signed area between γ and $T(\gamma)$, and this area does not depend on the choice of γ. We show that the projective billiard transformation in a circle, associated with a transverse field ξ, has zero flux if and only if ξ is exact. It follows that if this transformation has an invariant circle, then the corresponding transverse field is exact. We construct an example to show that the last claim may fail for plane curves, other than circles. We also interpret the projective billiard inside a circle as a dynamical system on the one-sheeted hyperboloid.

The mirror equation of geometric optics describes the following situation. Suppose that an infinitesimal beam of rays from a point A reflects at a point X of a smooth billiard curve M and focuses at a point B. Let a and b be the distances from A and B to X, k the curvature of M at X, and α the angle of incidence of the ray AX. The mirror equation reads

$$\frac{1}{a} + \frac{1}{b} = \frac{2k}{\sin \alpha}.$$

This equation proved very useful in the study of billiards; see [**W1,2**].

We find an analog of the mirror equation for the projective billiard in the unit circle M. Parameterize M by the usual angle parameter t. Denote by $\phi(t)$ the angle made by the line of the transverse field at point $M(t)$ with the tangent vector $M'(t)$, and let ϕ and ϕ' be its value and the value of its derivative at X. Let a and b have the same meaning as before, and α and β be the angles made by the rays XA and XB with the positive direction of M at X. Then

$$\frac{2\phi'}{\sin^2 \phi} + \frac{1}{\sin^2 \alpha} + \frac{1}{\sin^2 \beta} = \frac{1}{a \sin \alpha} + \frac{1}{b \sin \beta}.$$

A similar equation can be written for every plane projective billiard.

At the end of Section 3 we present a conjecture describing all the cases in which a plane projective billiard is integrable.

2. Exact transverse line fileds

2.1. We use the same notation as in the Introduction. Let $M \subset \mathbb{R}^n$ be a smooth cooriented hypersurface and v a vector field along M that defines an exact transverse line field. One has $vn = 1$ and $v\,dn|_M = df$. Consider the smooth map $g : M \to \mathbb{R}^n$ given by the formula: $g(x) = e^{f(x)} v(x)$. Let $N = g(M)$.

LEMMA 2.1.1. *At every smooth point of N the tangent hyperplane $T_{g(x)}N$ is parallel to $T_x M$.*

PROOF. One wants to show that $n(x)$ is orthogonal to $T_{g(x)}N$; that is, the 1-form $n(x)\,dg(x)$ vanishes on N. Since $vn = 1$, one has $v\,dn + n\,dv = 0$. Therefore,

$$n\,dg = e^{f(x)} n(v\,df + dv) = e^{f(x)}(df + n\,dv) = e^{f(x)}(df - v\,dn) = 0,$$

and we are done.

Geometrically this lemma means the following. Consider a homogeneous codimension one distribution in the punctured $\mathbb{R}^n$ whose hyperplanes along the ray, spanned by $v(x)$, are parallel to $T_x M$. Then this distribution is integrable, the leaves being homothetic to the hypersurface N.

In general, N may be singular. Assume, however, that M is a closed quadratically convex hypersurface and the Gauss map

$$G: M \to S^{n-1}, \quad G(x) = v(x)/|v(x)|,$$

is a diffeomorphism.

THEOREM 2.1.2. *There exists a (not necessarily symmetric) Minkowski metric in $\mathbb{R}^n$ such that the exact transverse field v is the field of Minkowski normals to M.*

PROOF. Our assumptions imply that N is a quadratically convex star-shaped closed hypersurface. The desired Minkowski metric has N as its indicatrix.

The support function of a convex star-shaped hypersurface in $\mathbb{R}^n$ is the distance from the origin to its tangent hyperplanes. The value of the support function of the indicatrix N at point $g(x) \in N$ equals $n(x) \cdot e^{f(x)} v(x) = e^{f(x)}$.

In general, the construction of this section gives a smooth map from M to the space of contact elements in $\mathbb{R}^n$: to $x \in M$ there corresponds the hyperplane $T_x M$ based at the point $e^{f(x)} v(x)$. The image of this map is Legendrian, but it may be singular.

2.2. Using the same notation as before, call a transverse line field along a hypersurface *closed* if the 1-form $v\,dn$ is closed on M. Unlike exactness, this is a local property: a field is closed iff its restriction to every open subset of M. For simply connected hypersurfaces, the two notions are equivalent.

In this subsection we discuss Finsler metrics in domains in $\mathbb{R}^n$ such that the field of the Finsler normals to every smooth hypersurface is closed.

A Finsler metric can be characterized by a positive function $H(q,p)$ in $T^*\mathbb{R}^n$, strictly convex and homogeneous of degree 2 in the momentum variable p; see [**Ar**]. The unit level hypersurface of H in each fiber of the cotangent bundle consists of Finsler unit covectors, and the Hamiltonian vector field sgrad H is the geodesic flow in the Finsler metric. Given a point q of a cooriented hypersurface $M \subset \mathbb{R}^n$, let $p \in T_q^*\mathbb{R}^n$ be a covector such that Ker $p = T_q M$ and p gives $T_q M$ the positive coorientation. The Finsler normal $\xi(q)$ to M at q is spanned by the projection to $\mathbb{R}^n$ of the vector sgrad $H(q,p)$; since H is homogeneous the line $\xi(q)$ does not depend on the choice of p.

THEOREM 2.2.1. *The field of Finsler normals ξ along every hypersurface $M \subset \mathbb{R}^n$ is closed if and only if the 2-form $\Omega = (\ln H)_{pq}\, dp \wedge dq$ belongs to the ideal $(p\,dp,\ p\,dq,\ dp \wedge dq)$ in the algebra of differential forms in $T^*\mathbb{R}^n$.*

Here we use the obvious notation

$$F_{pq}\,dp \wedge dq = \sum \frac{\partial^2 F}{\partial p_i \partial q_j}\,dp_i \wedge dq_j, \quad dp \wedge dq = \sum dp_i \wedge dq_i,$$

etc.

PROOF. Identify the tangent and cotangent spaces by the Euclidean structure. Then the unit normal vector $n(q)$, $q \in M$ is considered to be a covector p such that Ker $p = T_q M$. The projection of sgrad H to $\mathbb{R}^n$ is the vector H_p. Hence the field ξ is generated by the vectors $H_p(q,p)$, $q \in M$, $p = n(q)$. Let $v(q) = H_p/2H$; by the Euler formula, $vn = pH_p/2H = 1$. Thus, one wants the 2-form

$$d(v\,dn) = d((\ln H)_p\,dp)/2 = (\ln H)_{pq}\,dq \wedge dp/2$$

to vanish on every hypersurface M whenever p is the unit normal vector field along M.

Assume that $\Omega \in (p\,dp,\ p\,dq,\ dp \wedge dq)$. If p is the unit normal field along M, then $p^2 = 1$ and $p\,dq|_M = 0$ for $q \in M$. It follows that $p\,dp|_M = 0$ and $dp \wedge dq|_M = 0$; hence, $\Omega|_M = 0$.

Conversely, if $L^{n-1} \subset T^*\mathbb{R}^n$ is a submanifold such that $p\,dp|_L = 0$, $p\,dq|_L = 0$ and L projects diffeomorphically to a hypersurface $M \subset \mathbb{R}^n$, then L consists of the pairs (q, p) where $q \in M$ and p is a normal vector field to M of constant length. If $\Omega|_L = 0$ for every such L, then Ω belongs to the differential ideal generated by $p\,dp$ and $p\,dq$, that is, $\Omega \in (p\,dp,\ p\,dq,\ dp \wedge dq)$.

Notice that the condition of Theorem 2.2.1 is conformally invariant. If a Hamiltonian function $H(q, p)$ satisfies this condition, then so does $f(q)H(q, p)$ for every positive function $f(q)$.

COROLLARY 2.2.2. *Let g be a metric of constant curvature in a domain in $\mathbb{R}^n$ whose geodesics are straight lines. Then the field of g-normals to every hypersurface is exact.*

PROOF. Consider the case of the hyperbolic metric, the spherical case being analogous.

Recall the construction of the hyperbolic metric in the unit ball. Let H be the upper sheet of the hyperboloid $x^2 - y^2 = -1$ in $\mathbb{R}^n_x \times \mathbb{R}^1_y$ with the Lorentz metric $dx^2 - dy^2$. The restriction of the Lorentz metric to H is a metric of constant negative curvature. Project H from the origin to the hyperplane $y = 1$. Let q be the Euclidean coordinate in this hyperplane. The projection is given by the formula

$$x = \frac{q}{(1-q^2)^{1/2}}, \quad y = \frac{1}{(1-q^2)^{1/2}},$$

and the image of H is the open unit ball $q^2 < 1$. The hyperbolic metric g in the unit ball is given by the formula

$$g(u, v) = \frac{uv}{1 - q^2} + \frac{(uq)(vq)}{(1-q^2)^2},$$

where u and v are tangent vectors at q.

The Hamiltonian function $H(q, p)$ is the corresponding metric on the cotangent space. Lifting the indices yields the following formula:

$$H(q, p) = (1 - q^2)(p^2 - (pq)^2).$$

Therefore,

$$v\,dn = \frac{1}{2}(\ln H)_p dp = \frac{p\,dp - (pq)(q\,dp)}{2(p^2 - (pq)^2)} = \frac{1}{4}d(\ln(p^2 - (pq)^2)),$$

which is an exact 1-form.

2.3. This subsection concerns the number of lines from an exact transverse field passing through a fixed point in $\mathbb{R}^n$. Let $M \subset \mathbb{R}^n$ be a closed locally quadratically convex immersed hypersurface and ξ an exact transverse line field along M. Fix a point $p \in \mathbb{R}^n$.

THEOREM 2.3.1. *The number of points $x \in M$ such that the line $\xi(x)$ passes through p is greater than or equal to the least number of critical points of a smooth fucntion on M.*

PROOF. Without loss of generality, assume that p is the origin. Due to the local convexity, M is coorientable. As before, let v be the vector field along ξ such that $vn = 1$, $v\,dn = df$. Consider the function on M given by the formula $h(x) = e^{-f(x)}(x \cdot n(x))$.

We claim that if x is a critical point of this function, then $\xi(x)$ contains the origin. One has

$$e^f dh = d(xn) - (xn)df = x\,dn - (xn)(v\,dn) = (x - (xn)v)dn;$$

the second equality is due to the fact that $n\,dx = 0$ on M.

Suppose that $dh(x) = 0$; then $(x - (xn)v)dn = 0$. Notice that the vector $x - (xn)v$ is orthogonal to $n(x)$, thus $x - (xn)v \in T_x M$. The value of the second quadratic form of M on two tangent vectors $u, w \in T_x M$ is $(u\,dn(x))(w)$. Since M is quadratically nondegenerate, $udn = 0$ on $T_x M$ only if $u = 0$. Applying this to $u = x - (xn)v$, one concludes that $x = (x \cdot n(x))v(x)$. It follows that the line $\xi(x)$, spanned by $v(x)$, contains the origin.

The next construction associates a hypersurface N with a point p such that the lines $\xi(x)$, $x \in M$, passing through p, correspond to the perpendiculars from p to N. Consider the smooth map $F : M \to \mathbb{R}^n$ given by the formula

$$F(x) = e^{-f(x)}[x + (x \cdot n(x))(n(x) - v(x))],$$

and let N be its image.

LEMMA 2.3.2. a) *The tangent hyperplane to N at a smooth point $F(x)$ is parallel to $T_x M$.*

b) *A point $p \in \mathbb{R}^n$ lies on the line $\xi(x)$ if and only if p lies on the normal line to N at point $F(x)$ (if $F(x)$ is not a smooth point of N we mean the line, orthogonal to the hyperplane, parallel to $T_x M$ and based at $F(x)$).*

c) *The value of the support function of N at point $F(x)$ is $e^{-f(x)}(x \cdot n(x))$ (with the same convention if $F(x)$ is not a smooth point of N).*

PROOF. To prove a) one needs to show that the 1-form $n(x)dF(x)$ vanishes on M. One has

$$e^f dF = dx + (xn)(dn - dv) + (n - v)(n\,dx + x\,dn) - [x + (xn)(n - v)](v\,dn).$$

It follows that

$$e^f n dF = n\,dx + (xn)(n\,dn - n\,dv)) + (n \cdot (n - v))(n\,dx + x\,dn)$$
$$- (n \cdot [x + (xn)(n - v)])(v\,dn).$$

Since $n\,dx = 0$, $n\,dn = 0$, and $(n \cdot (n - v)) = 0$ one concludes that

$$e^f n dF = -(xn)(n\,dv) - (nx)(v\,dn) = -(xn)d(nv) = 0.$$

As before, let p be the origin. If $p \in \xi(x)$, then $x = cv(x)$ for some constant c. Scalar multiply by $n(x)$ to find that $c = x \cdot n(x)$. Thus $x = (xn)v$, and

$$F(x) = e^{-f(x)}[x + (xn)(n - v)] = e^{-f(x)}(xn)n(x).$$

It follows that the vector $F(x)$ is collinear with $n(x)$.

Conversely, if $F(x)$ is collinear with $n(x)$, then $x + (xn)(n - v) = cn$ for some constant c. One again finds that $c = x \cdot n(x)$. Thus $x + (xn)(n - v) = (xn)n$, and $x = (xn)v(x)$. Therefore the vector x is collinear with $v(x)$, and the claim b) follows. Finally the support function of N is $n(x) \cdot F(x) = e^{-f(x)}(x \cdot n(x))$. The lemma is proved.

Similar to subsection 2.1, the manifold N is, in general, singular. One may construct a Legendrian submanifold in the space of contact elements in $\mathbb{R}^n$ whose projection to $\mathbb{R}^n$ is N but this Legendrian manifold may have singularities as well.

2.4. In this subsection we discuss intrinsic definitions of exact transverse line fields along hypersurfaces in affine and projective spaces. Recall that the definition of the conormal bundle of a hypersurface M in an affine space V is

$$\nu(M) = \{(q,p) \in T^*V | \; q \in M, \text{ Ker } p = T_qM\}.$$

A section of $\nu(M)$ will be called a conormal field along M. A conormal field is considered as a covector-valued function $M \to V^*$.

LEMMA 2.4.1. *A transverse line field ξ along a coorientable hypersurface M is exact if and only if there exists a conormal field p along M such that for some, hence every, vector field u generating ξ one has $udp = 0$.*

Here udp is the 1-form on M resulting in pairing of a vector-valued function and a covector-valued 1-form.

PROOF. Introduce a Euclidean structure in V and identify V with V^*. Let ξ be exact; that is, ξ is generated by a vector field v such that $vn = 1$ and $v\,dn = df$. Consider the covector field $p = e^{-f}n$. One has

$$v\,dp = e^{-f}v(dn - n\,df) = e^{-f}(v\,dn - df) = 0.$$

Conversely, if p is a conormal field then $p = \phi n$ for some function ϕ. Given a vector field u along M such that $udp = 0$, assume, without loss of generality, that $up = 1$. Let $v = \phi u$. Then $vn = 1$ and

$$v\,dn = v\,d(\phi^{-1}p) = v(\phi^{-1}dp - \phi^{-2}p\,d\phi) = -\phi^{-1}d\phi = -d\ln\phi.$$

Thus the transverse line field generated by u is exact.

Recall the notion of polar duality. Let S be a star-shaped hypersurface in a linear space V. To a vector $x \in S$ there corresponds the covector $p_x \in V^*$ such that $xp_x = 1$ and Ker $p_x = T_xS$. The locus of p_x, $x \in S$, is a hypersurface $S^* \subset V^*$, polar dual to S. By construction, $p_x dx = 0$ on S.

The relation of exactness with polar duality is as follows. Let $p : M \to V^*$ be a conormal field along a hypersurface $M \subset V$. Generically, $p(M)$ is star shaped, and its polar dual hypersurface is given by the map $u : M \to V$, where $up = 1$ and $u\,dp = 0$. Thus u, considered as a vector field along M, generates an exact transverse line field. Polar duality also explains the construction of the Minkowski metric in subsection 2.1: the indicatrix N is polar dual to the hypersurface $p(M)$, where $p = e^{-f}n$ is the conormal field corresponding to the given exact transverse field along M. The hypersurface $p(M)$ is the unit level surface of the Hamiltonian defying the Minkowski metric (called the figuratrix).

Apply Lemma 2.4.1 to the affine normals of a convex hypersurface in an affine space defined in subsection 1.3.

COROLLARY 2.4.2. *The field of affine normals is exact.*

PROOF. Identify vectors and covectors by a Euclidean structure. Let n be the unit normal field to M^m and K its Gauss curvature. Consider the conormal field $p = K^{-1/(m+2)}n$, and let u be a vector field generating the affine normals to M. Then, according to [**Ca**], $udp = 0$, and the result follows from Lemma 2.4.1.

Consider the projective space $P(V)$, and let $\pi : V - \{O\} \to P(V)$ be the projection. Denote by E the Euler vector field in V; this field generates the fibers of π. Considered as a map $V \to V$, the field E is the identity.

Let N be a hypersurface in $P(V)$ and $M = \pi^{-1}(N) \subset V$. Given a transverse line field ξ along N, consider a line field η along M that projects to ξ.

DEFINITION. A field ξ is called exact if there exists a homogeneous degree zero conormal field p along M such that $vdp = 0$ for every vector field v generating η.

The next lemma justifies this definition.

LEMMA 2.4.3. a) *The definition is correct; that is, the exactness does not depend on the choice of η.*

b) *For a transverse line field along a hypersurface in an affine chart of $P(V)$, the affine and projective definitions of exactness are equivalent.*

PROOF. Let v be a vector field along M generating η, and p be a conormal field such that $v\,dp = 0$ on M. Any other lift η_1 is generated by the vector field $v_1 = v + \phi E$ where ϕ is a function. Notice that $Ep = 0$ since E is tangent to M. It follows that $E\,dp + p\,dE = 0$. Since p is conormal to M and $E(x) = x$ one has $p\,dx = p\,dE = 0$ on M. Thus $E\,dp = 0$. It follows that $v_1\,dp = 0$, and η_1 is exact. The claim a) is proved.

To prove b), choose coordinates $(x_0, x_1, \ldots, x_n)$ in V and identify the affine part of $P(V)$ with the hyperplane $x_0 = 1$. Decompose V into $\mathbb{R}^1_{x_0} \oplus \mathbb{R}^n_x$, $x = (x_1, \ldots, x_n)$; the vectors and covectors in V will be decomposed accordingly.

Assume that N belongs to the hyperplane $x_0 = 1$. Let u be a vector field along N generating a transverse line field, exact in the affine sense, and p be a conormal field along N such that $u\,dp = 0$. Consider the covector field

$$P(x_0, x) = \left(-\frac{x}{x_0} \cdot p\left(\frac{x}{x_0}\right), \; p\left(\frac{x}{x_0}\right)\right)$$

along $M = \pi^{-1}(N)$. Then $E(x_0, x)P(x_0, x) = 0$, so P is a conormal field along M. Lift u to the horizontal vector field $v(x_0, x) = (0, u(x/x_0))$. Then $vdP = u(x/x_0)dp(x/x_0) = 0$. Thus the transverse field generated by u is exact in the projective sense.

Conversely, assume that $v(x_0, x) = (0, u(x/x_0))$ generates a line field along M, exact in the projective sense. The respective conormal field along M is homogeneous and has degree zero. Therefore its $\mathbb{R}^n_x$-component $p(x/x_0)$ defines a conormal field along N, and $udp = 0$ on N. Thus u generates a line field, which is exact in the affine sense.

Note that Lemma 2.4.3 implies Theorem 1.2.1, the projective invariance of exactness. We add to it the following.

LEMMA 2.4.4. *If a local diffeomorphism f of $\mathbb{R}^n$ takes every exact transverse field along every hypersurface to an exact field, then f is a projective transformation.*

PROOF. The unit normal vector field along a hyperplane is constant, so every transverse line field along it is exact. Therefore f takes hyperplanes to hyperplanes, and it follows that f is projective.

Now we extend Theorem 2.3.1 to quadratically nondegenerate hypersurfaces in the projective space.

THEOREM 2.4.5. *Let N be a closed cooriented quadratically nondegenerate immersed hypersurface in the projective space $P(V)$ and ξ an exact transverse line field along N. Then for every point $a \in P(V)$ the number of points $x \in N$ such that the line $\xi(x)$ passes through a is greater than or equal to the least number of critical points of a smooth function on N.*

PROOF. We will repeat the arguments from the proof of Theorem 2.3.1. Introduce a Euclidean structure in V. Let η be a lift of ξ to $M = \pi^{-1}(N)$ and v a vector field along M normalized so that $vn = 1$, where n is the unit normal field along M. Let p be the homogeneous degree zero conormal field along M such that $v\,dp = 0$. Then, by Lemma 2.4.1, $p = e^{-f}n$, where f is a homogeneous function on M of degree zero, and $v\,dn = df$. Choose a point $A \in V$ that projects to $a \in P(V)$.

Consider the function $h(x) = e^{-f(x)}(A - x) \cdot n(x)$ on M. The same argument as in Theorem 2.3.1 shows that if $dh = 0$, then the vector $(A-x) - ((A-x) \cdot n)v(x)$ is in the kernel of the second quadratic form of M at x. This kernel is generated by the vector $E(x) = x$. Thus $A - x$ belongs to the span of the vectors $v(x)$ and $E(x)$. Projecting back to $P(V)$ one concludes that the line $\xi(\pi(x))$ passes through the point a.

Since M is conical, $xn(x) = 0$; thus the function $h(x) = e^{-f(x)}A \cdot n(x)$ is homogeneous of degree zero. Therefore this function descends to a function on N, and the result follows.

2.5. This subsection concerns exact transverse line fields along the unit sphere S^{n-1}.

We start with computing the canonical symplectic structure ω on the space of rays intersecting a closed convex hypersurface $M \subset \mathbb{R}^n$, associated with the Hilbert metric inside M. If $n > 2$, then the space of rays is simply connected, and the cohomology class of a 1-form λ such that $d\lambda = \omega$ is uniquely defined. If $n = 2$, then the space of rays is a cylinder, and we choose λ so that the curve consisting of the rays through a fixed point inside M is an exact Lagrangian manifold, which means that the restriction of λ on this manifold is exact. An oriented line will be characterized by its first and second intersection points x, y with M. Let $n(x)$ and $n(y)$ be the unit normal vectors to M at x and y.

LEMMA 2.5.1. *One has $\omega = d\lambda$, where*

$$\lambda = \frac{n(y)dx}{n(y) \cdot (y-x)} + \frac{n(x)dy}{n(x) \cdot (y-x)}.$$

PROOF. A Finsler metric is characterized by its Lagrangian L, a positive function on the tangent bundle, homogeneous of degree one in the velocity vectors (see [**Ar**]). We introduce the following notation: q, u and p will denote points, tangent vectors and covectors in $\mathbb{R}^n$. The symplectic form in question is obtained from the symplectic form in $T^*\mathbb{R}^n$ by symplectic reduction. Using the Legendre transform $p = L_u$, one writes the Liouville form $p\,dq$ as $L_u dq$. Restricting it to the space of oriented lines, one obtains the desired 1-form λ.

Let q be a point on the line xy and u a tangent vector at q along xy. Let s, t be the reals such that $q = (1-t)x + ty$, $u = s(y-x)$. Then x, y, t, s are coordinates in $T\mathbb{R}^n$.

First we compute the Lagrangian in these coordinates. Identify the segment (xy) with $(0, 1)$ by an affine transformation. Given a point $t \in (0, 1)$, consider the Hilbert geodesic $x(\tau)$ through t. Then $x(0) = t$, and the Hilbert distance $d(t, x(\tau)) = \tau$; that is,

$$\ln\left(\frac{(1-t)x(\tau)}{t(1-x(\tau))}\right) = \tau.$$

Solving for x, differentiating and evaluating at $\tau = 0$, one finds $x'(0) = t(1-t)$. It follows that

$$|u|_{\text{Hilbert}} = \frac{|u|_{\text{Euclidean}}}{t(1-t)}.$$

Returning to the segment (xy), one obtains the formula $L(x, y, t, s) = s/t(1-t)$.

Next we compute L_u. One has

$$dq = (1-t)dx + t\,dy + (y-x)dt, \qquad du = (y-x)ds + s(dy - dx).$$

Write the differential dL in two ways:

$$L_u du + L_q dq = L_u(y-x)ds + L_u s(dy - dx) + (1-t)L_q dx + tL_q dy + L_q(y-x)dt.$$

One obtains the equations

$$(y-x)L_u = \frac{1}{t(1-t)}, \qquad (y-x)L_q = -\frac{s(1-2t)}{t^2(1-t)^2},$$

$$(sL_u + tL_q)dy = 0, \qquad (sL_u - (1-t)L_q)dx = 0.$$

The latter two 1-forms vanish only if the vectors in front of dy and dx are proportional to $n(y)$ and $n(x)$, respectively

$$sL_u + tL_q = an(y), \qquad sL_u - (1-t)L_q = bn(x).$$

Therefore

$$sL_u = a(1-t)n(y) + btn(x), \qquad L_q = an(y) - bn(x).$$

Using the first two equations, one finds

$$a = \frac{s}{(1-t)^2 n(y) \cdot (y-x)}, \qquad b = \frac{s}{t^2 n(x) \cdot (y-x)}.$$

Thus

$$L_u = \frac{n(y)}{(1-t)n(y) \cdot (y-x)} + \frac{n(x)}{tn(x) \cdot (y-x)}.$$

Since $n(x)dx = 0$ and $n(y)dy = 0$, it follows that

$$L_u dq = \frac{n(y)dx}{n(y) \cdot (y-x)} + \frac{n(x)dy}{n(x) \cdot (y-x)} + d\ln\left(\frac{t}{1-t}\right),$$

and one may omit the last exact term to obtain the 1-form λ on the space of rays intersecting M.

Finally, consider the manifold C that consists of the rays through a fixed point inside M. Then C is obtained from a fiber of the cotangent bundle by symplectic reduction. The Liouville form vanishes on the fibers of $T^*\mathbb{R}^n$; therefore, the form λ on C is cohomologous to zero.

We apply Lemma 2.5.1 to the case where M is the unit sphere. The Hilbert metric is the hyperbolic metric in the Beltrami–Klein model of $\mathbb{H}^n$. As explained in the Introduction, a transverse line field ξ along the sphere determines a submanifold L^{n-1} of the space of rays in $\mathbb{H}^n$.

THEOREM 2.5.2. *The transverse line field ξ is exact if and only if L is an exact Lagrangian submanifold of the space of rays in $\mathbb{H}^n$.*

PROOF. One has $n(x) = x$, $n(y) = y$. Therefore $\lambda = (y\,dx - x\,dy)/(1-xy)$. Since $\lambda + d\ln(1-xy) = -2(x\,dy)/(1-xy)$ the form $\lambda|_L$ is exact if and only if so is $(x\,dy)/(1-xy)|_L$.

Let $v(y)$, $y \in M$, be a vector field generating a transverse line field along M, normalized by the condition $yv(y) = 1$. Then the other intersection point of the line $\xi(y)$ with M is
$$x = y - 2\frac{v(y)}{v(y) \cdot v(y)}.$$
It follows that $1 - xy = 2/v^2$. On L one has
$$-\frac{x\,dy}{1-xy} = \frac{v^2}{2}\left(-y\,dy + \frac{2v\,dy}{v^2}\right) = v(y)dy,$$
the second equality due to the fact that $y\,dy = 0$. Thus L is exact Lagrangian if and only if the field ξ is exact.

We conclude this section by formulating the following problem. Describe the closed convex hypersurfaces M such that a transverse line field along M is exact if and only if the corresponding submanifold in the space of rays intersecting M is exact Lagrangian with respect to the form λ from Lemma 2.5.1. We conjecture that only ellipsoids have this property.

3. Projective billiards

3.1. Start with an equation describing the projective billiard reflection in a closed convex hypersurface $M \subset \mathbb{R}^n$ equipped with a transverse line field ξ. Let v be a vector field along ξ such that $vn = 1$, and let the ray xy reflect to the ray yz; here $x, y, z \in M$.

LEMMA 3.1.1. *One has*
$$\frac{y-x}{n(y) \cdot (y-x)} + \frac{y-z}{n(y) \cdot (y-z)} = 2v(y).$$

PROOF. The vector
$$\frac{y-x}{n(y) \cdot (y-x)} - \frac{y-z}{n(y) \cdot (y-z)}$$

is normal to $n(y)$. Therefore this vector is tangent to M at y. Thus this vector spans the intersection line of the plane generated by $y - x$ and $z - y$ with the hyperplane $T_y M$. The fourth vector

$$\frac{y-x}{n(y)\cdot(y-x)} + \frac{y-z}{n(y)\cdot(y-z)}$$

constitutes a harmonic quadruple with the above three. Therefore

$$\frac{y-x}{n(y)\cdot(y-x)} + \frac{y-z}{n(y)\cdot(y-z)} = tv(y).$$

Take the scalar product with $n(y)$ to conclude that $t = 2$.

This lemma implies the following result.

THEOREM 3.1.2. *Let M be a closed quadratically convex hypersurface equipped with an exact transverse line field. For every two points A, B inside M there exist at least two rays through A that, after the projective billiard reflection in M, pass through B.*

PROOF. Let $v\,dn = df$. Consider the function

$$h(y) = \ln(n(y) \cdot (y - A)) + \ln(n(y) \cdot (y - B)) - 2f(y).$$

This function on M has at least two critical points. One has

$$dh = \frac{(y-A)dn + n\,dy}{n(y-A)} + \frac{(y-B)dn + n\,dy}{n(y-B)} - 2v\,dn$$

$$= \left(\frac{y-A}{n(y-A)} + \frac{y-B}{n(y-B)} - 2v\right)dn.$$

The vector in front of dn in the last formula is orthogonal to n, and M is quadratically convex. Therefore if $dh(y) = 0$, then

$$\frac{y-A}{n(y)\cdot(y-A)} + \frac{y-B}{n(y)\cdot(y-B)} = 2v(y).$$

According to Lemma 3.1.1, this implies that the ray Ay reflects to the ray yB.

We conjecture that the quadratic convexity condition may be relaxed, and the result will hold for the hypersurfaces, star shaped with respect to the points A and B.

3.2. It was mentioned in the Introduction that for every transverse line field, the projective billiard transformation in an ellipse preserves the symplectic form on the space of rays, associated with the hyperbolic metric. In this subsection we prove that this property characterizes the ellipses.

THEOREM 3.2.1. *Let M be a strictly convex closed plane curve equipped with a transverse line field. If the projective billiard transformation preserves the symplectic structure on the oriented lines associated with the Hilbert metric inside M, then M is an ellipse.*

PROOF. Introduce on M an affine parameterization. This means that for every t we have $[M'(t), M''(t)] = 1$, where $[\,,\,]$ is the cross-product of vectors in the plane. A ray is characterized by its first and second intersection points $M(t_1)$ and $M(t_2)$ with the curve. We use (t_1, t_2) as coordinates in the space of rays.

First, we rewrite the symplectic form ω from Lemma 2.5.1 in these coordinates. If $x = M(t_1)$, $y = M(t_2)$, then $dx = M'(t_1)dt_1$, $dy = M'(t_2)dt_2$. Since

$$\frac{n(y) \cdot M'(t_1)}{n(y) \cdot (y-x)} = \frac{[M'(t_1), M'(t_2)]}{[M(t_2) - M(t_1), M'(t_2)]},$$

$$\frac{n(x) \cdot M'(t_2)}{n(x) \cdot (y-x)} = \frac{[M'(t_2), M'(t_1)]}{[M(t_2) - M(t_1), M'(t_1)]},$$

the 1-form λ from Lemma 2.5.1 is as follows:

$$\lambda = \frac{[M'(t_1), M'(t_2)]}{[M(t_2) - M(t_1), M'(t_2)]} dt_1 + \frac{[M'(t_2), M'(t_1)]}{[M(t_2) - M(t_1), M'(t_1)]} dt_2.$$

Differentiating and using the identities

(1) $\qquad [a,b][c,d] = [a,c][b,d] + [a,d][c,b]\quad$ for all a, b, c, d

and $[M'(t), M''(t)] = 1$, one concludes that

(2) $\qquad \omega = \left(\frac{[M(t_2) - M(t_1), M'(t_2)]}{[M(t_2) - M(t_1), M'(t_1)]^2} - \frac{[M(t_2) - M(t_1), M'(t_1)]}{[M(t_2) - M(t_1), M'(t_2)]^2} \right) dt_1 \wedge dt_2.$

Let T be the projective billiard transformation, and let $T(t_1, t_2) = (t_2, t_3)$. Cross-multiply the equality of Lemma 3.1.1 by $v(y)$ where $y = M(t_2)$ and, as before, replace the dot product with $n(y)$ by the cross-product with $M'(t_2)$. One obtains the equality

(3) $\qquad \dfrac{[M(t_2) - M(t_1), v(y)]}{[M(t_2) - M(t_1), M'(t_2)]} + \dfrac{[M(t_3) - M(t_2), v(y)]}{[M(t_3) - M(t_2), M'(t_2)]} = 0.$

Take the exterior derivative, wedge multiply by dt_2, simplify using (1), and cancel the common factor $[v(y), M'(t_2)]$ to arrive at the equality

(4) $\qquad -\dfrac{[M(t_2) - M(t_1), M'(t_1)]}{[M(t_2) - M(t_1), M'(t_2)]^2} dt_1 \wedge dt_2 = \dfrac{[M(t_3) - M(t_2), M'(t_3)]}{[M(t_3) - M(t_2), M'(t_2)]^2} dt_2 \wedge dt_3.$

If $T^*\omega = \omega$, then, according to (2), one has

$$\left(\frac{[M(t_2) - M(t_1), M'(t_2)]}{[M(t_2) - M(t_1), M'(t_1)]^2} - \frac{[M(t_2) - M(t_1), M'(t_1)]}{[M(t_2) - M(t_1), M'(t_2)]^2} \right) dt_1 \wedge dt_2$$

$$= \left(\frac{[M(t_3) - M(t_2), M'(t_3)]}{[M(t_3) - M(t_2), M'(t_2)]^2} - \frac{[M(t_3) - M(t_2), M'(t_2)]}{[M(t_3) - M(t_2), M'(t_3)]^2} \right) dt_2 \wedge dt_3.$$

According to (4), the second term at the left-hand side equals the first one at the right. Thus,

(5) $\qquad \dfrac{[M(t_2) - M(t_1), M'(t_2)]}{[M(t_2) - M(t_1), M'(t_1)]^2} dt_1 \wedge dt_2 = -\dfrac{[M(t_3) - M(t_2), M'(t_2)]}{[M(t_3) - M(t_2), M'(t_3)]^2} dt_2 \wedge dt_3.$

Dividing (4) by (5) and taking the cubic root we arrive at the equality

(6) $\qquad \dfrac{[M(t_2) - M(t_1), M'(t_1)]}{[M(t_2) - M(t_1), M'(t_2)]} = \dfrac{[M(t_3) - M(t_2), M'(t_3)]}{[M(t_3) - M(t_2), M'(t_2)]}.$

Change the notation by letting $t_2 = t$, $t_3 = t + \epsilon$, and $t_1 = t - \delta$. Denote the right-hand side of (6) by $f(t, \epsilon)$; then (6) reads

$$f(t - \delta, \delta) = \frac{1}{f(t, \epsilon)}. \tag{7}$$

Differentiating the equality $[M'(t), M''(t)] = 1$, one has $[M'(t), M'''(t)] = 0$. Thus $M'''(t) = -k(t)M'(t)$ where the function $k(t)$ is called the affine curvature. Further differentiating, one finds

$$[M', M^{IV}] = -k, \quad [M'', M'''] = k, \quad [M', M^V] = -2k', \quad [M'', M^{IV}] = k'. \tag{8}$$

We will show that $k' = 0$ identically. This will imply the result since the only curves with constant affine curvature are the conics.

Using the Taylor expansion up to the fifth derivatives and taking (8) into account, one finds

$$f(t, \epsilon) = 1 + \frac{\epsilon^3 k'(t)}{60} + O(\epsilon^4), \tag{9}$$

and hence

$$f(t - \delta, \delta) = 1 + \frac{\delta^3 k'(t - \delta)}{60} + O(\delta^4). \tag{10}$$

Clearly $\delta = O(\epsilon)$. Consider the Taylor expansion of (3) up to ϵ^2,

$$\frac{[\delta M' - \frac{\delta^2}{2} M'', v]}{[\delta M' - \frac{\delta^2}{2} M'', M']} + \frac{[\epsilon M' + \frac{\epsilon^2}{2} M'', v]}{[\epsilon M' + \frac{\epsilon^2}{2} M'', M']} = 0.$$

It follows that $\delta = \epsilon + O(\epsilon^2)$. Then (10) and (9) imply that

$$f(t - \delta, \delta) = 1 + \frac{\epsilon^3 k'(t)}{60} + O(\epsilon^4) = f(t, \epsilon),$$

and one concludes from (7) that $k'(t) = 0$ for all t.

3.3. Let M be a circle in the plane and ξ a transverse line field along M. The phase space of the projective billiard map T is a cylinder C, and T preserves the area form $\omega = d\lambda$ from Theorem 2.5.2.

LEMMA 3.3.1. *The flux of T equals zero if and only if ξ is exact.*

PROOF. Let $L \subset C$ be the curve that consists of the lines $\xi(x)$, $x \in M$, oriented outward, and let τ be the involution of C that changes the orientation of a line to the opposite. Then $T(L) = \tau(L)$. The flux of T is the signed area between the curves L and $T(L)$, that is,

$$\int_L \lambda - \int_{\tau(L)} \lambda.$$

Clearly, $\tau^*(\lambda) = -\lambda$. Hence the flux of T equals $2 \int_L \lambda$. According to Theorem 2.5.2 this integral vanishes if and only if ξ is exact.

An invariant circle is a simple closed noncontractible T-invariant curve $\Gamma \subset C$. The map T is a twist map of the cylinder, the leaves of the vertical foliation consisting of the outward oriented lines through a fixed point on M. According to Birkhoff's theorem, an invariant circle intersects each vertical leaf once (see, e.g.,

[**H-K**]). Γ is a one parameter family of lines, and their envelope $\gamma \subset \mathbb{R}^2$ is called a projective billiard caustic. The caustic may have cusp singularities but from every point $x \in M$ there are exactly two tangents to γ. A ray, tangent to γ, remains tangent to it after the projective billiard reflection.

COROLLARY 3.3.2. *If a projective billiard map inside a circle has an invariant circle then the corresponding transverse line field is exact.*

In other words, let γ be a closed curve inside the circle M such that from every point $x \in M$ there are exactly two tangents to γ. Let $\xi(x)$ be the line that constitutes a harmonic quadruple with these two tangents and the tangent line to M at x. Then the transverse line field ξ is exact.

This property is specific to circles: we will construct a pair (M, γ) such that the corresponding transverse field ξ will fail to be exact.

Example. The caustic γ will be a triangle $A_1 A_2 A_3$ (if one wishes one may smoothly round the corners). Let M be a closed strictly convex curve enclosing the triangle. Denote by P_i and Q_i, $i = 1, 2, 3$, the intersections of the rays $A_i A_{i+1}$ and $A_{i+1} A_i$ with M (the indices are considered modulo 3). For example, a ray through the point A_3 reflects in the arc $Q_1 P_3$ to a ray through A_2. This determines the transverse line field ξ along $Q_1 P_3$, and a similar consideration applies to the other five arcs of M.

According to Theorem 1.2.4, the field ξ is exact if and only if it is generated by the acceleration vectors $M''(t)$ for some parameterization $M(t)$. Consider again the arc $Q_1 P_3$. The desired parameterization is determined by the equation (3) from the proof of Theorem 3.2.1. Namely, setting $v = M''(t)$, $M(t_1) = A_2$, $M(t_3) = A_3$, and $M(t_2) = M(t)$, this equation can be rewritten as

$$([M(t) - A_2, M'(t)][M(t) - A_3, M'(t)])' = 0.$$

Thus $[M - A_2, M'][M - A_3, M']$ is constant. Similar equations determine the parameterizations of the other five arcs. One can make a consistent choice of the constants involved if and only if

$$\prod_{i=1}^{3} \frac{[P_i - A_i, M'|_{P_i}][Q_i - A_i, M'|_{Q_i}]}{[P_i - A_{i+1}, M'|_{P_i}][Q_i - A_{i+1}, M'|_{Q_i}]} = 1.$$

This condition is equivalent to

$$\frac{|A_1 P_1||A_2 P_2||A_3 P_3||A_1 Q_1||A_2 Q_2||A_3 Q_3|}{|A_2 P_1||A_3 P_2||A_1 P_3||A_2 Q_1||A_3 Q_2||A_1 Q_3|} = 1.$$

Clearly this equality does not hold for a generic curve M (but it holds for a circle or an ellipse, as follows from elementary geometry). It is interesting to remark that a similar construction will not work for a "two-gon": it is shown in [**T2**] that for every two points A, B inside M there is an exact transverse line field along M such that every ray through A projectively reflects in M to a ray through B.

We conclude this section with a description of a dynamical system on the one-sheeted hyperboloid equivalent to the projective billiard in a circle.

Consider 3-space with the Lorentz quadratic form $Q = x^2 + y^2 - z^2$; denote by H_+ the upper sheet of the hyperboloid $Q = -1$, by H_0 the cone $Q = 0$ and by H_- the one-sheeted hyperboloid $Q = 1$. The Lorentz metric restricted to H_+ is a metric of constant negative curvature. The projection from the origin to the plane

$z = 1$ takes H_0 to the unit circle and identifies H_+ with the Beltrami–Klein model of the hyperbolic plane. In particular, the straight lines on H_+ are its intersections with the planes through the origin.

The Q-orthogonal complement to such a plane is a line intersecting H_-. Taking orientation into account, to every oriented line in H_+ there corresponds a point of H_-, and the orientation reversing involution on the lines corresponds to the central symmetry of H_-. This duality with respect to the quadratic form Q identifies H_- with the space of rays in the hyperbolic plane, that is, the phase space of the projective billiard in a circle.

The hyperboloid H_- carries two families of straight lines called the rulings; denote these families by U and V. Given a transverse line field ξ along the circle, consider it as a family of outward oriented lines; this determines a noncontractible simple closed curve $\gamma \subset H_-$. For example, if ξ consists of the lines through a fixed point, then γ is a plane section of the hyperboloid. A field ξ is exact if and only if the signed area between γ and the equator of H_- equals zero.

Let T be the projective billiard transformation considered as a map of H_-.

THEOREM 3.3.3. *The curve γ intersects each ruling from one of the families, say, U, at a unique point. Given a point $x \in S$, let $u \in U$ be the ruling through x. Then $T(x)$ is obtained from x by the composition of the two reflections: first, in the intersection point of u with γ, and then in the origin.*

PROOF. Let u be a ruling of H_-. Consider the plane π through u and the origin. Since H_0 and H_- are disjoint, the intersection of π with H_0 is a line l, parallel to u (and not a pair of lines). It follows that π is tangent to C. If l is a line on the cone H_0, then its Q-orthogonal complement is a plane tangent to H_0 along l; that is, it coincides with π. The line l projects to a point y of the unit circle. It follows that the duality takes a line through y to a point on u.

Taking orientations into account, we obtain the following result: the one parameter families of incoming and outgoing rays through a point y on the unit circle correspond on H_- to two parallel rulings $u \in U$ and $v \in V$. Then $\xi(y)$, oriented outward, corresponds to the unique intersection point of γ with u. The tangent line to the unit circle at y correponds to the point of u at infinity.

Consider the projective billiard reflection at y. The incoming and outgoing lines, both oriented outward, correspond to the points x, $x_1 \in u$. These points, along with the point $\gamma \cap u$ and the point of u at infinity, constitute a harmonic quadruple. Thus x and x_1 are symmetric with respect to the point $\gamma \cap u$. Reversing the orientation of the outgoing line amounts to reflecting x_1 in the origin, and the result follows.

REMARK. The canonical area form on the space of rays in the hyperbolic plane is the standard area form on H_- induced by the Euclidean metric. Equivalently, this area equals the volume of the layer between H_- and an infinitely close homothetic surface. The axial projection of H_- to the cylinder $x^2 + y^2 = 1$ is area preserving (a similar fact for the sphere is due to Archimedes).

3.4. In this subsection we will derive the mirror equation for the projective billiard in the unit circle. A similar approach works for an arbitrary billiard curve but the resulting formula is rather ugly.

Let $M(t) = (\cos t, \sin t)$ be the unit circle, and let $\phi(t)$ be the angle made by the line of the transverse line field at $M(t)$ with the tangent vector $M'(t)$. Let A, B be two points inside M such that the infinitesimal beam of rays from A reflects in $M(t_0)$ to the infinitesimal beam of rays through B. Let $\alpha(t)$ and $\beta(t)$ be the angles made by the rays $M(t)A$ and $M(t)B$ with $M'(t)$, and let $|M(t)A| = a(t)$, $|M(t)B| = b(t)$. Omit the argument for the values of these functions at t_0.

THEOREM 3.4.1. *One has*
$$\frac{2\phi'}{\sin^2 \phi} + \frac{1}{\sin^2 \alpha} + \frac{1}{\sin^2 \beta} = \frac{1}{a \sin \alpha} + \frac{1}{b \sin \beta}.$$

PROOF. It is easy to see that if the ray $AM(t)$ reflects to the ray $M(t)B$ then $2 \cot \phi(t) = \cot \alpha(t) + \cot \beta(t)$. Differentiating and evaluating at t_0 one obtains the equation:
$$\frac{2\phi'}{\sin^2 \phi} = \frac{\alpha'}{\sin^2 \alpha} + \frac{\beta'}{\sin^2 \beta}.$$
The result will follow once we show that
$$\alpha' = \frac{\sin \alpha}{a} - 1, \quad \beta' = \frac{\sin \beta}{b} - 1.$$
Indeed, $A - M(t) = a(-\sin(t + \alpha), \cos(t + \alpha))$. Differentiating, cross-multiplying by the vector $(-\sin(t + \alpha), \cos(t + \alpha))$ and evaluating at t_0 yields $\sin \alpha = a(1 + \alpha')$. A similar computation for β completes the proof.

We conclude the paper with a conjecture on the integrability of projective billiards. The projective billiard inside a plane curve M will be called integrable near the boundary if there is a neighbourhood of M foliated by the caustics. For example, this holds for the usual billiard inside an ellipse, the caustics being the confocal ellipses. Birkhoff's conjecture states that if a plane billiard is integrable near the boundary, then the billiard curve is an ellipse; as far as I know, this conjecture is still open. A related theorem by Bialy states that if the whole phase space of a plane billiard is foliated by invariant circles then the billiard curve is a circle; see [**Bi**].

Recall that a pencil of conics consists of the conics passing through four fixed points (in what follows these points will be complex).

CONJECTURE 3.4.2. a) *If the projective billiard inside a closed convex smooth plane curve M is integrable near the boundary, then M and the caustics are the ellipses belonging to the family, projectively dual to a pencil of conics.*

b) *If the whole phase space of a plane projective billiard is foliated by invariant circles, then M is an ellipse and the transverse field consists of the lines through a fixed point inside M.*

As a justification, we prove the converse statements. The second one is obvious: modulo a projective transformation, M is a circle, and the point is its center. Then the concentric circles are the caustics.

To prove the statement converse to a), consider the curve M^*, projectively dual to M. A point $x \in M$ becomes a tangent line l to M^*, and the incoming and outgoing rays at x, tangent to a caustic C, become the intersection points of l with the dual curve C^*. The curves C^* constitute a pencil of conics, and one wants to show that the (local) involution of l that interchanges its intersection points with each of these conics is a projective transformation. According to the Desargues theorem this holds for every line intersecting a nondegenerate pencil of conics (see [**Be**]), and we are done.

Acknowledgments. I am grateful to J.-C. Alvarez, E. Ferrand, E. Gutkin, A. McRae and P. Tantalo for stimulating discussions. Part of this work was done while the author participated in the "Geometry and Dynamics" program at the Tel Aviv University; I am grateful to the organizers, M. Bialy and L. Polterovich, for their hospitality. Research was supported in part by NSF grant DMS-9402732.

References

[Ar] V. Arnold, *mathematical methods of classical mechanics*, Springer–Verlag, Berlin, 1978.
[A-G] V. Arnold and A. Givental, *Symplectic geometry*, Dynamical Systems–4, Encyclopedia of Mathematical Sciences, Springer–Verlag, Berlin, 1990, pp. 1–136.
[Be] M. Berger, *Geometry*, Springer–Verlag, Berlin, 1987.
[Bi] M. Bialy, *Convex billiards and a theorem by E. Hopf*, Math. Z. **214** (1993), 147–154.
[Ca] E. Calabi, *Géometrie différentielle affine des hypersurfaces*, Sém. Bourbaki, 1980/81, Lecture Notes in Math., vol. 901, Springer–Verlag, Berlin–New York, 1981.
[D]   G. Darboux, *Leçon sur le théorie générale des surfaces*, Chelsea, New York, 1972.
[Fe1] E. Ferrand, *On a theorem of Chekanov*, Banach Center Publ. (1997), no. 39, Warszawa.
[Fe2] _____, *Sur la structure symplectique de l'espace des géodésiques d'une variété de Hadamard*, Geom. Dedicata (to appear).
[H-K] B. Hasselblatt and A. Katok, *Introduction to the modern theory of dynamical systems*, Cambridge University Press, Cambridge, 1995.
[Pu] P. Pushkar', *A generalization of Chekanov's theorem. Diameters of immersed manifolds and wave fronts*, Preprint, 1997.
[T1] S. Tabachnikov, *Billiards*, Panoramas et Syntheses, vol. 1, Soc. Math. France, Paris, 1995.
[T2] _____, *Introducing projective billiards*, Ergodic Theory Dynamical Systems **17** (1997), 957–976.
[T3] _____, *Exact transverse line fields and projective billiards in a ball*, Geom. Anal. and Functional Anal. **7** (1997), 594–608.
[T4] _____, *Parameterized curves, Minkowski caustics, Minkowski vertices and conservative line fields*, L'Enseign. Math. **43** (1997), 3–26.
[T5] _____, *The four vertex theorem revisited—two variations on the old theme*, Amer. Math. Monthly **102** (1995), no. 10, 912–916.
[W1] M. Wojtkowski, *Principles for the design of billiards with nonvanishing Lyapunov exponents*, Comm. Math. Phys. **105** (1986), 391–414.
[W2] _____, *Two applications of Jacobi fields in the billiard ball problem*, J. Diff. Geom. **40** (1994), 155–164.

DEPARTMENT OF MATHEMATICS, UNIVERSITY OF ARKANSAS, FAYETTEVILLE, ARKANSAS 72701 USA

E-mail address: serge@comp.uark.edu

Shadows of Wave Fronts and Arnold–Bennequin Type Invariants of Fronts on Surfaces and Orbifolds

Vladimir Tchernov

ABSTRACT. A first-order Vassiliev invariant of an oriented knot in an S^1-fibration and a Seifert fibration over a surface is constructed. It takes values in a quotient of the group ring of the first homology group of the total space of the fibration. It gives rise to an invariant of wave fronts on surfaces and orbifolds related to the Bennequin-type invariants of the Legendrian curves studied by F. Aicardi, V. Arnold, M. Polyak, and S. Tabachnikov. Formulas expressing these relations are presented.

We also calculate Turaev's shadow for the Legendrian lifting of a wave front. This allows us to use all invariants known for shadows in the case of wave fronts.

Most of the proofs in this paper are postponed until the last section.

Everywhere in this text an S^1-fibration is a locally trivial fibration with fibers homeomorphic to S^1.

In this paper the multiplicative notation for the operation in the first homology group is used. The zero homology class is denoted by e. The reason for this is that we have to deal with the integer group ring of the first homology group. For a group G the group of all formal half-integer linear combinations of elements of G is denoted by $\frac{1}{2}\mathbb{Z}[G]$.

We work in the differential category.

I am deeply grateful to Oleg Viro for the inspiration of this work and many enlightening discussions. I am thankful to Francesca Aicardi, Thomas Fiedler, and Michael Polyak for our valuable discussions.

1. Introduction

In [5] Polyak suggested a quantization

$$l_q(L) \in \frac{1}{2}\mathbb{Z}[q, q^{-1}]$$

of the Bennequin invariant of a generic cooriented oriented wave front $L \subset \mathbb{R}^2$. In this paper we construct an invariant $S(L)$ which is, in a sense, a generalization of $l_q(L)$ to the case of a wave front on an arbitrary surface F.

1991 *Mathematics Subject Classification.* Primary 57M25.

©1999 American Mathematical Society

In the same paper [5] Polyak introduced Arnold's [3] J^+-type invariant of a front L on an oriented surface F. It takes values in $H_1(ST^*F, \frac{1}{2}\mathbb{Z})$. We show that $S(L) \in \frac{1}{2}\mathbb{Z}[H_1(ST^*F)]$ is a refinement of this invariant in the sense that it is taken to Polyak's invariant under the natural mapping $\frac{1}{2}\mathbb{Z}[H_1(ST^*F)] \to H_1(ST^*F, \frac{1}{2}\mathbb{Z})$.

Further we generalize $S(L)$ to the case where L is a wave front on an orbifold.

Invariant $S(L)$ is constructed in two steps. The first consists of lifting L to the Legendrian knot λ in the S^1-fibration $\pi : ST^*F \to F$. The second step can be applied to any knot in an S^1-fibration, and it involves the structure of the fibration in a crucial way. This step allows us to define the S_K invariant of a knot K in the total space N of an S^1-fibration. Since ordinary knots are considered up to a rougher equivalence relation (ordinary isotopy versus Legendrian isotopy), in order for S_K to be well defined it has to take values in a quotient of $\mathbb{Z}[H_1(N)]$. This invariant is generalized to the case of a knot in a Seifert fibration, and this allows us to define $S(L)$ for wave fronts on orbifolds.

All these invariants are Vassiliev invariants of order one in an appropriate sense.

For each of these invariants we introduce its version with values in the group of formal linear combinations of the free homotopy classes of oriented curves in the total space of the corresponding fibration.

The first invariants of this kind were constructed by Fiedler [4] in the case of a knot K in an $\mathbb{R}^1$-fibration over a surface and by Aicardi in the case of a generic oriented cooriented wave front $L \subset \mathbb{R}^2$. The connection between these invariants and S_K is discussed in [9].

The space ST^*F is naturally fibered over a surface F with a fiber S^1. In [10] Turaev introduced a shadow description of a knot K in an oriented three-dimensional manifold N fibered over an oriented surface with a fiber S^1. A shadow presentation of a knot K is a generic projection of K, together with an assignment of numbers to regions. It describes a knot type modulo a natural action of $H_1(F)$. It appeared to be a very useful tool. Many invariants of knots in S^1-fibrations, particularly quantum state sums, can be expressed as state sums for their shadows. In this paper we construct shadows of Legendrian liftings of wave fronts. This allows one to use any invariant already known for shadows in the case of wave fronts.

However, in this paper shadows are used mainly for the purpose of depicting knots in S^1-fibrations.

2. Shadows

2.1. Preliminary constructions. We say that a one-dimensional submanifold L of a total space N^3 of a fibration $\pi : N^3 \to F^2$ is *generic with respect to* π if $\pi|_L$ is a generic immersion. Recall that an immersion of 1-manifold into a surface is said to be *generic* if it has neither self-intersection points with multiplicity greater than 2 nor self-tangency points, and at each double point its branches are transversal to each other. An immersion of (a circle) S^1 to a surface is called a *curve*.

Let π be an oriented S^1-fibration of N over an oriented closed surface F. Then N admits a fixed point free involution which preserves fibers. Let $\tilde{N}$ be the quotient of N by this involution, and let $p : N \to \tilde{N}$ be the corresponding double covering. Each fiber of p (a pair of antipodal points) is contained in a fiber of π. Therefore, π factorizes through p and we have a fibration $\tilde{\pi} : \tilde{N} \to F$. Fibers of $\tilde{\pi}$ are projective lines. They are homeomorphic to circles.

An isotopy of a link $L \subset N$ is said to be vertical with respect to π if each point of L moves along a fiber of π. It is clear that if two links are vertically isotopic, then their projections coincide. Using vertical isotopy we can modify each generic link L in such a way that any two points of L belonging to the same fiber lie in the same orbit of the involution. Denote the obtained generic link by L'.

Let $\tilde{L} = p(L')$. It is obtained from L' by gluing together points lying over the same point of F. Hence $\tilde{\pi}$ maps $\tilde{L}$ bijectively to $\pi(L) = \pi(L')$. Let $r : \pi(L) \to \tilde{L}$ be an inverse bijection. It is a section of $\tilde{\pi}$ over $\pi(L)$.

For a generic nonempty collection of curves on a surface by a *region* we mean the closure of a connected component of the complement of this collection. Let X be a region for $\pi(L)$ on F. Then $\tilde{\pi}|_X$ is a trivial fibration. Hence we can identify it with the projection $S^1 \times X \to X$. Let ϕ be a composition of the section $r|_{\partial X}$ with the projection to S^1. It maps ∂X to S^1. Denote by α_X the degree of ϕ. (This is actually an obstruction to an extension of $r|_{\partial X}$ to X.) One can see that α_X does not depend on the choice of the trivialization of $\tilde{\pi}$ and on the choice of L'.

2.2. Basic definitions and properties.

DEFINITION 2.2.1. The number $\frac{1}{2}\alpha_X$ corresponding to a region X is called the *gleam* of X and is denoted by $\mathrm{gl}(X)$. A *shadow* $s(L)$ of a generic link $L \subset N$ is a (generic) collection of curves $\pi(L) \subset F$ with the gleams assigned to each region X. The sum of gleams over all regions is said to be the *total gleam* of the shadow.

2.2.2. One can check that for any region X the integer α_X is congruent modulo 2 to the number of corners of X. Therefore, $\mathrm{gl}(X)$ is an integer if the region X has even number of corners, and a half-integer otherwise.

2.2.3. The total gleam of the shadow is equal to the Euler number of π.

DEFINITION 2.2.4. A *shadow* on F is a generic collection of curves together with the numbers $\mathrm{gl}(X)$ assigned to each region X. These numbers can be either integers or half-integers, and they should satisfy the conditions of 2.2.2 and 2.2.3.

There are three local moves S_1, S_2, and S_3 of shadows shown in Figure 1. They are similar to the well-known Riedemeister moves of planar knot diagrams.

DEFINITION 2.2.5. Two shadows are said to be *shadow equivalent* if they can be transformed to each other by a finite sequence of moves S_1, S_2, S_3, and their inverses.

2.2.6. There are two other important shadow moves $\bar{S}_1$ and $\bar{S}_3$ shown in Figure 2. They are similar to the previous versions of the first and the third Riedemeister moves and can be expressed in terms of S_1, S_2, S_3, and their inverses.

2.2.7. In [10] the action of $H_1(F)$ on the set of all isotopy types of links in N is constructed as follows. Let L be a generic link in N and β an oriented (possibly self-intersecting) curve on F presenting a homology class $[\beta] \in H_1(F)$. Deforming β we can assume that β intersects $\pi(K)$ transversally at a finite number of points different from the self-intersection points of $\pi(K)$. Denote by $\alpha = [a,b]$ a small segment of L such that $\pi(\alpha)$ contains exactly one intersection point c of $\pi(L)$ and β. Assume that $\pi(a)$ lies to the left, and $\pi(b)$ to the right of β. Replace α by the arc α' shown in Figure 3. We will call this transformation of L a *fiber fusion* over the point c. After we apply fiber fusion to L over all points of $\pi(L) \cap \beta$ we get a

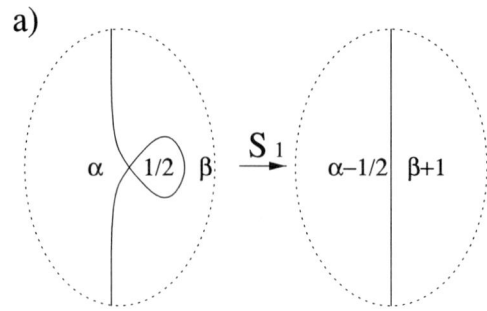

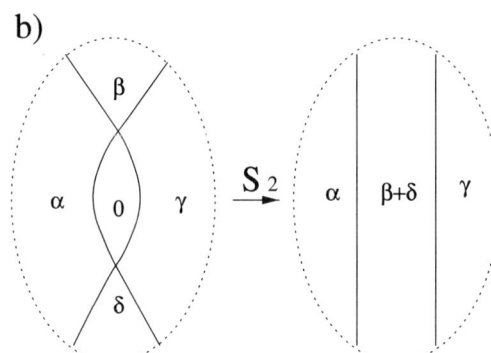

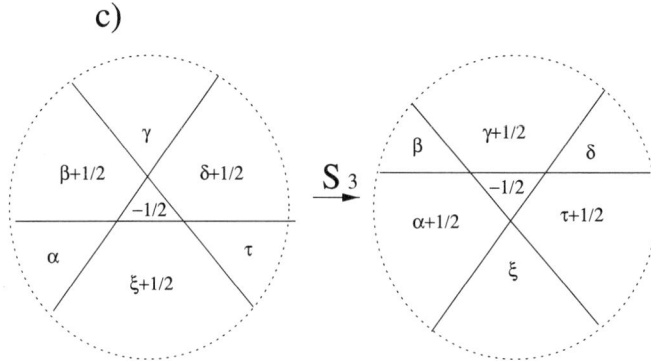

FIGURE 1

new generic link L' with $\pi(L) = \pi(L')$. One can notice that the shadows of K and K' coincide. Indeed, each time β enters a region X of $s(L)$, it must leave it. Hence the contributions of the newly inserted arcs to the gleam of X cancel out. Thus links belonging to one $H_1(F)$-orbit always produce the same shadow-link on F.

THEOREM 2.2.8 (Turaev [10]). *Let N be an oriented closed manifold, F an oriented surface, and $\pi : N \to F$ an S^1-fibration with the Euler number $\chi(\pi)$. The mapping that associates to each link $L \subset N$ its shadow equivalence class on F*

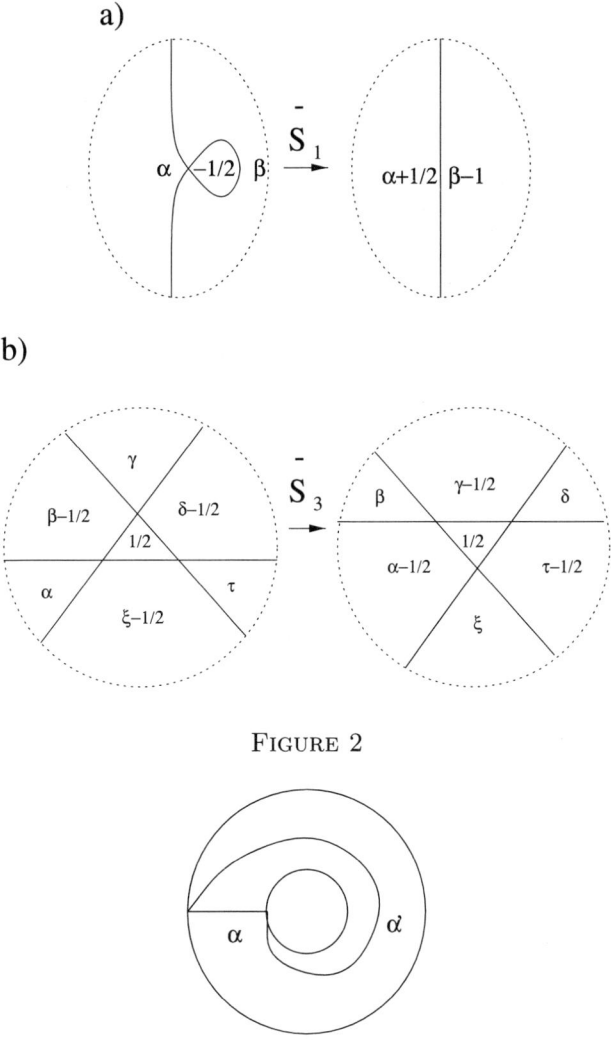

FIGURE 2

FIGURE 3

establishes a bijective correspondence between the set of isotopy types of links in N modulo the action of $H_1(F)$ and the set of all shadow equivalence classes on F with the total gleam $\chi(\pi)$.

2.2.9. It is easy to see that all links whose projections represent $0 \in H_1(F)$ and whose shadows coincide are homologous to each other. To prove this, one looks at the description of a fiber fusion and notices that to each fiber fusion where we add a positive fiber there corresponds another where we add a negative one. Thus the numbers of positively and negatively oriented fibers we add are equal, and they cancel out.

2.2.10. As was remarked in [**10**], it is easy to transfer the construction of shadows and Theorem 2.2.8 to the case where F is a nonclosed oriented surface and N is an oriented manifold. In order to define the gleams of the regions that have a noncompact closure or contain components of ∂F, we have to choose a section of the fibration over all boundary components and ends of F. In the case of nonclosed F the total gleam of the shadow is equal to the obstruction to the extension of the section to the entire surface.

3. Invariants of knots in S^1-fibrations

3.1. Main constructions. In this section we deal with knots in an S^1-fibration π of an oriented three-dimensional manifold N over an oriented surface F. We do not assume F and N to be closed. As was said in 2.2.10, all theorems from the previous section are applicable in this case.

DEFINITION 3.1.1 (of S_K). Orientations of N and F determine an orientation of a fiber of the fibration. Denote by $f \in H_1(N)$ the homology class of a positively oriented fiber.

Let $K \subset N$ be an oriented knot which is generic with respect to π. Let v be a double point of $\pi(K)$. The fiber $\pi^{-1}(v)$ divides K into two arcs that inherit the orientation from K. Complete each arc of K to an oriented knot by adding the arc of $\pi^{-1}(v)$ such that the orientations of these two arcs define an orientation of their union. The orientations of F and $\pi(K)$ allow one to identify a small neighborhood of v in F with a model picture shown in Figure 4a. Denote the knots obtained by the operation above by μ_v^+ and μ_v^- as shown in Figure 4. We will often call this construction a *splitting* of K (with respect to the orientation of K).

This splitting can be described in terms of shadows as follows. Note that μ_v^+ and μ_v^- are not in general position. We slightly deform them in a neighborhood of $\pi^{-1}(v)$, so that $\pi(\mu_v^+)$ and $\pi(\mu_v^-)$ do not have double points in the neighborhood of v. Let P be a neighborhood of v in F homeomorphic to a closed disk. Fix a section over ∂P such that the intersection points of $K \cap \pi^{-1}(\partial P)$ belong to the section. Inside P we can construct Turaev's shadow (see 2.2.10). The action of $H_1(\text{Int } P) = e$ on the set of the isotopy types of links is trivial. Thus the part of K can be reconstructed in the unique way (up to an isotopy fixed on ∂P) from the shadow over P (see 2.2.10). The shadows for μ_v^+ and μ_v^- are shown in Figures 4a and 4b, respectively.

Regions for the shadows $s(\mu_v^+)$ and $s(\mu_v^-)$ are, in fact, unions of regions for $s(K)$. One should think of gleams as of measure, so that the gleam of a region is the sum of all numbers inside.

Let H be the quotient of the group ring $\mathbb{Z}[H_1(N)]$ (viewed as a $\mathbb{Z}$-module) by the submodule generated by $\{[K] - f, [K]f - e\}$. Here by $[K] \in H_1(N)$ we denote the homology class represented by the image of K.

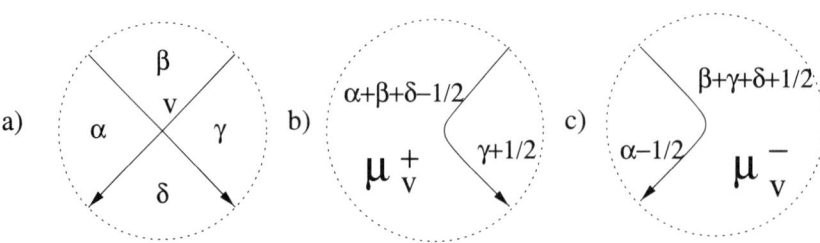

FIGURE 4

Finally define $S_K \in H$ by the following formula, where the summation is taken over all double points v of $\pi(K)$:

$$S_K = \sum_v ([\mu_v^+] - [\mu_v^-]). \tag{1}$$

3.1.2. Since $\mu_v^+ \cup \mu_v^- = K \cup \pi^{-1}(v)$ we have

$$[\mu_v^+][\mu_v^-] = [K]f. \tag{2}$$

THEOREM 3.1.3. S_K is an isotopy invariant of the knot K.

For the proof of Theorem 3.1.3 see subsection 7.1.

3.1.4. It follows from 3.1.2 that S_K can also be described as an element of $\mathbb{Z}[H_1(N)]$ equal to the sum of $([\mu_v^+] - [\mu_v^-])$ over all double points for which the sets $\{[\mu_v^+], [\mu_v^-]\}$ and $\{e, f\}$ are disjoint. Note that in this case we do not need to factorize $\mathbb{Z}[H_1(N)]$ to make S_K well defined.

3.1.5. One can obtain an invariant similar to S_K with values in the free $\mathbb{Z}$-module generated by the set of all free homotopy classes of oriented curves in N. To do this one substitutes the homology classes of μ_v^+ and μ_v^- in (1) with their free homotopy classes and takes the summation over the set of all double points v of $\pi(K)$ such that neither one of the knots μ_v^+ and μ_v^- is homotopic to a trivial loop and neither one of them is homotopic to a positively oriented fiber (see 3.1.4).

To prove that this is indeed an invariant of K one can easily modify the proof of Theorem 3.1.3.

3.2. S_K is a Vassiliev invariant of order one.

3.2.1. Let $\pi : N \to F$ be an S^1-fibration over a surface. Let $K \subset N$ be a generic knot with respect to π and v a double point of $\pi(K)$. A modification of pushing of one branch of K through the other along the fiber $\pi^{-1}(v)$ is called a *modification of K along the fiber $\pi^{-1}(v)$*.

3.2.2. If a fiber fusion increases by one the gleam γ in Figure 4b, then $[\mu_v^+]$ is multiplied by f. If a fiber fusion increases by one the gleam α in Figure 4c, then $[\mu_v^-]$ is multiplied by f^{-1}. These facts are easy to verify.

3.2.3. Let us find out how S_K changes under the modification along a fiber over a double point v. Consider a singular knot K' (whose only singularity is a point v of transverse self-intersection). Let ξ_1 and ξ_2 be the homology classes of the two loops of K' adjacent to v. The two resolutions of this double point correspond to adding $\pm\frac{1}{2}$ to the gleams of the regions adjacent to v in two ways shown in Figures 5b and 5c.

Using 3.2.2, one verifies that under the corresponding modification S_K changes by

$$(f - e)(\xi_1 + \xi_2). \tag{3}$$

This means that the first derivative of S_K depends only on the homology classes of the two loops adjacent to the singular point. Hence the second derivative of S_K is 0. Thus it is a Vassiliev invariant of order one in the usual sense.

For similar reasons the version of S_K with values in the free $\mathbb{Z}$-module generated by all free homotopy classes of oriented curves in N is also a Vassiliev invariant of order one.

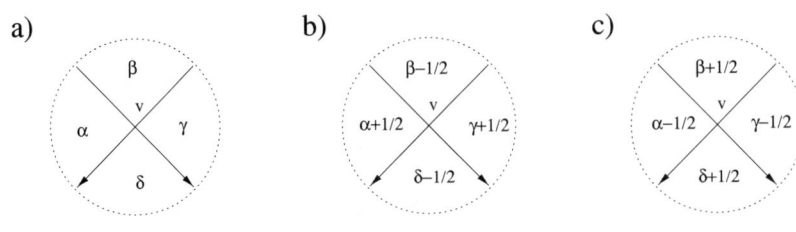

FIGURE 5

THEOREM 3.2.4. I. *If K and K' are two knots representing the same free homotopy class, then S_K and $S_{K'}$ are congruent modulo the submodule generated by elements of form*

(4) $$(f-e)(j+[K]j^{-1})$$

for $j \in H_1(N)$.

II. *If K is a knot, and $S \in H$ is congruent to S_K modulo the submodule generated by elements of form (4) (for $j \in H_1(N)$), then there exists a knot K' such that*

  a. K *and* K' *represent the same free homotopy class;*
  b. $S_{K'} = S$.

For the proof of Theorem 3.2.4 see subsection 7.2.

3.3. Example. If N is a solid torus T fibered over a disk, then we can calculate the value of S_K directly from the shadow of K.

DEFINITION 3.3.1. Let C be an oriented closed curve in $\mathbb{R}^2$ and X a region for C. Take a point $x \in \text{Int } X$ and connect it to a point near infinity by a generic oriented path D. Define the sign of an intersection point of C and D as shown in Figure 6. Let $\text{ind}_C X$ be the sum over all intersection points of C and D of the signs of these points.

It is easy to see that $\text{ind}_C(X)$ is independent on the choices of x and D.

DEFINITION 3.3.2. Let $K \subset T$ be an oriented knot which is generic with respect to π, and let $s(K)$ be its shadow. Define $\sigma(s(K)) \in \mathbb{Z}$ as the following sum over all regions X for $\pi(K)$:

(5) $$\sigma(s(K)) = \sum_X \text{ind}_{\pi(K)}(X) \,\text{gl}(X).$$

Denote by $h \in \mathbb{Z}$ the image of $[K]$ under the natural identification of $H_1(T)$ with $\mathbb{Z}$.

FIGURE 6

LEMMA 3.3.3. $\sigma(s(K)) = h$.

3.3.4. Put

(6) $$S'_K = \sum t^{\sigma(s(\mu_v^+))} - t^{\sigma(s(\mu_v^-))},$$

where the sum is taken over all double points v of $\pi(K)$ such that $\{0,1\}$ and $\{\sigma(s(\mu_v^+)), \sigma(s(\mu_v^-))\}$ are disjoint (see 3.1.4).

Lemma 3.3.3 implies that S'_K is the image of S_K under the natural identification of $\mathbb{Z}[H_1(T)]$ with the ring of finite Laurent polynomials (see 3.1.4).

One can show [9] that S'_K and Aicardi's partial linking polynomial of K (which was introduced in [1]) can be expressed explicitly in terms of each other.

3.4. Further generalizations of the S_K invariant. One can show that an invariant similar to S_K can be introduced in the case where N is oriented and F is nonorientable.

DEFINITION 3.4.1 (of $\tilde{S}_K$). Let N be oriented and F nonorientable. Let $K \subset N$ be an oriented knot generic with respect to π, and let v be a double point of $\pi(K)$. Fix an orientation of a small neighborhood of v in F. Since N is oriented this induces an orientation of the fiber $\pi^{-1}(v)$. Similar to the definition of S_K (see 3.1.1), we split our knot with respect to the orientation and obtain two knots $\mu_1^+(v)$ and $\mu_1^-(v)$. Then we take the other orientation of the neighborhood of v in F, and in the same way we obtain another pair of knots $\mu_2^+(v)$ and $\mu_2^-(v)$. The element $\bigl([\mu_1^+(v)] - [\mu_1^-(v)] + [\mu_2^+(v)] - [\mu_2^-(v)]\bigr) \in \mathbb{Z}[H_1(N)]$ does not depend on which orientation of the neighborhood of v we choose first.

Similar to the definition of S_K, we can describe all this in terms of shadows as it is shown in Figure 7. These shadows are constructed with respect to the same orientation of the neighborhood of v.

Let f be the homology class of a fiber of π oriented in some way. As one can easily prove $f^2 = e$, so it does not matter which orientation we choose to define f. Let $\tilde{H}$ be the quotient of $\mathbb{Z}[H_1(N)]$ (viewed as a $\mathbb{Z}$-module) by the $\mathbb{Z}$-submodule generated by $\{[K] - f + e - [K]f = (e-f)([K]+e)\}$. Finally define $\tilde{S}_K \in \tilde{H}$ by the following formula, where the summation is taken over all double points v of $\pi(K)$:

(7) $$\tilde{S}_K = \sum_v \Bigl([\mu_1^+(v)] - [\mu_1^-(v)] + [\mu_2^+(v)] - [\mu_2^-(v)]\Bigr).$$

THEOREM 3.4.2. $\tilde{S}_K$ is an isotopy invariant of the knot K.

The proof is essentially the same as the proof of Theorem 3.1.3.

3.4.3. One can easily prove that the $\tilde{S}_K$ invariant satisfies relations similar to (3). In particular, $\tilde{S}_K$ is also a Vassiliev invariant of order one.

One can introduce a version of this invariant with values in the free $\mathbb{Z}$-module generated by all free homotopy classes of oriented curves in N. To do this, we substitute the homology classes of $\mu_1^+(v)$, $\mu_1^-(v)$, $\mu_2^+(v)$, and $\mu_2^-(v)$ with the corresponding free homotopy classes. The summation should be taken over the set of all double points of $\pi(K)$ for which not one of $\mu_1^+(v), \mu_1^-(v), \mu_2^+(v)$, and $\mu_2^-(v)$ is homotopic to a trivial loop and not one of them is homotopic to a fiber of π. To prove that this is indeed an invariant of K, one easily modifies the proof of Theorem 3.1.3.

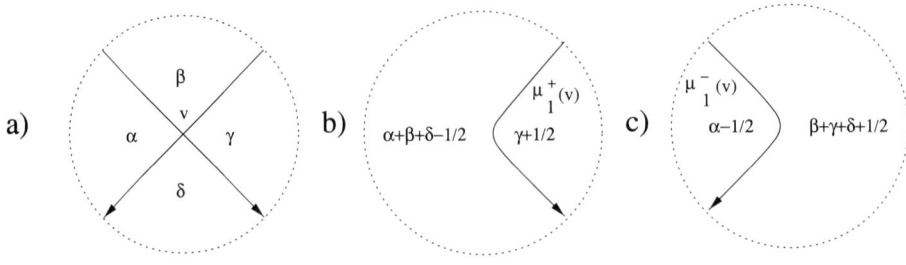

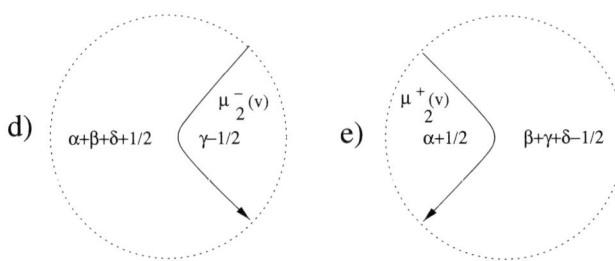

Figure 7

4. Invariants of knots in Seifert fibered spaces

Let (μ, ν) be a pair of relatively prime integers. Let
$$D^2 = \{(r, \theta); 0 \leq r \leq 1, 0 \leq \theta \leq 2\pi\} \subset \mathbb{R}^2$$
be the unit disk defined in polar coordinates. A fibered solid torus of type (μ, ν) is the quotient space of the cylinder $D^2 \times I$ via the identification $\big((r, \theta), 1\big) = \big((r, \theta + \frac{2\pi\nu}{\mu}), 0\big)$. The fibers are the images of the curves $x \times I$. The integer μ is called the index or the multiplicity. For $|\mu| > 1$ the fibered solid torus is said to be *exceptionally fibered,* and the fiber that is the image of $0 \times I$ is called the *exceptional fiber.* Otherwise the fibered solid torus is said to be *regularly fibered,* and each fiber is a *regular fiber.*

DEFINITION 4.1.1. An orientable three manifold S is said to be a *Seifert fibered manifold* if it is a union of pairwise disjoint closed curves, called fibers, such that each one has a closed neighborhood which is a union of fibers and is homeomorphic to a fibered solid torus by a fiber preserving homeomorphism.

A fiber h is called *exceptional* if h has a neighborhood homeomorphic to an exceptionally fibered solid torus (by a fiber preserving homeomorphism), and h corresponds via the homeomorphism to the exceptional fiber of the solid torus. If $\partial S \neq \emptyset$, then ∂S should be a union of regular fibers.

The quotient space obtained from a Seifert fibered manifold S by identifying each fiber to a point is a 2-manifold. It is called the orbit space and the images of the exceptional fibers are called the *cone points.*

4.1.2. For an exceptional fiber a of an oriented Seifert fibered manifold there is a unique pair of relatively prime integers (μ_a, ν_a) such that $\mu_a > 0$, $|\nu_a| < \mu_a$, and a neighborhood of a is homeomorphic (by a fiber preserving homeomorphism) to a fibered solid torus of type (μ_a, ν_a). We call the pair (μ_a, ν_a) the *type of the exceptional fiber* a. We also call this pair the type of the corresponding cone point.

We can define an invariant of an oriented knot in a Seifert fibered manifold that is similar to the S_K invariant.

Clearly any S^1-fibration can be viewed as a Seifert fibration without cone points. This justifies the notation in the definition below.

DEFINITION 4.1.3 (of S_K). Let N be an oriented Seifert fibered manifold with an oriented orbit space F. Let $\pi : N \to F$ be the corresponding fibration and $K \subset N$ an oriented knot in general position with respect to π. Assume also that K does not intersect the exceptional fibers. For each double point v of $\pi(K)$ we split K into μ_v^+ and μ_v^- (see 3.1.1). Let A be the set of all exceptional fibers. Since N and F are oriented, we have an induced orientation of each exceptional fiber $a \in A$. For $a \in A$ set f_a to be the homology class of the fiber with this orientation. For $a \in A$ of type (μ_a, ν_a) (see 4.1.2) set $N_1(a) = \{k \in \{1, \ldots, \mu_a\} | \frac{2\pi k \nu_a}{\mu_a} \bmod 2\pi \in (0, \pi]\}$, $N_2(a) = \{k \in \{1, \ldots, \mu_a\} | \frac{2\pi k \nu_a}{\mu_a} \bmod 2\pi \in (0, \pi)\}$. Define $R_a^1, R_a^2 \in \mathbb{Z}[H_1(N)]$ by the following formulas:

$$(8) \quad R_a^1 = \sum_{k \in N_1(a)} \left([K] f_a^{\mu_a - k} - f_a^k\right) - \sum_{k \in N_2(a)} \left(f_a^{\mu_a - k} - [K] f_a^k\right),$$

$$(9) \quad R_a^2 = \sum_{k \in N_1(a)} \left(f_a^{\mu_a - k} - [K] f_a^k\right) - \sum_{k \in N_2(a)} \left([K] f_a^{\mu_a - k} - f_a^k\right).$$

Let H be the quotient of $\mathbb{Z}[(H_1(N)]$ (viewed as a $\mathbb{Z}$-module) by the free $\mathbb{Z}$-submodule generated by $\{[K]f - e, [K] - f, \{R_a^1, R_a^2\}_{a \in A}\}$. Finally, define $S_K \in H$ by the following formula, where the summation is taken over all double points v of $\pi(K)$

$$(10) \quad S_K = \sum_v \left([\mu_v^+] - [\mu_v^-]\right).$$

THEOREM 4.1.4. S_K *is an isotopy invariant of the knot* K.

For the proof of Theorem 4.1.4 see subsection 7.4.

We introduce a similar invariant in the case where N is oriented and F is nonorientable.

DEFINITION 4.1.5 (of $\tilde{S}_K$). Let N be an oriented Seifert fibered manifold with a nonorientable orbit space F. Let $\pi : N \to F$ be the corresponding fibration and $K \subset N$ an oriented knot in general position with respect to π. Assume also that K does not intersect the exceptional fibers. For each double point v of $\pi(K)$ we split K into $\mu_1^+(v)$, $\mu_1^-(v)$, $\mu_2^+(v)$, and $\mu_2^-(v)$ as in Definition 3.4.1. The element $\left([\mu_1^+(v)] - [\mu_1^-(v)] + [\mu_2^+(v)] - [\mu_2^-(v)]\right) \in \mathbb{Z}[H_1(N)]$ is well defined.

Denote by f the homology class of a regular fiber oriented in some way. Note that $f^2 = e$, so the orientation we use to define f does not matter. For a cone point a denote by f_a the homology class of the fiber $\pi^{-1}(a)$ oriented in some way.

For $a \in A$ of type (μ_a, ν_a) set $N_1(a) = \{k \in \{1, \ldots, \mu_a\} | \frac{2\pi k \nu_a}{\mu_a} \bmod 2\pi \in (0, \pi]\}$, $N_2(a) = \{k \in \{1, \ldots, \mu_a\} | \frac{2\pi k \nu_a}{\mu_a} \bmod 2\pi \in (0, \pi)\}$.

Define $R_a \in \mathbb{Z}[H_1(N)]$ by the following formula:

$$R_a = \sum_{k \in N_1(a)} \left([K]f_a^{\mu_a-k} - f_a^k + f_a^{k-\mu_a} - [K]f_a^{-k} \right) \tag{11}$$

$$- \sum_{k \in N_2(a)} \left(f_a^{\mu_a-k} - [K]f_a^k + [K]f_a^{k-\mu_a} - f_a^{-k} \right).$$

Let $\tilde{H}$ be the quotient of $\mathbb{Z}[H_1(N)]$ (viewed as a $\mathbb{Z}$-module) by the free $\mathbb{Z}$-submodule generated by $\left\{ (e-f)([K]+e), \{R_a\}_{a \in A} \right\}$.

One can prove that under the change of the orientation of $\pi^{-1}(a)$ (used to define f_a) R_a goes to $-R_a$. Thus $\tilde{H}$ is well defined. To show this, one verifies that if μ_a is odd, then $N_1(a) = N_2(a)$. Under this change each term from the first sum (used to define R_a) goes to minus the corresponding term from the second sum and vice versa. (Note that $f^2 = e$.) If $\mu_a = 2l$ is even, then $N_1(a) \setminus \{l\} = N_2(a)$. Under this change each term with $k \in N_1(a) \setminus \{l\}$ goes to minus the corresponding term with $k \in N_2(a)$ and vice versa. The term in the first sum that corresponds to $k = l$ goes to minus itself.

Finally define $\tilde{S}_K \in \tilde{H}$ as the sum over all double points v of $\pi(K)$:

$$\tilde{S}_K = \sum_v \left([\mu_1^+(v)] - [\mu_1^-(v)] + [\mu_2^+(v)] - [\mu_2^-(v)] \right). \tag{12}$$

THEOREM 4.1.6. $\tilde{S}_K$ is an isotopy invariant of K.

The proof is a straightforward generalization of the proof of Theorem 4.1.4.

4.1.7. One can easily verify that S_K and $\tilde{S}_K$ satisfy relations similar to (3); hence, both of them are Vassiliev invariants of order one (see 3.2.3).

5. Wave fronts on surfaces

5.1. Definitions. Let F be a two-dimensional manifold. A *contact element* at a point in F is a one-dimensional vector subspace of the tangent plane. This subspace divides the tangent plane into two half-planes. A choice of one of them is called a *coorientation* of a contact element. The space of all cooriented contact elements of F is a spherical cotangent bundle ST^*F. We will also denote it by N. It is an S^1-fibration over F. The natural contact structure on ST^*F is a distribution of hyperplanes given by the condition that a velocity vector of an incidence point of a contact element belongs to the element. A *Legendrian* curve λ in N is an immersion of S^1 into N such that for each $p \in S^1$ the velocity vector of λ at $\lambda(p)$ lies in the contact plane. The naturally cooriented projection $L \subset F$ of a Legendrian curve $\lambda \subset N$ is called the *wave front* of λ. A cooriented wave front may be uniquely lifted to a Legendrian curve $\lambda \subset N$ by taking a coorienting normal direction as a contact element at each point of the front. A wave front is said to be generic if it is an immersion everywhere except a finite number of points, where it has cusp singularities, and all multiple points are double points with transversal self-intersection. A cusp is the projection of a point where the corresponding Legendrian curve is tangent to the fiber of the bundle.

5.2. Shadows of wave fronts.

5.2.1. For any surface F the space ST^*F is canonically oriented. The orientation is constructed as follows. For a point $x \in F$ fix an orientation of T_xF. It induces an orientation of the fiber over x. These two orientations determine an orientation of three-dimensional planes tangent to the points of the fiber over x. A straightforward verification shows that this orientation is independent on the orientation we chose of T_xF. Hence the orientation of ST^*F is well defined. Thus for oriented F the shadow of a generic knot in ST^*F is well defined (see 2.1 and 2.2.10). Theorem 5.2.3 describes the shadow of a Legendrian lifting of a generic cooriented wave front $L \subset F$.

DEFINITION 5.2.2. Let X be a connected component of $F \setminus L$. We denote by $\chi \operatorname{Int}(X)$ the Euler characteristic of $\operatorname{Int}(X)$, by C_X^i the number of cusps in the boundary of the region X pointing inside X (as in Figure 8a), by C_X^o the number of cusps in the boundary of X pointing outside (as in Figure 8b), and by V_X the number of corners of X where locally the picture looks in one of the two ways shown in Figure 8c. It can happen that a cusp point is pointing both inside and outside of X. In this case it contributes both in C_X^i and in C_X^o. If the corner of the type shown in Figure 8c enters twice in ∂X, then it should be counted twice.

THEOREM 5.2.3. *Let F be an oriented surface and L a generic cooriented wave front on F corresponding to a Legendrian curve λ. There exists a small deformation of λ in the class of all smooth (not only Legendrian) curves such that the resulting curve is generic with respect to the projection, and the shadow of this curve can be constructed in the following way. We replace a small neighborhood of each cusp of L with a smooth simple arc. The gleam of an arbitrary region X that has a compact closure and does not contain boundary components of F is calculated by the following formula:*

$$\operatorname{gl}_X = \chi \operatorname{Int}(X) + \frac{1}{2}(C_X^i - C_X^o - V_X). \tag{13}$$

For the proof of Theorem 5.2.3 see subsection 7.5.

REMARK. The surface F in the statement of Theorem 5.2.3 is not assumed to be compact.

Note that as we mentioned in 2.2.10, the gleam of a region X that does not have compact closure or contains boundary components is not well defined unless we fix a section over all ends of X and components of ∂F in X.

This theorem first appeared in [**9**]. A similar result was independently obtained by Polyak [**6**].

5.2.4. A self-tangency point p of a wave front is said to be a point of a *dangerous self-tangency* if the coorienting normals of the two branches coincide at p (see

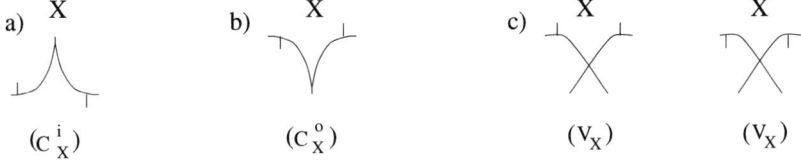

FIGURE 8

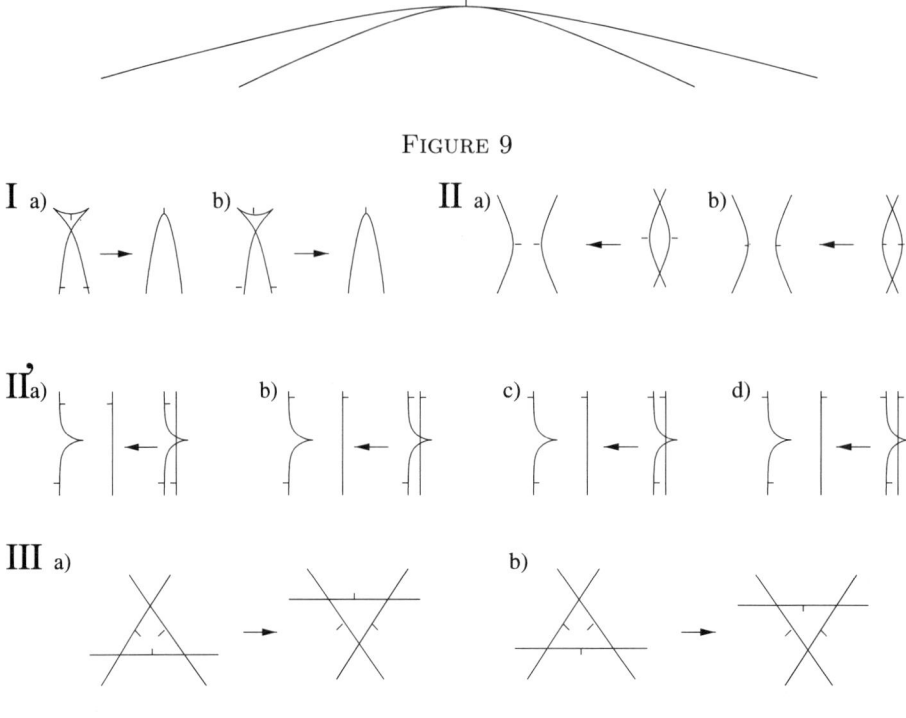

Figure 9

Figure 10

Figure 9). Dangerous self-tangency points correspond to the self-intersection of the Legendrian curve. Hence a generic deformation of the front L not involving *dangerous* self-tangencies corresponds to an isotopy of the Legendrian knot λ.

Any generic deformation of a wave front L corresponding to an isotopy in the class of the Legendrian knots can be split into a sequence of modifications shown in Figure 10. The construction of Theorem 5.2.3 transforms these generic modifications of wave fronts to shadow moves: Ia and Ib in Figure 10 are transformed to the $\bar{S}_1$ move for shadow diagrams IIa, IIb; II'a, II'b, II'c, and II'd are transformed to the S_2 move; and finally IIIa and IIIb are transformed to S_3 and $\bar{S}_3$, respectively.

5.2.5. Thus for the Legendrian lifting of a wave front we are able to calculate all invariants that we can calculate for shadows. This includes the analogue of the linking number for the fronts on $\mathbb{R}^2$ (see [**10**]), the second order Vassiliev invariant (see [**7**]), and quantum state sums (see [**10**]).

5.3. Invariants of wave fronts on surfaces. In particular, the S_K invariant gives rise to an invariant of a generic wave front. This invariant appears to be related to the formula for the Bennequin invariant of a wave front introduced by Polyak in [**5**].

Let us recall the corresponding results and definitions of [**5**].

Let L be a generic cooriented oriented wave front on an oriented surface F. A branch of a wave front is said to be positive (resp. negative) if the frame of coorienting and orienting vectors defines positive (resp. negative) orientation of the surface F. Define the *sign* $\epsilon(v)$ of a double point v of L to be $+1$ if the signs of both branches of the front intersecting at v coincide, and -1 otherwise. Similarly,

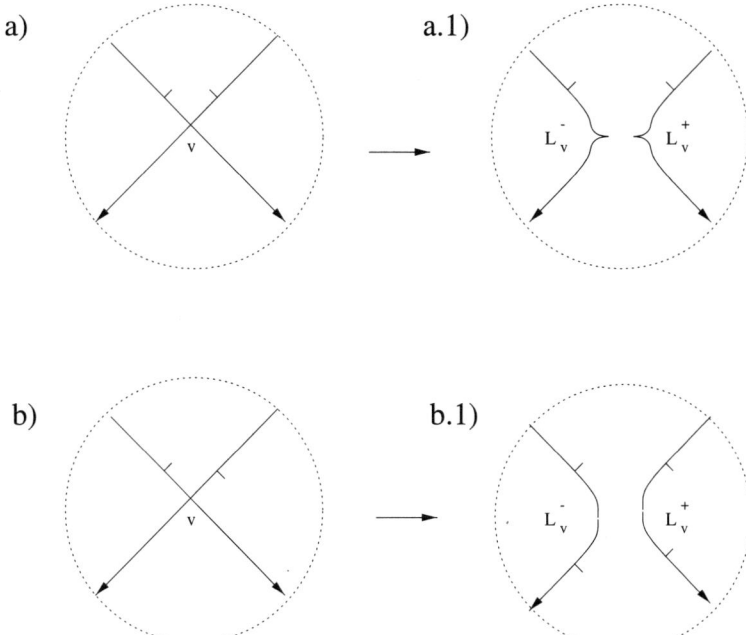

FIGURE 11

we assign a positive (resp. negative) sign to a cusp point if the coorienting vector turns in a positive (resp. negative) direction while traversing a small neighborhood of the cusp point along the orientation. We denote half of the number of positive and negative cusp points by C^+ and C^-, respectively.

Let v be a double point of L. The orientations of F and L allow one to distinguish the two wave fronts L_v^+ and L_v^- obtained by splitting of L in v with respect to orientation and coorientation (see Figures 11a.1 and 11b.1). (Locally one of the two fronts lies to the left and another to the right of v.)

For a Legendrian curve λ in $ST^*\mathbb{R}^2$ denote by $l(\lambda)$ its Bennequin invariant described in the works of Tabachnikov [8] and Arnold [3] with the sign convention of [3] and [5].

THEOREM 5.3.1 (Polyak [5]). *Let L be a generic oriented cooriented wave front on $\mathbb{R}^2$, and λ the corresponding Legendrian curve. Denote by $\mathrm{ind}(L)$ the degree of the mapping taking a point $p \in S^1$ to the point of S^1 corresponding to the direction of the coorienting normal of L at $L(p)$. Define S as the following sum over all double points of L:*

$$(14) \qquad S = \sum_v (\mathrm{ind}(L_v^+) - \mathrm{ind}(L_v^-) - \epsilon(v)).$$

Then

$$(15) \qquad l(\lambda) = S + (1 - \mathrm{ind}(L))C^+ + (\mathrm{ind}(L) + 1)C^- + \mathrm{ind}(L)^2.$$

In [5] it is shown that the Bennequin invariant of a wave front on the $\mathbb{R}^2$ plane admits quantization. Consider a formal quantum parameter q. Recall that for any $n \in \mathbb{Z}$ the corresponding quantum number $[n]_q \in \mathbb{Z}[q, q^{-1}]$ is a finite Laurent polynomial in q defined by

(16) $$[n]_q = \frac{q^n - q^{-n}}{q - q^{-1}}.$$

Substituting quantum integers instead of integers in Theorem 5.3.1 we get the following theorem.

THEOREM 5.3.2 (Polyak [5]). *Let L be a generic cooriented oriented wave front on $\mathbb{R}^2$, and λ the corresponding Legendrian curve. Define S_q by the following formula, where the sum is taken over the set of all double points of L:*

(17) $$S_q = \sum_v [\text{ind}(L_v^+) - \text{ind}(L_v^-) - \epsilon(v)]_q.$$

Put

(18)
$$l_q(L) = S_q + [1 - \text{ind}(L)]_q C^+ + [\text{ind}(L) + 1]_q C^- + [\text{ind}(L)]_q \text{ind}(L).$$

Then $l_q(\lambda) = l_q(L) \in \frac{1}{2}\mathbb{Z}[q, q^{-1}]$ is invariant under isotopy in the class of the Legendrian knots.

The $l_q(\lambda)$ invariant can be expressed [2] in terms of the partial linking polynomial of a generic cooriented oriented wave front introduced by Aicardi [1].

The reason why this invariant takes values in $\frac{1}{2}\mathbb{Z}[q, q^{-1}]$ and not in $\mathbb{Z}[q, q^{-1}]$ is that the number of positive (or negative) cusps can be odd. This makes C^+ (C^-) a half-integer.

Let λ_v^ϵ with $\epsilon = \pm$ be the Legendrian lifting of the front L_v^ϵ. Let $f \in H_1(ST^*F)$ be the homology class of a positively oriented fiber.

THEOREM 5.3.3 (Polyak [5]). *Let L be a generic oriented cooriented wave front on an oriented surface F. Let λ be the corresponding Legendrian curve. Define $l_F(\lambda) \in H_1(ST^*F, \frac{1}{2}\mathbb{Z})$ by the following formula:*

(19) $$l_F(\lambda) = \left(\prod_v [\lambda_v^+][\lambda_v^-]^{-1} f^{-\epsilon(v)}\right) (f[\lambda]^{-1})^{C^+} ([\lambda]f)^{C^-}.$$

(We use the multiplicative notation for the group operation in $H_1(ST^*F)$.)

Then $l_F(\lambda)$ is invariant under isotopy in the class of the Legendrian knots.

The proof is straightforward. One verifies that $l_F(\lambda)$ is invariant under all oriented versions of nondangerous self-tangency, triple point, cusp crossing, and cusp birth moves of the wave front.

In [5] this invariant is denoted by $I_\Sigma^+(\lambda)$ and, in a sense, it appears to be a natural generalization of Arnold's J^+ invariant [3] to the case of an oriented cooriented wave front on an oriented surface.

Note that in the situation of Theorem 5.3.1, the indices of all the fronts involved are the images of the homology classes of their Legendrian liftings under the natural identification of $H_1(ST^*\mathbb{R}^2)$ with $\mathbb{Z}$. If one replaces indices everywhere in Theorem 5.3.1 with the corresponding homology classes and puts f instead of 1, then the only difference between the two formulas is the term $\text{ind}^2(L)$. (One has to remember that we use the multiplicative notation for the operation in $H_1(ST^*F)$.)

5.3.4. The splitting of a Legendrian knot K into μ_v^+ and μ_v^- (see 3.1.1) can be done up to an isotopy in the class of the Legendrian knots. Although this can be done in many ways, there exists a simplest way. The projections $\tilde{L}_v^+$ and $\tilde{L}_v^-$ of

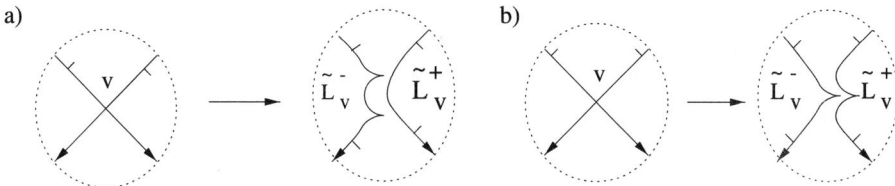

Figure 12

the Legendrian curves created by the splitting are shown in Figure 12. (This fact follows from Theorem 5.2.3.)

Let $\tilde{\lambda}_v^\epsilon$ with $\epsilon = \pm$ be the Legendrian lifting of the front $\tilde{L}_v^\epsilon$.

THEOREM 5.3.5. *Let L be a generic oriented cooriented wave front on an oriented surface F. Let λ be the corresponding Legendrian curve. Define $S(\lambda) \in \frac{1}{2}\mathbb{Z}[H_1(ST^*F)]$ by the following formula:*

$$(20) \quad S(\lambda) = \sum_v \left([\tilde{\lambda}_v^+] - [\tilde{\lambda}_v^-]\right) + (f - [\lambda])C^+ + ([\lambda]f - e)C^-.$$

Then $S(\lambda)$ is invariant under isotopy in the class of the Legendrian knots.

The proof is straightforward. One verifies that $S(\lambda)$ is indeed invariant under all oriented versions of nondangerous self-tangency, triple point, cusp crossing, and cusp birth moves of the wave front.

5.3.6. By taking the free homotopy classes of $\tilde{\lambda}_v^+$ and $\tilde{\lambda}_v^-$ instead of the homology classes, one obtains a different version of the $S(\lambda)$ invariant. It takes values in the group of formal half-integer linear combinations of the free homotopy classes of oriented curves in ST^*F. In this case the terms $[\lambda]$ and f in (20) should be substituted with the free homotopy classes of λ and of a positively oriented fiber, respectively. The terms $[\lambda]f$ and e in (20) should be substituted with the free homotopy classes of λ with a positive fiber added to it and the class of a contractible curve, respectively. Note that f lies in the center of $\pi_1(ST^*F)$, so that the class of λ with a fiber added to it is well defined.

A straightforward verification shows that this version of $S(\lambda)$ is also invariant under isotopy in the class of the Legendrian knots.

THEOREM 5.3.7. *Let L be a generic oriented cooriented wave front on an oriented surface F. Let λ be the corresponding Legendrian curve. Let $S(\lambda)$ and $l_F(\lambda)$ be the invariants introduced in Theorems 5.3.5 and 5.3.3, respectively. Let*

$$(21) \quad \mathrm{pr} : \frac{1}{2}\mathbb{Z}[H_1(ST^*F)] \to H_1(ST^*F, \frac{1}{2}\mathbb{Z})$$

be the mapping defined as follows: for any $n_i \in \frac{1}{2}\mathbb{Z}$ and $g_i \in H_1(ST^*F)$,

$$(22) \quad \sum n_i g_i \mapsto \prod g_i^{n_i}.$$

Then $\mathrm{pr}(S(\lambda)) = l_F(\lambda)$.

The proof is straightforward: one must verify that

$$(23) \quad [\lambda_v^+][\lambda_v^-]^{-1} f^{-\epsilon(v)} = [\tilde{\lambda}_v^+][\tilde{\lambda}_v^-]^{-1} \quad \text{in } H_1(ST^*F).$$

(Recall that we use a multiplicative notation for the group operation in $H_1(ST^*F)$.) This means that $S_F(\lambda)$ is a refinement of Polyak's invariant $l_F(\lambda)$.

5.3.8. One can verify that there is a unique linear combination $\sum_{m\in\mathbb{Z}} n_m[m]_q = l_q(\lambda)$ with n_m being nonnegative half-integers such that $n_0 = 0$, and if $n_m > 0$, then $n_{-m} = 0$. To prove this one must verify that $\{\frac{1}{2}[n]_q | 0 < n\}$ is a basis for the $\mathbb{Z}$-submodule of $\frac{1}{2}\mathbb{Z}[q, q^{-1}]$ generated by the quantum numbers and use the identity $n[m]_q = -n[-m]_q$.

The following theorem shows that if $L \subset \mathbb{R}^2$, then $S(\lambda)$ and Polyak's quantization $l_q(\lambda)$ (see Theorem 5.3.2) of the Bennequin invariant can be expressed explicitly in terms of each other.

THEOREM 5.3.9. *Let $f \in H_1(ST^*\mathbb{R}^2)$ be the class of a positively oriented fiber. Let L be a generic oriented cooriented wave front on $\mathbb{R}^2$, λ the corresponding Legendrian curve, and f^h the homology class realized by it. Let $l_q(\lambda) - [h]_q h = \sum_{m\in\mathbb{Z}} n_m[m]_q$ be written in the form described in 5.3.8 and $S(\lambda) = \sum_{l\in\mathbb{Z}} k_l f^l$. Then*

(24) $$l_q(\lambda) = [h]_q h + \sum_{k_l > 0} k_l [2l - h - 1]_q,$$

and

(25) $$S(\lambda) = \sum_{n_m > 0} n_m (f^{\frac{h+1+m}{2}} - f^{\frac{h+1-m}{2}}).$$

For the proof of Theorem 5.3.9 see subsection 7.6.

One can show that for $n_m > 0$ both $\frac{h+1+m}{2}$ and $\frac{h+1-m}{2}$ are integers, so that the sum (25) takes values in $\frac{1}{2}\mathbb{Z}[H_1(ST^*\mathbb{R}^2)]$.

Note that the $l_q(\lambda)$ invariant was defined only for fronts on the plane $\mathbb{R}^2$. Thus $S(\lambda)$ is, in a sense, a generalization of Polyak's $l_q(\lambda)$ to the case of wave fronts on an arbitrary oriented surface F.

5.3.10. The splitting of the Legendrian knot K into $\mu_1^+(v)$, $\mu_1^-(v)$, $\mu_2^+(v)$, and $\mu_2^-(v)$, which was used to define $\tilde{S}(K)$ (see 3.4.1), can be done up to an isotopy in the class of the Legendrian knots. Although this can be done in many ways, there is the simplest one. The projections $\tilde{L}_1^+(v)$, $\tilde{L}_1^-(v)$, $\tilde{L}_2^+(v)$, and $\tilde{L}_2^-(v)$ are shown in Figure 13. (This fact follows from Theorem 5.2.3.)

This allows us to introduce an invariant similar to $S(\lambda)$ for generic oriented cooriented wave fronts on a nonorientable surface F in the following way.

Let L be a generic wave front on a nonorientable surface F. Let v be a double point of L. Fix some orientation of a small neighborhood of v in F. The orientations of the neighborhood and L allow one to distinguish the wave fronts L_1^+, L_1^-, L_2^+, and L_2^- obtained by the two splittings of L with respect to the orientation and coorientation (see Figure 13). Locally the fronts with the upper indices plus and minus are located to the right and to the left of v, respectively. To each double point v of L we associate an element $([\tilde{\lambda}_1^+(v)] - [\tilde{\lambda}_1^-(v)] + [\tilde{\lambda}_2^+(v)] - [\tilde{\lambda}_2^-(v)]) \in \mathbb{Z}[H_1(ST^*(F))]$. Here we denote by lambdas the Legendrian curves corresponding to the wave fronts appearing under the splitting. Clearly this element does not depend on the orientation of the neighborhood of v we have chosen.

For a wave front L let C be half of the number of cusps of L. Denote by f the homology class of the fiber of ST^*F oriented in some way. Note that $f^2 = e$, so it does not matter which orientation of the fiber we use to define f.

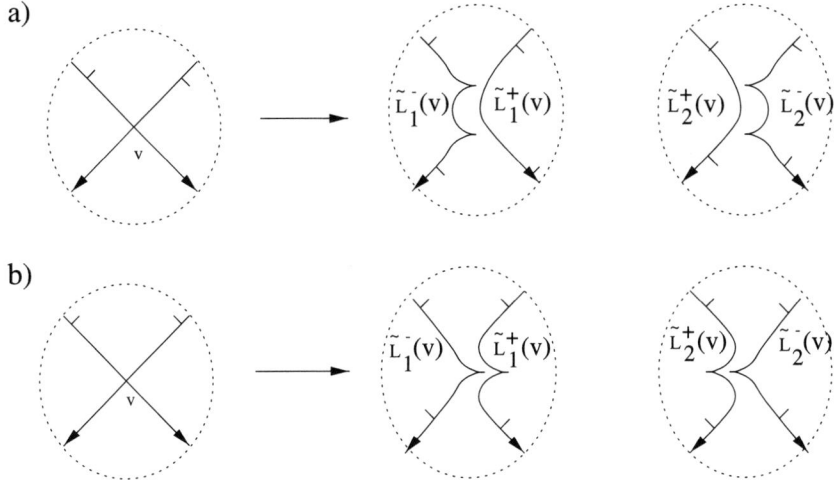

FIGURE 13

THEOREM 5.3.11. *Let L be a generic cooriented oriented wave front on a nonorientable surface F and λ the corresponding Legendrian curve. Define $\tilde{S}(\lambda) \in \frac{1}{2}\mathbb{Z}[H_1(ST^*(F))]$ by the following formula, where the summation is taken over the set of all double points of L:*

(26)
$$\tilde{S}(\lambda) = \sum_v \left([\tilde{\lambda}_1^+(v)] - [\tilde{\lambda}_1^-(v)] + [\tilde{\lambda}_2^+(v)] - [\tilde{\lambda}_2^-(v)] \right) + C\big([\lambda]f - e + f - [\lambda]\big).$$

Then $\tilde{S}(\lambda)$ is invariant under isotopy in the class of the Legendrian knots.

The proof is straightforward. One verifies that $\tilde{S}(\lambda)$ is indeed invariant under all oriented versions of nondangerous self-tangency, triple-point passing, cusp crossing, and cusp birth moves of the wave front.

The reason we have $\tilde{S}(\lambda) \in \frac{1}{2}\mathbb{Z}[H_1(ST^*F)]$ is that if L is an orientation reversing curve, then the number of cusps of L is odd. In this case C is a half-integer.

6. Wave fronts on orbifolds

6.1. Definitions.

DEFINITION 6.1.1. An *orbifold* is a surface F with the additional structure consisting of

1) set $A \subset F$;
2) smooth structure on $F \setminus A$;
3) set of homeomorphisms ϕ_a of neighborhoods U_a of a in F onto $\mathbb{R}^2/G_a$ such that $\phi_a(a) = 0$ and $\phi_a|_{U_a \setminus a}$ is a diffeomorphism. Here $G_a = \{e^{\frac{2\pi k}{\mu_a}} | k \in \{1, \ldots, \mu_a\}\}$ is a group acting on $\mathbb{R}^2 = \mathbb{C}$ by multiplication ($\mu_a \in \mathbb{Z}$ is assumed to be positive).

The points $a \in A$ are called *cone points*.

The action of G on $\mathbb{R}^2$ induces the action of G on $ST^*\mathbb{R}^2$. This makes $ST^*\mathbb{R}^2/G$ a Seifert fibration over $\mathbb{R}^2/G$. Gluing together the pieces over neighborhoods of F

we obtain a Seifert fibration $\pi : N \to F$. The fiber over a cone point a is an exceptional fiber of type $(\mu_a, -1)$ (see 4.1.2).

The natural contact structure on $ST^*\mathbb{R}^2$ is invariant under the induced action of G. Since G acts freely on $ST^*\mathbb{R}^2$, this implies that N has an induced contact structure. As before, the naturally cooriented projection $L \subset F$ of a generic Legendrian curve λ is called the *front of* λ.

6.2. Invariants for fronts on orbifolds. For oriented F we construct an invariant similar to $S(\lambda)$. It corresponds to the S_K invariant of a knot in a Seifert fibered space. For a nonorientable surface F we construct an analogue of $\tilde{S}(\lambda)$. It corresponds to the $\tilde{S}_K$ invariant of a knot in a Seifert fibered space.

Note that any surface F can be viewed as an orbifold without cone points. This justifies the notation below.

Let F be an oriented surface. The orientation of F induces an orientation of all fibers. Denote by f the homology class of a positively oriented fiber. For a cone point a denote by f_a the homology class of a positively oriented fiber $\pi^{-1}(a)$. For a generic oriented cooriented wave front $L \subset F$ denote by C^+ (resp. C^-) half of the number of positive (resp. negative) cusps of L. Note that for a double point v of a generic front L, the splitting into $\tilde{L}_v^+$ and $\tilde{L}_v^-$ is well defined. The corresponding Legendrian curves $\tilde{\lambda}_v^+$ and $\tilde{\lambda}_v^-$ in N are also well defined.

For $a \in A$ of type $(\mu_a, -1)$ put $N_1(a) = \{k \in \{1, \ldots, \mu_a\} | \frac{-2k\pi}{\mu_a} \bmod 2\pi \in (0, \pi]\}$, $N_2(a) = \{k \in \{1, \ldots, \mu_a\} | \frac{2k\pi}{\mu_a} \bmod 2\pi \in (0, \pi)\}$. Define $R_a^1, R_a^2 \in \mathbb{Z}[H_1(N)]$ by the following formulas:

$$R_a^1 = \sum_{k \in N_1(a)} \left([\lambda] f_a^{\mu_a - k} - f_a^k\right) - \sum_{k \in N_2(a)} \left(f_a^{\mu_a - k} - [\lambda] f_a^k\right), \tag{27}$$

$$R_a^2 = \sum_{k \in N_1(a)} \left(f_a^{\mu_a - k} - [\lambda] f_a^k\right) - \sum_{k \in N_2(a)} \left([\lambda] f_a^{\mu_a - k} - f_a^k\right). \tag{28}$$

Set J to be the quotient of $\frac{1}{2}\mathbb{Z}[H_1(N)]$ by the free Abelian subgroup generated by $\{\{\frac{1}{2} R_1(a), \frac{1}{2} R_2(a)\}_{a \in A}\}$.

THEOREM 6.2.1. *Let L be a generic cooriented oriented wave front on F and λ the corresponding Legendrian curve.*

Then $S(\lambda) \in J$ defined by the sum over all double points of L,

$$S(\lambda) = \sum \left([\tilde{\lambda}^+(v)] - [\tilde{\lambda}^-(v)]\right) + (f - [\lambda])C^+ + ([\lambda]f - e)C^-, \tag{29}$$

is invariant under isotopy in the class of the Legendrian knots.

For the proof of Theorem 6.2.1 see subsection 7.7.

Let F be a nonorientable surface. Denote by f the homology class of a regular fiber oriented in some way. Note that $f^2 = e$, so the orientation we use to define f does not matter. For a cone point a denote by f_a the homology class of the fiber $\pi^{-1}(a)$ oriented in some way. For a generic oriented cooriented wave front $L \subset F$ denote by C half of the number of cusps of L. Note that for a double point v of a generic front L the element $\left([\tilde{\lambda}_1^+(v)] - [\tilde{\lambda}_1^-(v)] + [\tilde{\lambda}_2^+(v)] - [\tilde{\lambda}_2^-(v)]\right) \in \mathbb{Z}[H_1(N)]$ used to introduce $\tilde{S}(\lambda)$ is well defined.

For $a \in A$ of type $(\mu_a, -1)$ put $N_1(a) = \{k \in \{1, \ldots, \mu_a\} | \frac{-2k\pi}{\mu_a} \bmod 2\pi \in (0, \pi]\}$, and $N_2(a) = \{k \in \{1, \ldots, \mu_a\} | \frac{-2k\pi}{\mu_a} \bmod 2\pi \in (0, \pi)\}$. Define $R_a \in$

$\mathbb{Z}[H_1(N)]$ by the following formula:

(30) $$R_a = \sum_{k \in N_1(a)} \left([\lambda] f_a^{\mu_a - k} - f_a^k + f_a^{k - \mu_a} - [\lambda] f_a^{-k} \right)$$
$$- \sum_{k \in N_2(a)} \left(f_a^{\mu_a - k} - [\lambda] f_a^k + [\lambda] f_a^{k - \mu_a} - f_a^{-k} \right).$$

Put $\tilde{J}$ to be the quotient of $\frac{1}{2}\mathbb{Z}[H_1(N)]$ by a free Abelian subgroup generated by $\{\{\frac{1}{2} R_a\}_{a \in A}\}$.

Similar to Definition 4.1.5, one can prove that under the change of the orientation of $\pi^{-1}(a)$ (used to define f_a) R_a goes to $-R_a$. Thus $\tilde{J}$ is well defined.

THEOREM 6.2.2. *Let L be a generic cooriented oriented wave front on F and λ the corresponding Legendrian curve. Then $\tilde{S}(\lambda) \in \tilde{J}$ defined by the summation over all double points of L,*

(31)
$$\tilde{S}(\lambda) = \sum \left([\tilde{\lambda}_1^+(v)] - [\tilde{\lambda}_1^-(v)] + [\tilde{\lambda}_2^+(v)] - [\tilde{\lambda}_2^-(v)] \right) + ([\tilde{\lambda}]f - e + f - [\lambda])C,$$

is invariant under isotopy in the class of the Legendrian knots.

The proof is a straightforward generalization of the proof of Theorem 6.2.1.

7. Proofs

7.1. Proof of Theorem 3.1.3. To prove the theorem, it suffices to show that S_K is invariant under the elementary isotopies. They project to a death of a double point, the cancellation of two double points, and passing through a triple point.

To prove the invariance, we fix a part P of F homeomorphic to a closed disk and containing the projection of one of the elementary isotopies. Fix a section over the boundary of P such that the points of $K \cap \pi^{-1}(\partial P)$ belong to the section. Inside P we can construct the Turaev shadow (see 2.2.10). The action of $H_1(\text{Int } P) = e$ on the set of isotopy types of links is trivial (see Theorem 2.2.8). Thus the part of K can be reconstructed in the unique way from the shadow over P. In particular, one can compare the homology classes of the curves created by splitting at a double point inside P. Hence to prove the theorem, it suffices to verify the invariance under the oriented versions of the moves S_1, S_2, and S_3.

There are two versions of the oriented move S_1 shown in Figures 14a and 14b. For Figure 14a the term $[\mu_v^+]$ appears to be equal to f. From 3.1.2 we know that $[\mu_v^+][\mu_v^-] = [K]f$, so that $[\mu_v^-] = [K]$. Hence $[\mu_v^+] - [\mu_v^-] = f - [K]$ and is equal to zero in H. In the same way we verify that $[\mu_v^+] - [\mu_v^-]$ (for v shown in Figure 14b) is equal to $[K]f - e$. It is also zero in H. The summands corresponding to other double points do not change under this move, since it does not change the homology classes of the knots created by the splittings.

There are three oriented versions of the S_2 move. We show that S_K does not change under one of them. The proof for the other two is the same or easier. We choose the version corresponding to the upper part of Figure 15. The summands corresponding to the double points not in this figure are preserved under the move, since it does not change the homology classes of the corresponding knots. So it suffices to show that the terms produced by the double points u and v in this figure cancel out. Note that the shadow μ_v^- is transformed to μ_u^+ by the $\bar{S}_1$ move. It is known that $\bar{S}_1$ can be expressed in terms of S_1, S_2, and S_3, thus it also does not

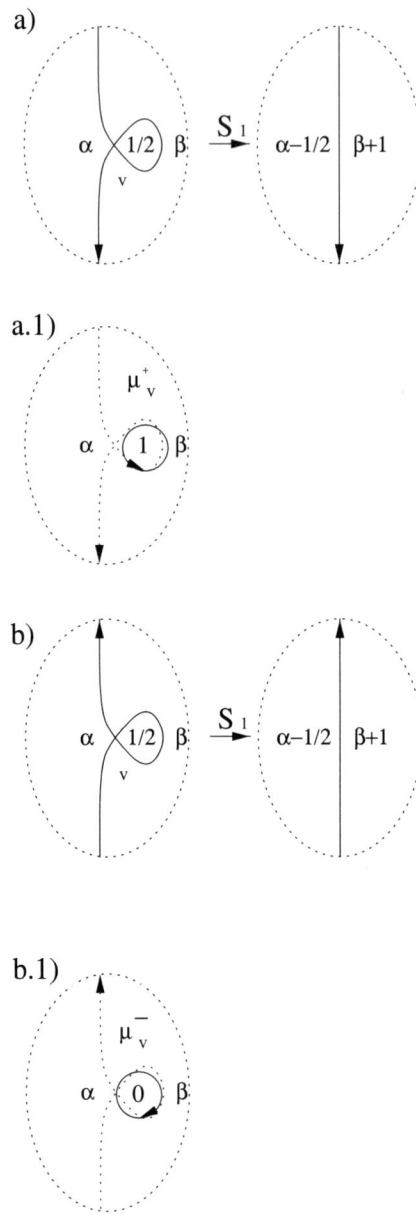

FIGURE 14

change the homology classes of the knots created by the splittings. Hence $[\mu_u^+]$ and $[\mu_u^-]$ cancel out. In the same way one proves that $[\mu_v^-]$ and $[\mu_v^+]$ also cancel out.

There are two oriented versions of the S_3 move: S_3' and S_3'', shown in Figures 16a and 16b, respectively. The S_3'' move can be expressed in terms of S_3' and oriented versions of S_2 and S_2^{-1}. To prove this we use Figure 17. There are two ways to get from Figure 17a to Figure 17b. One is to apply S_3''. Another way is to apply three times the oriented version of S_2 to obtain Figure 17c, then apply S_3' to get

INVARIANTS OF FRONTS ON SURFACES

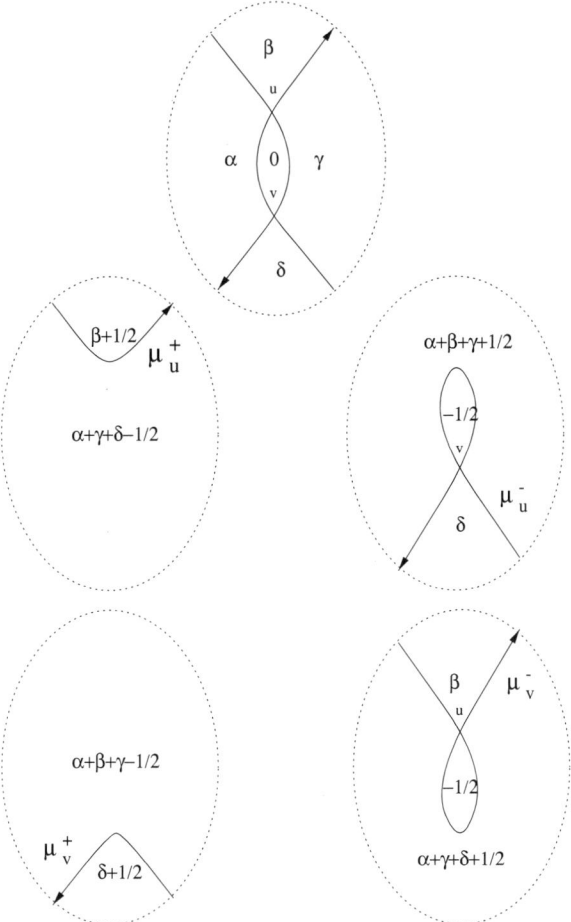

FIGURE 15

Figure 17d, and finally use three times the oriented version of S_2^{-1} to end up at Figure 17b.

Thus it suffices to verify the invariance under S_3'. The terms corresponding to the double points not in Figures 18a and 18b are preserved for the same reasons as above. The terms coming from double points u in Figure 18a and u in Figure 18b are the same. This holds also for the v- and w-pairs of double points in these two figures. We prove this statement only for the u-pair of double points. For v- and w-pairs the proof is the same or simpler. There is only one possibility: either the dashed line belongs to both $\pi(\mu_u^+)$ in Figures 18a.1 and 18b.1, respectively, or to both $\pi(\mu_u^-)$ in Figures 18a.2 and 18b.2, respectively. We choose the one to which it does not belong. Summing up gleams on each of the two sides of it we immediately see that the corresponding shadows are the same on both pictures. Thus the homology classes of the corresponding knots are equal. But $[\mu_u^+][\mu_u^-] = [K]f$ (see 3.1.2), thus the homology classes of the knots represented by the other shadows are also equal. This completes the proof of Theorem 3.1.3. □

a)

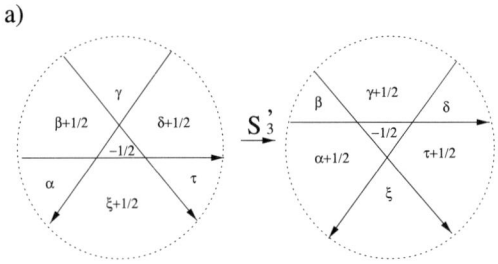

b)

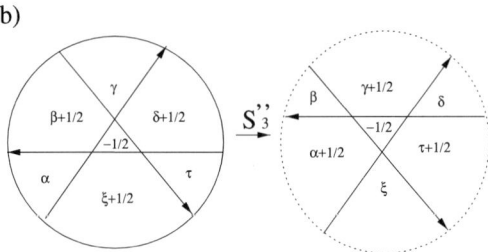

FIGURE 16

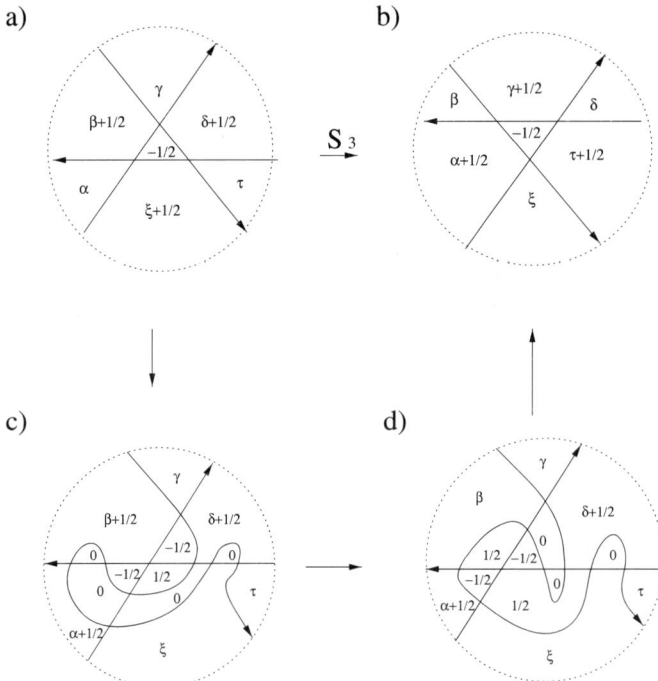

FIGURE 17

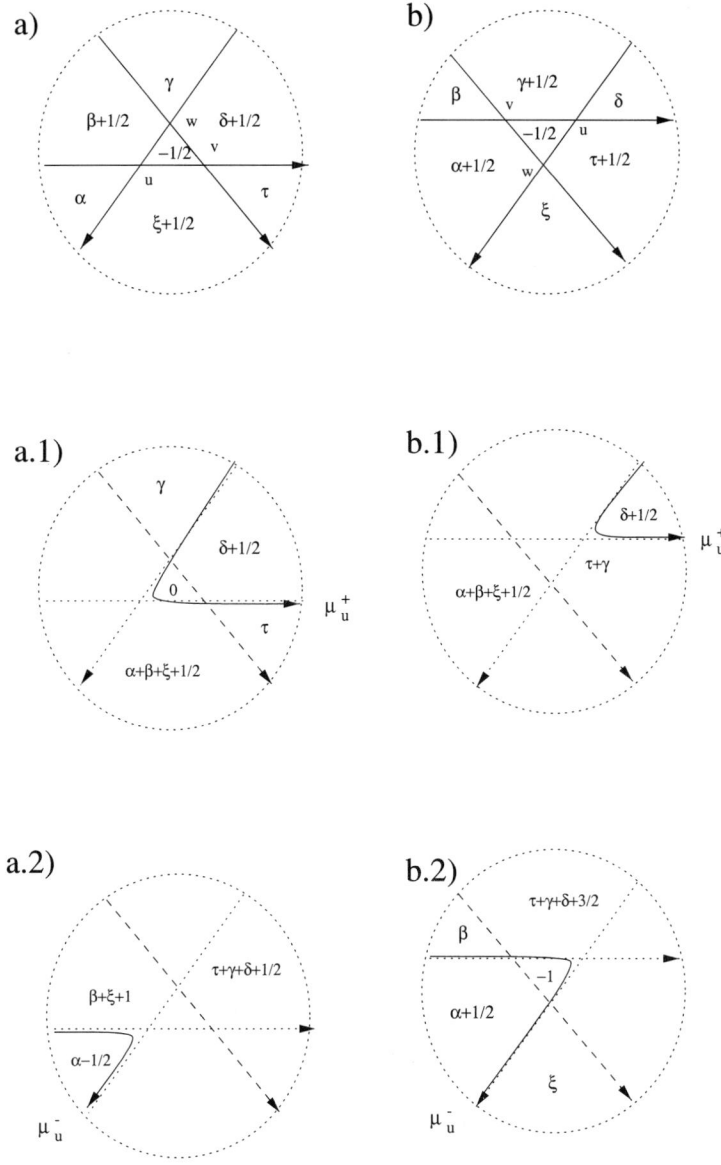

FIGURE 18

7.2. Proof of Theorem 3.2.4. I: K' can be obtained from K by a sequence of isotopies and modifications along fibers. Isotopies do not change S. The modifications change S by elements of type (3). To complete the proof, we use the identity $\xi_1 \xi_2 = [K]$.

II: We prove that for any $i \in H_1(N)$ there exist two knots K_1 and K_2 such that they represent the same free homotopy class as K,

(32) $$S_{K_1} = S_K + (f - e)([K]i^{-1} + i),$$

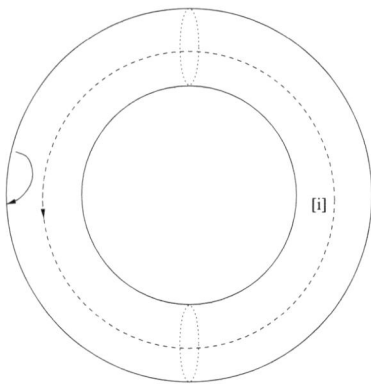

FIGURE 19

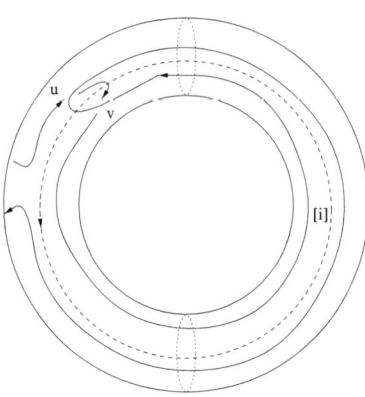

FIGURE 20

and

(33) $$S_{K_2} = S_K - (f - e)([K]i^{-1} + i).$$

Clearly this would imply the second statement of the theorem.

Take $i \in H_1(N)$. Let K_i be an oriented knot in N such that $[K_i] = i$. The space N is oriented, hence the tubular neighborhood T_{K_i} of K_i is homeomorphic to an oriented solid torus T. Deform K_i so that $K_i \cap T_{K_i}$ is a small arc (see Figure 19). Pull one part of the arc along K_i in T_{K_i} under the other part of the arc (see Figure 20). This isotopy creates two new double points u and v of $\pi(K)$. (Since T_{K_i} may be knotted, it might happen that there are other new double points, but we do not need them for our construction.) Making a fiber modification along the part of $\pi^{-1}(u)$ that lies in T one obtains K_2. Making a fiber modification along the part of $\pi^{-1}(v)$ that lies in T one obtains K_1.

This completes the proof of Theorem 3.2.4. □

7.3. Proof of Lemma 3.3.3. It is easy to verify that any two shadows with the same projection can be transformed to each other by a sequence of fiber fusions. One can easily create a trivial knot with an ascending diagram such that its projection is any desired curve. This implies that any two shadows on $\mathbb{R}^2$ can be transformed to each other by a sequence of fiber fusions, movements S_1, S_2, S_3, and

FIGURE 21

their inverses. A straightforward verification shows that $\sigma(s(K))$ does not change under the moves S_1, S_2, S_3, and their inverses. Under fiber fusions, the homology class of the knot and the element σ change in the same way. To prove this, we use Figure 21, where Figure 21a shows the shadow before the application of the fiber fusion (that adds 1 to the homology class of the knot) and Figure 21b after. In these diagrams the indices of the regions are denoted by Latin letters. Now one easily verifies that σ also increases by one. Finally, for the trivial knot with a trivial shadow diagram its homology class and $\sigma(s(K))$ are both equal to 0. This completes the proof of Lemma 3.3.3. □

7.4. Proof of Theorem 4.1.4. It suffices to show that S_K does not change under the elementary isotopies of the knot. Three of them correspond in the projection to a birth of a small loop, passing through a point of self-tangency, and passing through a triple point. The fourth one is passing through an exceptional fiber.

From the proof of Theorem 3.1.3 one gets that S_K is invariant under the first three of the elementary isotopies described above. Thus it suffices to prove invariance under passing through an exceptional fiber a.

Let a be a singular fiber of type (μ_a, ν_a) (see 4.1.2). Let T_a be a neighborhood of a which is fiber-wise isomorphic to the standardly fibered solid torus with an exceptional fiber of type (μ_a, ν_a).

We can assume that the move proceeds as follows. At the start K and T_a intersect along a curve lying in the meridianal disk D of T_a. The part of K close to a in D is an arc C of a circle of a very large radius. This arc is symmetric with respect to the y axis passing through a in D. During the move this arc slides along the y axis through the fiber a (see Figure 22).

Clearly two points u and v of C after this move are in the same fiber if and only if they are symmetric with respect to the y axis, and the angle formed by v, a, u in D is less or equal to π and is equal to $\frac{2l\pi}{\mu_a}$ for some $l \in \{1, \ldots, \mu_a\}$ (see Figure 22). They are in the same fiber before the move if and only if the angle formed by u, a, v in D is less than π and is equal to $\frac{2l\pi}{\mu_a}$ for some $l \in \{1, \ldots, \mu_a\}$ (see Figure 22).

Consider a double point v of $\pi\big|_D(K)$ that appears after the move and corresponds to the angle $\frac{2l\pi}{\mu_a}$. There is a unique $k \in N_1(a)$ such that $\frac{2\pi\nu_a k}{\mu_a} \bmod 2\pi = \frac{2l\pi}{\mu}$. Note that to make the splitting of $[K]$ into $[\mu_v^+]$ and $[\mu_v^-]$ well defined, we do not need the two points of K projecting to v to be antipodal in $\pi^{-1}(v)$. This allows one to compare these homology classes with f_a. For the orientation of C shown in Figure 22 one verifies that connecting v to u along the orientation of the fiber we are adding k fibers f_a. Note that the factorization we used to define the exceptionally

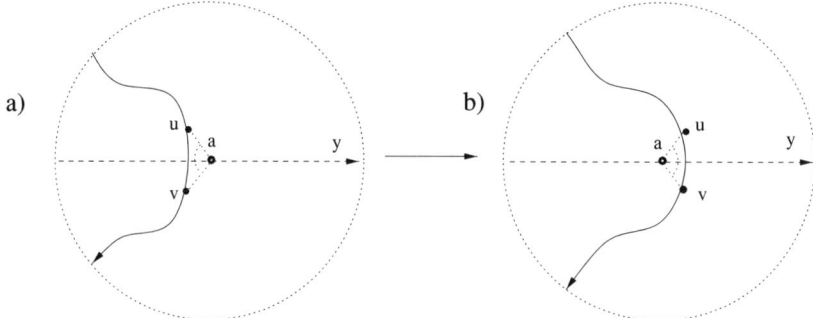

FIGURE 22

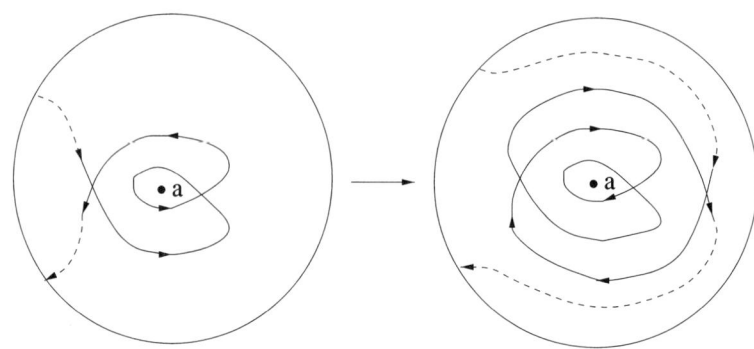

FIGURE 23

fibered torus was

$$((r,\theta),1) = ((r,\theta + \frac{2\pi\nu}{\mu}),0).$$

Thus $[\mu_v^-] = f_a^k$ (see Figure 23). From 3.1.2 we know that $[\mu_v^+][\mu_v^-] = [K]f$. Hence $[\mu_v^+] = [K]f_a^{\mu_a - k}$.

As above, to each double point v of $\pi\big|_D(K)$ before this move there corresponds $k \in N_2(a)$. For this double point $[\mu_v^+] = f_a^{\mu_a - k}$ and $[\mu_v^-] = [K]f_a^k$.

Summing up over the corresponding values of k we see that S_K changes by R_a^1 under this move. Recall that $R_a^1 = 0 \in H$. Thus S_K is invariant under the move.

For the other choice of the orientation of C the value of S_K changes by $R_a^2 = 0 \in H$. Thus S_K is invariant under all elementary isotopies, and this proves Theorem 4.1.4. □

7.5. Proof of Theorem 5.2.3. Deform L in the neighborhoods of all double points of L (see Figure 24), so that the two points of the Legendrian knot corresponding to the double point of L are antipodal in the fiber. After we make the quotient of the fibration by the $\mathbb{Z}_2$-action, the projection of the deformed λ is not a cooriented front anymore but a front equipped with a normal field of lines. (This corresponds to the factorization $S^1 \to \mathbb{R}P^1$.) Using Figure 25 one calculates the contributions of different cusps and double points to the total rotation number of the line field under traversing the boundary in the counterclockwise direction.

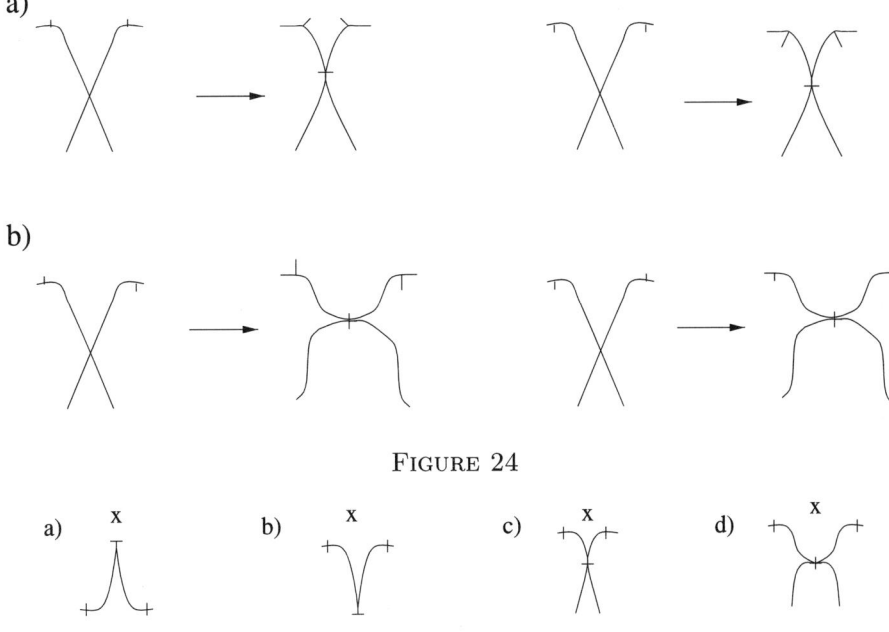

FIGURE 24

FIGURE 25

These contributions are as follows:

(34) $\begin{cases} 1 & \text{for every cusp point pointing inside } X; \\ -1 & \text{for every cusp point pointing outside } X; \\ -1 & \text{for every double point of the type shown in Figure 8c}; \\ 0 & \text{for the other types of double points.} \end{cases}$

To get the contributions to gleams, we divide these numbers by 2 (as we do in the construction of shadows, see subsection 2.1).

If the region does not have cusps and double points in its boundary, then the obstruction to an extension of the section over ∂X to X is equal to $\chi(\text{Int } X)$.

This completes the proof of Theorem 5.2.3. □

7.6. Proof of Theorem 5.3.9. A straightforward verification shows that

(35) $$\text{ind } \tilde{L}_u^+ - \text{ind } \tilde{L}_u^- = \text{ind } L_u^+ - \text{ind } L_u^- - \epsilon(u),$$

(36) $$\text{ind } \tilde{L}_u^+ + \text{ind } \tilde{L}_u^- = \text{ind } L + 1,$$

and

(37) $$\text{ind } L_u^+ + \text{ind } L_u^- = \text{ind } L$$

for any double point u of L.

Let us prove (24). We write down the formal sums used to define $S(\lambda)$ and $l_q(\lambda)$ and start to reduce them in a parallel way as described below.

We say that a double point is essential if $[\tilde{\lambda}_u^+] \neq [\tilde{\lambda}_u^-]$.

For nonessential u we see that the term $([\tilde{\lambda}_u^+] - [\tilde{\lambda}_u^-])$ in $S(\lambda)$ is zero. Using (35) we get that the term $[\text{ind } L_v^+ - \text{ind } L_v^- - \epsilon(v)]_q$ in $l_q(\lambda)$ is also zero.

The index of a wave front coincides with the homology class of its lifting under the natural identification of $H_1(ST^*F)$ with $\mathbb{Z}$. This fact and (36) imply that if we have $[\tilde{\lambda}_u^+] = [\tilde{\lambda}_v^-]$ for two double points u and v, then $[\tilde{\lambda}_u^-] = [\tilde{\lambda}_v^+]$. Hence $([\tilde{\lambda}_u^+] - [\tilde{\lambda}_u^-]) = -([\tilde{\lambda}_v^+] - [\tilde{\lambda}_v^-])$, and these two terms cancel out. Identity (35) implies that the terms $[\text{ind } L_u^+ - \text{ind } L_u^- - \epsilon(u)]_q$ and $[\text{ind } L_v^+ - \text{ind } L_v^- - \epsilon(v)]_q$ also cancel out.

For similar reasons, if for a double point u the term $([\tilde{\lambda}_u^+] - [\tilde{\lambda}_u^-])$ is equal to $([\lambda] - f)$, so that we can simplify $S(\lambda)$ by crossing out the term and decreasing the coefficient C^+ by one. Then $[\text{ind } L_u^+ - \text{ind } L_u^- - \epsilon(u)]_q = [\text{ind } L - 1]_q$, and we can simplify $l_q(\lambda)$ by crossing out the term and decreasing the coefficient C^+ by one.

Similarly, if the input of double point u into $S(\lambda)$ is $(e - [\lambda]f)$, then we reduce the two sums in the parallel way by crossing out the corresponding terms and decreasing by one the coefficients C^-.

We make the cancellations described above in both $S(\lambda)$ and $(l_q(\lambda) - [h]_q h)$ in a parallel way until we cannot reduce $S(\lambda)$ any more. In this reduced form the terms of the form $k_l f^l$ with $k_l > 0$ corresponds to the terms of type $[\tilde{\lambda}_u^+]$ for some double points u of L. (The case where a term of this type corresponds to cusps is treated separately below.) Identities (35) and (36) imply that the contribution of the corresponding double points into $l_q(\lambda)$ is $k_l [2l - h - 1]_q$.

In the case where the $k_l f^l$ term comes from the cusps and not from the double points of L, one can easily verify that the corresponding input of cusps into $(l_q(\lambda) - [h]_q h)$ can still be written as $k_l [2l-h-1]q$. Thus $l_q(\lambda) = [h]_q h + \sum_{k_l > 0} k_l [2l-h-1]_q$, and we have proved (24).

Let us prove (25). As above we reduce $S(\lambda)$ and $(l_q(\lambda) - [h]_q h)$ in a parallel way. Note that the coefficient at each $[m]_q$ was positive from the very beginning by the definition of $l_q(\lambda)$, and it stays positive under the cancellations described above. After this reduction each term $n_m [m]_q$ is a contribution of n_m double points. (The case where it is a contribution of cusps is treated separately as in the proof of (25).) Let u be one of these double points. Then from (35) and (36) we get the following system of two equations in variables $\text{ind } \tilde{L}_u^+$ and $\text{ind } \tilde{L}_u^-$:

$$(38) \quad \begin{cases} \text{ind } \tilde{L}_u^+ - \text{ind } \tilde{L}_u^- = m, \\ \text{ind } \tilde{L}_u^+ + \text{ind } \tilde{L}_u^- = \text{ind } L + 1. \end{cases}$$

Solving the system we get that $[\tilde{\lambda}_u^+] = f^{\frac{m+h+1}{2}}$ and $[\tilde{\lambda}_u^-] = f^{\frac{h+1-m}{2}}$.

This proves identity (25) and Theorem 5.3.9. □

7.7. Proof of Theorem 6.2.1.

There are five elementary isotopies of a generic front L on an orbifold F. Four of them are the birth of two cusps, passing through a nondangerous self-tangency point, passing through a triple point, and passing of a branch through a cusp point. For all possible oriented versions of these moves a straightforward calculation shows that $S(\lambda) \in \frac{1}{2}\mathbb{Z}[H_1(N)]$ is preserved.

The fifth move is more complicated. It corresponds to a generic passing of a wave front lifted to $\mathbb{R}^2$ through the preimage of a cone point a. We can assume that this move is a symmetrization by G_a of the following move. The lifted front in the neighborhood of a is an arc C of a circle of large radius with center at the y axis, and during this move this arc slides through a along y (see Figure 26).

Clearly after this move points u and v on the arc C turn out to be in the same fiber if and only if they are symmetric with respect to the y axis, and the angle

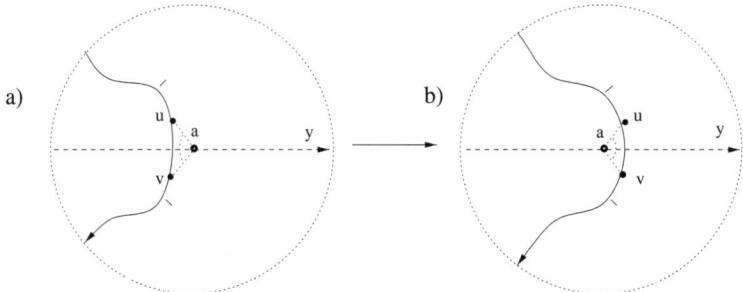

Figure 26

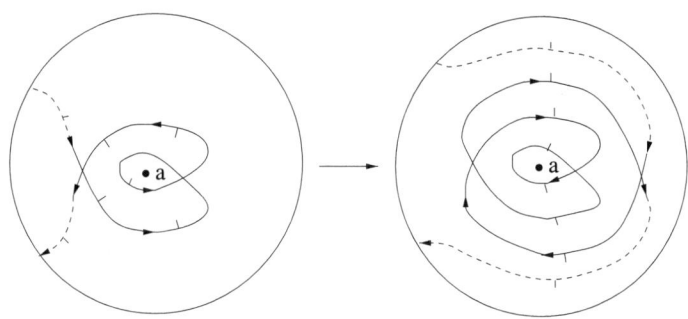

Figure 27

formed by v, a, u is less or equal to π and is equal to $\frac{2k\pi}{\mu_a}$ for some $k \in \{1, \ldots, \mu_a\}$ (see Figure 26). We denote the set of such numbers k by $\bar{N}_1(a) = \{k \in \{1, \ldots, \mu_a\} \big| \frac{2k\pi}{\mu_a} \in (0, \pi]\}$.

Two points u and v on the arc C are in the same fiber before the move if and only if they are symmetric with respect to the y axis, and the angle formed by u, a, v is less than π and is equal $\frac{2k\pi}{\mu_a}$ for some $k \in \{1, \ldots, \mu_a\}$ (see Figure 26). We denote the set of such numbers k by $\bar{N}_2(a) = \{k \in \{1, \ldots, \mu_a\} \big| \frac{2k\pi}{\mu_a} \in (0, \pi)\}$.

The projection of this move for the orientation of L' drawn in Figure 26 is shown in Figure 27.

Split the wave front in Figure 27 at the double point v (appearing after the move) that corresponds to some $k \in \bar{N}_1(a)$. Then $\tilde{\lambda}_v^-$ is a front with two positive cusps that rotates k times around a in the clockwise direction. Hence $[\tilde{\lambda}_v^-] = ff_a^{-k} = f_a^{\mu_a - k}$. We know that $[\tilde{\lambda}_v^+][\tilde{\lambda}_v^-] = [\lambda] f$ and that $f_a^{\mu_a} = f$. Thus $[\tilde{\lambda}_v^+] = [\lambda] f_a^k$.

In the same way we verify that if we split the front at the double point v (existing before the move) that corresponds to some $k \in \bar{N}_2(a)$, then $[\tilde{\lambda}_v^+] = f_a^k$ and $[\tilde{\lambda}_v^-] = [K] f_a^{|\mu_a| - k}$.

Now making sums over all corresponding numbers $k \in \{1, \ldots, \mu_a\}$ we get that under this move $S(\lambda)$ changes by

$$(39) \qquad \bar{R}_a^1 = \sum_{k \in \bar{N}_1(a)} \left([\lambda] f_a^k - f_a^{\mu_a - k} \right) - \sum_{k \in \bar{N}_2(a)} \left(f_a^k - [\lambda] f_a^{\mu_a - k} \right).$$

A straightforward verification shows that $R_a^1 = \bar{R}_a^1$. (Note that the sets $N_1(a)$ and $N_2(a)$ are different from $\bar{N}_1(a)$ and $\bar{N}_2(a)$.)

Recall that $R_a^1 = 0 \in J$. Thus $S(\lambda)$ is invariant under the move.

For the other choice of the orientation of C the value of $S(\lambda)$ changes by $R_a^2 = 0 \in J$. Hence, $S(\lambda)$ is invariant under all elementary isotopies, and we have proved Theorem 6.2.1. □

References

[1] F. Aicardi, *Invariant polynomial of framed knots in the solid torus and its application to wave fronts and Legendrian knots*, J. Knot Theory Ramifications **15** (1996), no. 6, 743–778.

[2] _____, Private communication.

[3] V. I. Arnold, *Topological invariants of plane curves and caustics*, Univ. Lecture Series, vol. 5, Amer. Math. Soc., Providence, RI, 1994.

[4] T. Fiedler, *A small state sum for knots*, Topology **30** (1993), no. 2, 281–294.

[5] M. Polyak, *On the Bennequin invariant of Legendrian curves and its quantization*, C. R. Acad. Sci. Paris Ser. I Math. **322** (1996), No. 1, 77–82.

[6] _____, *Shadows of Legendrian links and J-theory of curves*, Brieskorn's Jubilee Volume on Singularities (V. I. Arnold, G.-M. Greuel, J. H. M. Steenbrink, eds.), Birkhäuser, Boston. (to appear)

[7] A. Shumakovitch, *Shadow formulas for the Vassiliev invariant of degree two*, Topology **36** (1997), no. 2, 449–469.

[8] S. L. Tabachnikov, *Computation of the Bennequin invariant of a Legendrian curve from the geometry of its front*, Funktsional. Anal. i Prilozhen. **22** (1988). no. 3, 89–90; English transl., Functional Anal. Appl. **22** (1988). no. 3, 246–248.

[9] V. Tchernov, *First degree Vassiliev invariants of knots in $\mathbb{R}^1$- and S^1-fibrations*, preprint, Uppsala Univ., U.U.D.M. report 1996:11.

[10] V. G Turaev, *Shadow links and face models of statistical mechanics*, J. Diff. Geometry **36** (1992), 35–74.

D-MATH, ETH ZENTRUM, CH-8092, ZÜRICH, SWITZERLAND
E-mail address: `Chernov@math.ethz.ch`

A Unified Approach to the Four Vertex Theorems. I

Masaaki Umehara

Introduction

In 1932, Bose [**Bo**] established the following formula for a given noncircular simple closed convex plane curve γ

$$(0.1) \qquad s^\bullet - t^\bullet = 2,$$

where $s^\bullet$ is the number of enclosed osculating circles and $t^\bullet$ is the number of triple tangent enclosed circles in γ. Haupt [**Hu**] (1969) extended it to simple closed curves in the category of Ordnungscharakteristiken(=OCh) mit der Grundzahl $k = 3$, which is defined in Haupt and Künneth [**HK1**].

Roughly speaking, the formula for generic simple closed curves can be obtained by the following simple observations. Let γ be a generic C^∞-regular simple closed curve and D the domain bounded by γ. The *cut locus* K ($\subset D$) of γ is the closure of the set of points which have more than one minimizing line segment from γ. Then K has a structure of a tree and each boundary point corresponds to the center of an enclosed osculating circle; see Thom [**Tm1, Tm2**] and Wegner [**W1, W2**]. Moreover, it can be observed that the branch points of K are the centers of triple tangent enclosed circles. Hence $s^\bullet$ is the number of the boundary points of K and $t^\bullet$ is the total branching number at the branch points. Since K is contractible, the formula $s^\bullet - t^\bullet = 2$ follows immediately. This observation is justified for any C^1-regular simple closed curves with $s^\bullet < \infty$; see the last remark in Section 2. Recently, Shiohama and Tanaka [**ST**] investigated the structure of a cut locus of an arbitrary compact subset A of an Alexandrov surface X with curvature bounded below. Applying their results, one can easily deduce the formula $s - t = \chi(X \setminus A)$ under a suitable regularity of the boundary ∂A, where s (resp. t) is the number of single (resp. triple) tangent maximal inscribed circle in $X \setminus A$;[1] see Remark 2 of Corollary 3.4.

We give here a brief history of the four vertex theorems for simple closed curves. In 1909, Mukhopadhyaya [**Mu1**] proved for convex closed curves. A. Kneser [**KA**] (1912) proved it for simple closed curves. However a vertex (that is, a critical point of the curvature function) may not be a point where the osculating circle is

---

1991 *Mathematics Subject Classification.* Primary 53A04; Secondary 57M25.

[1]The author wishes to thank J. Ito for letting him know about the reference [**ST**] and for helpful discussions.

completely inside and outside the curve. The inequality $s^\bullet \geq 2$ for simple closed curves was proved by H. Kneser [**KH**] (1922–1923) who is the son of A. Kneser. The Bose formula and its generalization by Haupt [**Hu**] is a refinement of it. Jackson [**J1**] (1944) gave many other fundamental tools for the study of vertices on plane curves.

On the other hand, the four vertex theorem was extended to simple closed curves on closed convex surfaces by Mohrmann [**Mo**] (1917) without details and its complete proof was given by Barner and Flohr [**BF**] in 1958. To generalize the four vertex theorem for simple closed convex space curves (that is, curves lying on the boundary of their convex hulls) with nonvanishing curvature, Romero-Fuster [**R**] proved a Bose type formula

$$(0.2) \qquad s - t = 4$$

for convexly generic convex curves γ in $\mathbf{R}^3$, where s is the number of supporting osculating planes and t is the number of tritangent supporting planes. (Various approaches for the same problem are found in [**Bi1, Bi2**], [**RCN2**] and [**BR1, BR2**].) After that, Sedykh [**Sd5**] showed that (0.2) is true for generic simple closed strictly convex space curves. (Moreover, he gave a generalization of (0.2) for strictly convex manifolds M^k in the Euclidean space $\mathbf{R}^n$ ($k < n - 1$); see [**Sd3–Sd7**].) The four vertex theorem for simple closed convex space curves with nonvanishing curvature itself was proved by Sedykh in [**Sd2**] using a different approach. Recently, Kazarian [**Ka**] established some formulas similar to (0.1) representing the Chern–Euler class of a circle bundle over a Riemann surface in terms of global singularities of restrictions of a generic function to the fibers.

There are interesting connections between vertices and integral geometry (e.g., [**Bl2, Hy, Ba, Gu1, Gu2, He5**]) or contact geometry. The author was inspired by them, especially recent papers [**A1–A4, GMO, OT, Ta1–Ta3**] in which several variants of the four vertex theorem are given from the view of contact geometry or proved by using the technique of disconjugate operators on S^1.

The purpose of the paper is to give a unified treatment of the formulas (0.1) and (0.2). More precisely, we will introduce the notion of an "intrinsic circle system" as a certain multivalued function on the unit circle without referring to ambient spaces, which characterizes the cut loci of plane curves intrinsically and enables us to prove the formula (0.1) abstractly.[2] Consequently, (0.1) or (0.2) is proved under much weaker assumptions for the following three cases:

(1) Piecewise C^1-regular simple closed curves on the Euclidean or Minkowski plane;
(2) Piecewise C^1-regular simple closed curves on an embedded surface with positive Gaussian curvature in $\mathbf{R}^3$;
(3) Convex simple closed space curves in $\mathbf{R}^3$ with some additional conditions. As an application, the Sedykh's 4-vertex theorem [**Sd2**] is obtained; see Corollary 4.18.

The formula similar to (0.1) will be shown for these three cases (see Theorem 2.7 and Theorem 3.3). However, the formula similar to (0.2) requires piecewise C^2-regularity of curves (see Corollary 3.4 and Theorem 4.11). Haupt's proof partially

---

[2]E. Heil pointed out to the author that quite different intrinsic approaches to a four vertex theorem or other ordered geometry have been given in Valette [**V**] and Nöbeling [**N1, N2**]. Though the Bose type formula and intrinsic definition of inscribed circles are not treated in them, their approaches have independent interest, especially toward higher order geometry.

covers the cases (1)–(2) but not (3). (In his paper, the existence of osculating circles and the finiteness of vertices are assumed.) Here number $s^\bullet$ on curves defined for the cases (1)–(2) counts not only the number of inscribed circles but also singular points of curves. This gives a new interpretation for the existence of the unique inscribed circle in a triangle. (In this case, $s^\bullet = 3$ is the number of vertices and $t^\bullet = 1$ is the number of inscribed circles and they satisfy the relation $s^\bullet - t^\bullet = 2$ trivially.) Though it is not directly concerned with the Bose-type formulas, several generalization of four vertex theorems without differentiability have been investigated by [**LSC, J2, LSp, Sp1–Sp4**], etc. It should also be remarked that vertices for polygons have been studied by several authors (see [**Sa, W2, Sd8**]), but their definition of vertex is different from ours. In our setting, the vertices of polygons are considered as vertices of curves.

Finally, we remark here that this paper is prepared for the ensuing paper by Thorbergsson and Umehara [**TU**], in which we shall prove in the same axiomatic setting that *for any C^2-regular simple closed curve $\gamma : [a, b] \to \mathbf{R}^2$, there exist four points t_1, t_2, t_3, t_4 ($t_1 < t_2 < t_3 < t_4$) such that the osculating circles at t_1 and t_3 are enclosed in γ and the osculating circles at t_2 and t_4 enclose γ.* (Here the the order of the osculating circles is important. The corresponding version for convex simple closed space curves also holds.) The statement looks obvious at the first glance, but it is one of the deepest versions of the four vertex theorems, and provides many applications.

1. Intrinsic circle systems

We fix an oriented unit circle S^1. Let $\succ$ denote the order induced by the orientation on the complement of any interval in S^1. Any two distinct points $p, q \in S^1$ divide S^1 into two closed arcs $[p, q]$ and $[q, p]$, such that on $[p, q]$ we have $q \succ p$ and on $[q, p]$ we have $p \succ q$. We let (p, q) and (q, p) denote the corresponding open arcs. We also use the notation $p \succeq q$, which means $p = q$ or $p \succ q$. Let A be a subset of S^1 and $p \in A$. We denote by $Z_p(A)$ the connected component of A containing p.

DEFINITION 1.1. A family of nonempty closed subsets $F := (F_p)_{p \in S^1}$ of S^1 is called an *intrinsic circle system* on S^1 if it satisfies the following three conditions for any $p \in S^1$:
  (I1) $p \in F_p$ for each $p \in S^1$. If $q \in F_p$, then $F_p = F_q$.
  (I2) If $q \in S^1 \setminus F_p$, then $F_q \subset Z_q(S^1 \setminus F_p)$. Or equivalently, if $p' \in F_p$, $q' \in F_q$ and $q \succeq p' \succeq q' \succeq p(\succeq q)$, then $F_p = F_q$ holds.
  (I3) Let $(p_n)_{n \in \mathbf{N}}$ and $(q_n)_{n \in \mathbf{N}}$ be two sequences in S^1 such that $\lim_{n \to \infty} p_n = p$ and $\lim_{n \to \infty} q_n = q$, respectively. Suppose that $q_n \in F_{p_n}$ ($n = 1, 2, 3, \dots$). Then $q \in F_p$ holds.

REMARK. Let γ be a piecewise C^1-regular simple closed curve in $\mathbf{R}^2$. Let $C_p^\bullet$ be the maximal inscribed circle which is tangent to γ at p. Then $F_p := \gamma \cap C_p^\bullet$ satisfies the above three conditions; see Proposition 3.1. The definition of the intrinsic circle system characterizes the properties of maximal inscribed circles of a curve without referring to an ambient space, which enables us to generalize the Bose type formula to convex simple closed space curves. This is the reason for the terminology "intrinsic circle system". By (I1), F induces an equivalence relation. Later (see the last remark in Section 3), we will show that the quotient topological space S^1/F is homeomorphic to 1the cut locus K of γ. In this sense, the intrinsic

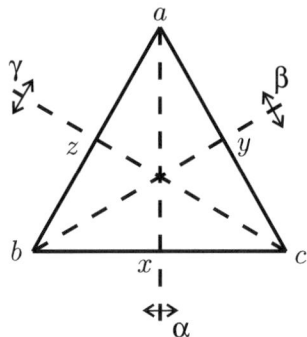

FIGURE 1.1

circle system can also be interpreted as an abstract characterization of the cut loci of plane curves. We give here two elementary examples.

Let $\gamma : x^2/a^2 + y^2/b^2 = 1$ $(a > b)$ be an ellipse in $\mathbf{R}^2$. Then the maximal inscribed circle $C_p^\bullet$ at each point $p = (x, y)$ on γ has two contact points at p and $\bar{p} = (x, -y)$ unless $y \neq 0$. So if we set $F_p := C_p^\bullet \cap \gamma$, then

$$F_p := \begin{cases} \{p, \bar{p}\} & \text{if } p \neq (\pm a, 0), \\ \{p\} & \text{if } p = (\pm a, 0). \end{cases}$$

One can easily verify that $(F_p)_{p \in \gamma}$ is an intrinsic circle system.

Another typical example is the triangle $\triangle abc$ as in Figure 1.1, which is invariant under the reflections α, β and γ. We consider the maximal circle $C_p^\bullet$ at each point on the triangle. Then $C_p^\bullet$ has two contact points to the triangle unless $p = a, b, c, x, y, z$, where $x := (a+b)/2$, $y := (b+c)/2$, and $z := (c+a)/2$. So if we set $F_p := C_p^\bullet \cap \gamma$, then

$$F_p := \begin{cases} \{p, \alpha(p)\} & \text{if } p \in \overline{ay} \cup \overline{az} \text{ and } p \neq a, y, z, \\ \{p, \beta(p)\} & \text{if } p \in \overline{bz} \cup \overline{bx} \text{ and } p \neq b, z, x, \\ \{p, \gamma(p)\} & \text{if } p \in \overline{cx} \cup \overline{cy} \text{ and } p \neq c, x, y, \\ \{p\} & \text{if } p = a, b, c, \\ \{x, y, z\} & \text{if } p = x, y, z. \end{cases}$$

One can also easily verify that $(F_p)_{p \in \triangle abc}$ is an intrinsic circle system. We will give further examples of intrinsic circle systems in Sections 3 and 4.

Let A be a subset of S^1. The number of connected components of A is called the *rank* of A and is denoted by $\mathrm{rank}(A)$. For a family of nonempty closed subsets $(F_p)_{p \in S^1}$, we set

$$\mathrm{rank}(p) := \mathrm{rank}(F_p).$$

The next lemma, which plays a fundamental role in this paper, is a generalization of the main argument in [**KH**].

LEMMA 1.1. *Let $(F_p)_{p \in S^1}$ be a family of nonempty closed subsets satisfying* (I2). *Let p, q be points on S^1 such that $q \in F_p$. Suppose that $(p, q) \not\subset F_p$. Then there exists a point $x \in (p, q)$ such that $\mathrm{rank}(x) = 1$.*

PROOF. If necessary, taking a subarc in (p,q), we may assume that $F_p \cap (p,q)$ is empty. We fix a metric $d(\,,\,)$ on S^1. Let x be the middle point of $[p,q]$ with respect to the distance function. If $\text{rank}(x) = 1$, the proof is finished. So we may assume that $\text{rank}(x) > 1$. By (I2), $F_x \subset (p,q)$. Since $S^1 \setminus F_x$ is an open subset, we can choose a connected component (p_1, q_1) of $S^1 \setminus F_x$ such that $(p_1, q_1) \subset [p,q]$. Then $p_1, q_1 \in F_x$. Instead of p and q, we apply the above argument for p_1 and q_1. Let x_1 be the middle point of the arc $[p_1, q_1]$. Then we find a subarc $[p_2, q_2]$ such that $p_2, q_2 \in F_{x_1}$ and $(p_2, q_2) \subset S^1 \setminus F_{x_1}$. Continuing this argument, we get a sequence of arcs $\{[p_n, q_n]\}_{n \in \mathbf{N}}$ such that

$$d(p_n, q_n) < \frac{1}{2} d(p_{n-1}, q_{n-1}).$$

Thus, there exists a point $y \in (p,q)$ such that

$$y = \lim_{n \to \infty} p_n = \lim_{n \to \infty} q_n.$$

If $\text{rank}(y) \neq 1$, then there exists an element $z \in F_y$ different from y. Then $z \notin (p_n, q_n) = Z_y(S^1 \setminus F_{p_n})$ for a sufficiently large n. This contradicts (I2). Thus we have $\text{rank}(y) = 1$. □

REMARK. Suppose that γ is a simple closed curve in $\mathbf{R}^2$. Let $C_p^\bullet$ be a maximal circle and $F_p = C_p^\bullet \cap \gamma$. Then the argument above was applied to show the existence of two distinct enclosed osculating circles in [**KH**]. In fact, using Lemma 1.1, one can easily get the existence of two distinct maximal circles $C_x^\bullet$ and $C_y^\bullet$ ($x \neq y$), which are tangent to γ with only one connected component. If the curve γ is C^2-differentiable, then $C_x^\bullet$ and $C_y^\bullet$ must coincide with the osculating circles at $x, y \in \gamma$, respectively; see Corollary A.5 in Appendix A. We remark that Thorbergsson [**Tr**] generalized this argument for a certain class of simple closed curves in any complete Riemannian 2-manifold.

From now on, we fix an intrinsic circle system $F = (F_p)_{p \in S^1}$ on S^1.

DEFINITION 1.2. $p \in S^1$ is called *regular* (resp. *weakly regular*) if $\text{rank}(p) = 2$ (resp. $2 \leq \text{rank}(p) \leq \infty$). A subarc I of S^1 whose elements are all regular (resp. weakly regular) is called a *regular arc* (resp. *weakly regular arc*).

The following corollary follows immediately from Lemma 1.1.

COROLLARY 1.2. *Let I be an open weakly regular arc. Then for each $p \in I$, the set*

$$Y_p := F_p \setminus Z_p(F_p)$$

is contained in $S^1 \setminus \overline{I}$. In particular, the closure $\overline{Y_p}$ lies in $S^1 \setminus I$.

DEFINITION 1.3. Let I be a closed arc on S^1 and A be a subset in I. Then the points $\sup_I(A)$ and $\inf_I(A)$ on I which are called the least upper bound and the greatest lower bound of A, are defined as the smallest (resp. greatest) points satisfying

$$\sup_I(A) \succeq x \quad (\text{for all } x \in A),$$
$$x \succeq \inf_I(A) \quad (\text{for all } x \in A).$$

DEFINITION 1.4. Let $I = (x_1, x_2)$ be a weakly regular arc. For any $p \in I$, we set
$$\mu_+(p) := \sup_{S^1 \setminus I}(Y_p), \qquad \mu_-(p) := \inf_{S^1 \setminus I}(Y_p),$$
where $Y_p := F_p \setminus Z_p(F_p)$. Moreover, we extend the definition of $\mu_\pm$ to the boundary of I as follows. If x_j ($j = 1, 2$) is weakly regular, we set

(1.1) $$\mu_+(x_j) := \sup_{S^1 \setminus I}(Y_{x_j}), \qquad \mu_-(x_j) := \inf_{S^1 \setminus I}(Y_{x_j}).$$

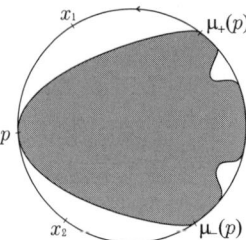

FIGURE 1.2

On the other hand, if x_j is of rank 1, we set

(1.2) $$\mu_+(x_j) := \sup_{S^1 \setminus I}(F_{x_j}), \qquad \mu_-(x_j) := \inf_{S^1 \setminus I}(F_{x_j}).$$

Obviously, we have $\mu_+(x_1) = x_1$ and $\mu_-(x_2) = x_2$. We will call $\mu_\pm$ antipodal maps. By definition, $\mu_\pm(\overline{I}) \subset S^1 \setminus I$ holds.

The following lemma is a simple consequence of the properties (I1) and (I2).

LEMMA 1.3. *Let $I = (x_1, x_2)$ be an open weakly regular arc and $p, q \in \overline{I}$ two points such that $p \succ q$ on $\overline{I}$. Then the following relations hold:*
$$\mu_+(q) \succeq \mu_+(p), \qquad \mu_-(q) \succeq \mu_-(p) \qquad (on\ S^1 \setminus I).$$
Moreover if $F_p \neq F_q$, then $\mu_-(q) \succ \mu_+(p)$ holds on $S^1 \setminus I$.

PROOF. We only prove the first relation. (The second relation is obtained if one reverses the orientation of S^1 and replaces p by q.) Suppose that $\mu_+(p) \succ \mu_+(q)$ on $S^1 \setminus I$. Then we have
$$q \succeq x_1 \succeq \mu_+(p) \succ \mu_+(q) \succeq x_2 \succeq p \qquad \text{on } [p, q].$$
By (I2), we have $F_p = F_q$. Since I contains no points of rank 1, Lemma 1.1 yields that $Z_p(F_p) = Z_q(F_q)$. Hence $\mu_+(p) = \mu_+(q)$ but it is a contradiction. Thus we have $\mu_+(q) \succeq \mu_+(p)$.

Next we suppose that $\mu_+(p) \succeq \mu_-(q)$ holds. Then we have
$$\mu_+(q) \succeq \mu_+(p) \succeq \mu_-(q)(\succeq p).$$
Since F_p and F_q are closed subsets of S^1, we have $\mu_\pm(q) \in F_q$ and $\mu_+(p) \in F_p$. Thus (I2) yields that $F_p = F_q$, which proves the second assertion. □

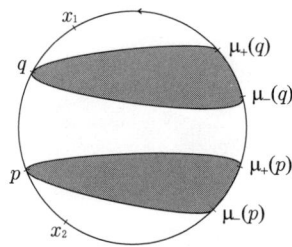

Figure 1.3

Theorem 1.4. *Let $I = (x_1, x_2)$ be an open weakly regular arc. Then the following two formulas hold*:

$$\lim_{x \to p-0} \mu_+(x) = \mu_+(p) + 0 \quad \text{(for } p \in (x_1, x_2]),$$
$$\lim_{x \to p+0} \mu_-(x) = \mu_-(p) - 0 \quad \text{(for } p \in [x_1, x_2)).$$

Proof. We shall prove the first formula; the second formula follows by the same arguments. We take a sequence $(p_n)_{n \in \mathbf{N}}$ such that $p_n \to p - 0$. Without loss of generality, we may assume that $p_{n+1} \succ p_n$ for any $n \in \mathbf{N}$. Since S^1 is compact, $(\mu_+(p_n))_{n \in \mathbf{N}}$ contains a convergent subsequence. Thus, we may assume that there exists a point $q \in S^1 \setminus I$ such that $\mu_+(p_n) \to q$. Since $p_{n+1} \succ p_n$, it holds that $\mu_+(p_n) \succeq \mu_+(p_{n+1})$ by Lemma 1.3. So we have $\mu_+(p_n) \to q + 0$. Then the proof of the formula follows from Lemma 1.5. □

Lemma 1.5. *Let $(p_n)_{n \in \mathbf{N}}$ be a sequence in an open weakly regular arc $I = (x_1, x_2)$ such that $p_n \to p - 0$, where $p \in (x_1, x_2]$. Suppose there exists $q \in S^1 \setminus I$ such that $\mu_+(p_n) \to q + 0$. Then $q = \mu_+(p)$.*

Proof. First, we consider the case that rank$(p) \geq 2$. By (I3), we have $p, q \in F_p$. Since $\mu_+(I) \subset S^1 \setminus I$, Lemma 1.3 yields

$$x_1 \succeq \mu_+(p_n) \succeq \mu_+(p) \quad \text{on } S^1 \setminus I.$$

By taking the limit $\mu_+(p_n) \to q$, we have

(1.3) $$x_1 \succeq q \succeq \mu_+(p) \quad \text{on } S^1 \setminus I.$$

In particular $p \neq q$. Suppose that $q \in Z_p(F_p)$. Then $[q, p] \subset F_p$. Since $\mu_+(p_n) \to q + 0$, we have $p_n \in Z_p(F_p)$ and thus $\mu_+(p_n) = \mu_+(p)$ for sufficiently large n. Hence we have $q = \mu_+(p)$. However, because $\mu_+(p) \notin Z_p(F_p)$, it is impossible. So we conclude that $q \in \overline{Y_p}$. Since $\mu_+(p) = \sup_{S^1 \setminus I}(Y_p)$, we have $q = \mu_+(p)$ by (1.3).

Next we consider the case that rank$(p) = 1$. This case happens only if $p = x_2$. By (I3) we have $q \in F_{x_2}$. If $F_{x_2} = \{x_2\}$, then we have $q = x_2 = \mu_+(x_2)$. So we may assume that F_{x_2} consists of more than two points. Then F_{x_2} is written as

$$F_{x_2} = [x_2, y] \quad (y \in S^1 \setminus \overline{I}).$$

Suppose that $q \in [x_2, y)$. Since $\mu_+(p_n) \to q + 0$, we have $\mu_+(p_n) \in (x_2, y)$. Then by (I1), $F_{p_n} = F_{\mu_+(p_n)} = F_{x_2}$. But this contradicts the fact rank$(p_n) \geq 2$. Hence we have $q = y = \mu_+(x_2)$ because of $q \in F_{x_2}$. □

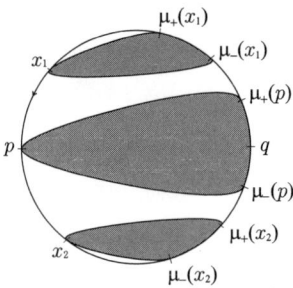

FIGURE 1.4

THEOREM 1.6. *Let $I = (x_1, x_2)$ be an open weakly regular arc. Then $\mu_-(x_1) \succ \mu_+(x_2)$ holds on the arc $S^1 \setminus I$. Moreover, for any $q \in (\mu_+(x_2), \mu_-(x_1))$, there exists a point $p \in I$ such that*

(1.4) $$\mu_+(p) \succeq q \succeq \mu_-(p) \qquad (on\ S^1 \setminus I).$$

PROOF. We divide the proof into three steps.

Step 1. First prove the relation $\mu_-(x_1) \succ \mu_+(x_2)$ on $S^1 \setminus I$. Suppose $F_{x_2} = F_{x_1}$. Then there is a point of rank 1 on I by Lemma 1.1. But this contradicts the weak regularity of I. So we have $F_{x_2} \neq F_{x_1}$. Then $\mu_-(x_1) \succ \mu_+(x_2)$ holds by Lemma 1.3.

Step 2. Next we prove that $p \in I$. We set

$$p := \inf_{\overline{I}}(B_q),$$

where B_q is the set defined by

$$B_q := \{x \in \overline{I}\,;\, q \succeq \mu_+(y) \text{ for all } x_2 \succeq y \succeq x\}.$$

For any $z \in I$ which is sufficiently close to x_2, it holds that $q \succ \mu_+(z)$ by Theorem 1.4. This implies $z \in B_q$, and thus B_q is nonempty. Moreover, the definition of p yields that

$$x_2 \succ z \succ p.$$

In particular $p \neq x_2$. Next we suppose that $p = x_1$. By Theorem 1.4, we have $\lim_{w \to x_1 + 0} \mu_-(w) = \mu_-(x_1)$. In particular, it holds that $\mu_-(w) \succ q$ on $S^1 \setminus I$ for $w \in I$ sufficiently close to x_1. On the other hand, the definition of p yields $q \succeq \mu_+(w)$ on $S^1 \setminus I$, since $x_2 \succeq w \succ p$. Thus we have $\mu_-(w) \succ \mu_+(w)$, a contradiction. Hence $p \in I$.

Step 3. By Theorem 1.4, we have

(1.5) $$\mu_+(p) = \lim_{x \to p-0} \mu_+(x) + 0,$$

(1.6) $$\mu_-(p) = \lim_{x \to p+0} \mu_-(x) - 0.$$

Suppose that $q \succ \mu_+(p)$ on $S^1 \setminus I$. Then (1.5) implies that there exists $u(\prec p)$ such that $q \succ \mu_+(x)$ for $x \in (u, p)$. On the other hand, $q \succeq \mu_+(y)$ holds for $y \succ p$. Thus $q \succeq \mu_+(z)$ holds for $z \in (u, x_2)$. Hence $u \in B_q$. But this contradicts that $p = \inf_{\overline{I}}(B_q)$. So we have $\mu_+(p) \succeq q$ on $S^1 \setminus I$. Next we suppose that $\mu_-(p) \succ q$ on $S^1 \setminus I$. Then (1.6) implies that there exists $v(\succ p)$ such that $\mu_-(v) \succ q$. Since

$\mu_+(v) \succeq \mu_-(v)$, we have $\mu_+(v) \succ q$. On the other hand, since $v \succ p$, we have $v \in B_q$. This contradicts the relation $\mu_+(v) \succ q$. So we have $q \in [\mu_-(p), \mu_+(p)]$. □

If the arc I is regular, the following stronger assertion follows immediately.

COROLLARY 1.7. *Let $I = (x_1, x_2)$ be a regular arc. Then $\mu_-(x_1) \succ \mu_+(x_2)$ holds on the arc $S^1 \setminus I$. Moreover, for any $q \in (\mu_+(x_2), \mu_-(x_1))$, there exists a point $p \in I$ such that $F_p = F_q$. In particular, $(\mu_+(x_2), \mu_-(x_1))$ is also a regular arc.*

2. A generalization of the Bose formula

In this section, we fix an intrinsic circle system $F = (F_p)_{p \in S^1}$. We define a relation $\sim$ on S^1 as follows. For $p, q \in S^1$, we denote $p \sim q$ if $F_p = F_q$. Then by (I1), this is an equivalence relation on S^1. We denote by S^1/F the quotient space of S^1 by the relation. The equivalence class containing $p \in S^1$ is denoted by $[p]$. Then $\mathrm{rank}([p]) := \mathrm{rank}(p)$ is well defined on S^1/F by (I1).

DEFINITION 2.1. We set
$$S(F) := \{[p] \in S^1/F \,;\, \mathrm{rank}([p]) = 1\},$$
$$T(F) := \{[p] \in S^1/F \,;\, \mathrm{rank}([p]) \geq 3\}.$$

The set $S(F)$ is called the *single tangent set* and $T(F)$ is called the *tritangent set*. Moreover, we set
$$s(F) := \text{the cardinality of the set } S(F),$$
$$t(F) := \sum_{[p] \in T(F)} (\mathrm{rank}(p) - 2).$$

DEFINITION 2.2. The single tangent set $S(F)$ is said to be *supported by a continuous function* $\tau : S^1 \to \mathbf{R}$ if for each $[p] \in S(F)$, F_p is a connected component of the zero set of τ.

In Section 3, we will give several examples of intrinsic circle systems whose single tangent sets are supported by continuous functions (see the Remark to Theorem 3.2).

LEMMA 2.1. *Suppose that $3 \leq s(F) < \infty$. Let $p, q \in S^1$ be points such that $\mathrm{rank}(p) = \mathrm{rank}(q) = 1$ and $F_p \neq F_q$. Then there is a point $x \in (p, q)$ such that $\mathrm{rank}(x) \geq 3$. Moreover, if the single tangent set $S(F)$ is supported by a continuous function τ, the assumption $s(F) < \infty$ is not needed.*

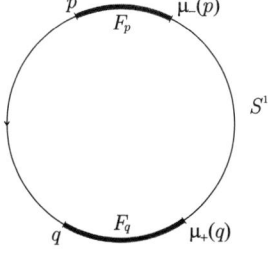

FIGURE 2.1

PROOF. Suppose that there are no points $x \in (p,q)$ such that $\operatorname{rank}(x) \geq 3$. Since $s(F) < \infty$, we may assume that there are no points of rank $= 1$ on (p,q). Then (p,q) is a regular arc. By Corollary 1.7, the open arc $(\mu_+(q), \mu_-(p))$ is also a regular arc. On the other hand, we have $\mu_+(p) = p$ and $\mu_-(q) = q$ by (1.2). So all the elements in $[\mu_-(p), p] \cup [q, \mu_+(q)]$ are of rank one. Since γ is expressed as

$$\gamma = (p,q) \cup [q, \mu_+(q)] \cup (\mu_+(q), \mu_-(p)) \cup [\mu_-(p), p],$$

there are no elements of rank(≥ 3) and $s(F) = 2$. But this contradicts $s(F) \geq 3$. This proves the first assertion. When $S(F)$ is supported by τ, we do not need the assumption $s(F) < \infty$. In fact, we get the same contradiction if we can take an open subarc (p', q') of (p,q) satisfying the following three properties:
1. $[p'], [q'] \in S(F)$;
2. $F_{p'} \neq F_{q'}$;
3. (p', q') is a regular arc.

If there are no such p' and q', then the subset

$$\{x \in (p,q)\,;\, [x] \in S(F)\}$$

is dense in (p,q). This implies that the function τ vanishes identically on (p,q) and thus $F_p = F_q$, which is a contradiction. □

THEOREM 2.2. *If $s(F) < \infty$, then $t(F) < \infty$. The converse is also true if the single tangent set $S(F)$ is supported by a continuous function $\tau : S^1 \to \mathbf{R}$.*

REMARK. In general, $t(F) < \infty$ does not imply $s(F) < \infty$. For example, we set $F_p := \{p\}$ ($p \in S^1$). Then $F = (F_p)_{p \in S^1}$ is an intrinsic circle, which satisfies $s(F) = \infty$ but $t(F) = 0$.

The theorem follows from the following three lemmas.

LEMMA 2.3. *If there exists a point $p \in S^1$ such that $\operatorname{rank}(p) = \infty$, then $s(F) = \infty$.*

PROOF. Let O be the open subset of S^1 given by $O := S^1 \setminus F_p$. We take a sequence $(x_n)_{n \in \mathbf{N}}$ in O such that x_i and x_j are in mutually different components of O unless $i = j$. Let (p_n, q_n) be the maximal open interval in O containing x_n. Then $p_n, q_n \in F_p$. By Lemma 1.1, there exists $[y_n] \in S(F)$ on (p_n, q_n). By (I2), we have $F_{y_n} \subset (p_n, q_n)$. Thus $(F_{y_n})_{n \in \mathbf{N}}$ are all disjoint. Hence $s(F) = \infty$. □

LEMMA 2.4. *Suppose that $S(F)$ is supported by a continuous function $\tau : S^1 \to \mathbf{R}$. If $s(F) = \infty$, then $t(F) = \infty$.*

PROOF. Let $n \geq 3$ be a fixed integer. We assume that $s(F) = \infty$. Then there exists a mutually distinct equivalence classes $[x_1], \cdots, [x_n] \in S(F)$. We set

$$M := \bigcup_{j=1}^{n} F_{x_j}.$$

Then $S^1 \setminus M$ is a union of disjoint open subsets $\{(p_j, q_j)\}_{j=1,\ldots,n}$. By Lemma 2.1, there exists a point y_j ($j = 1, \ldots, n$) on (p_j, q_j) such that $\operatorname{rank}(y_j) \geq 3$. This implies that $t(F) \geq n$. Since n is an arbitrary integer, we have $t(F) = \infty$. □

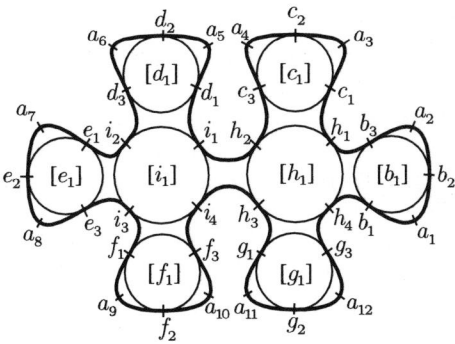

FIGURE 2.2

DEFINITION 2.3. Let Δ be a subset of $T(F)$ such that rank($[x]$) $< \infty$ for all $[x] \in \Delta$. Then for each $x \in \Delta$, $S^1 \setminus F_x$ is a union of disjoint open arcs $I_x^1, \ldots, I_x^{r_x}$, where $r_x := \operatorname{rank}(x)$. Such an open arc I_x^l is called a *primitive arc* with respect to the subset Δ if $I_x^l \cap F_y$ is empty for all $[y] \in \Delta$. If Δ is a finite subset and given by $\Delta := \{[x_1], \ldots, [x_n]\}$, then we set
(2.1)
$$N(\Delta) := \#\{I_{x_j}^{l_j} \; ; \; 1 \leq j \leq n, 1 \leq l_j \leq r_{x_j} \text{ and } I_{x_j}^{l_j} \cap F_{x_k} = \emptyset \text{ for all } k = 1, \ldots, n\},$$

that is $N(\Delta)$ is the total number of primitive arcs with respect to Δ among $\{I_{x_j}^{l_j}\}$.

We give an example which will be helpful for the arguments below.

EXAMPLE. Let γ be the smooth curve as shown in Figure 2.2 and $C_p^\bullet$ the maximal circle $C_p^\bullet$ at each point $p \in \gamma$. We set $F_p := C_p^\bullet \cap \gamma$. Then it can be easily checked that $(F_p)_{p \in \gamma}$ is an intrinsic circle system. The points $a_1, \ldots, a_{12}$ are of rank one and the points $b_1, b_2, b_3, c_1, c_2, c_3, d_1, d_2, d_3, e_1, e_2, e_3, f_1, f_2, f_3, g_1, g_2, g_3$ are of rank three. Finally, h_1, h_2, h_3, h_4 and i_1, i_2, i_3, i_4 are of rank four. Other points of γ are all regular. In this case,
$$S(F) := \{[a_1], \ldots, [a_{12}]\},$$
$$T(F) := \{[b_1], [c_1], [d_1], [e_1], [f_1], [g_1], [h_1], [i_1]\}.$$

For example, $\gamma \setminus F_{b_1}$ has three components $J_1 := \gamma_{(b_1, b_2)}$, $J_2 := \gamma_{(b_2, b_3)}$ and $J_3 := \gamma_{(b_3, b_1)}$. In this case J_1 and J_2 are primitive with respect to $T(F)$, but J_3 is not.

DEFINITION 2.4. Let Δ be a subset of $T(F)$. An element $[x] \in \Delta$ ($x \in S^1$) is called *totally primitive* if there exists a nonprimitive arc I_x^l such that all the other arcs
$$I_x^i (\subset S^1 \setminus F_x) \qquad (i \neq l, 1 \leq i \leq r_x)$$
are primitive with respect to Δ.

Let γ be the curve as in Figure 2.2 and F the intrinsic circle system defined in the example above. Then $[b_1], [c_1], [d_1], [e_1], [f_1], [g_1]$ are totally primitive with respect to $T(F)$, but $[h_1], [i_1]$ are not.

LEMMA 2.5. *If $s(F) < \infty$, then $t(F) < \infty$.*

PROOF. We prove this lemma by induction. If $t(F) \geq 1$, then by Lemma 1.1, we have $s(F) \geq 3$. Thus the lemma holds for $s(F) \leq 2$. So we assume that $t(F) < \infty$ holds if $s(F) < n$ ($n \geq 3$) and prove the assertion in the case $s(F) = n$. Although the set $T(F)$ need not be finite, the rank of each element is finite by Lemma 2.3.

Step 1. Suppose that there is a totally primitive element $[x] \in T(F)$ with respect to $T(F)$. Without loss of generality, we may assume that I_x^1 is not a primitive arc and the other arcs $I_x^2, \ldots, I_x^{r_x}$ are all primitive. We consider the quotient topological space $S^1/(S^1 \setminus I_x^1)$ and $\pi : S^1 \to S^1/(S^1 \setminus I_x^1)$ by the canonical projection. Then $S^1/(S^1 \setminus I_x^1)$ is also homeomorphic to S^1. For each $p \in S^1$, we set

$$\hat{F}_{\pi(p)} := \begin{cases} \pi(F_p) & \text{if } p \in I_x^1, \\ \pi(F_x) & \text{if } p \notin I_x^1. \end{cases}$$

Then it can be easily checked that $\hat{F}$ is an intrinsic circle system on $S^1/(S^1 \setminus I_x^1)$. By Lemma 1.1, each I_x^l contains at least one component of rank one points. On the other hand, I_x^l ($l \neq 1$) has at most one component of rank one points by Lemma 2.1, since it is a primitive arc with respect to $T(F)$. Thus each I_x^l ($l \neq 1$) contains exactly one component of rank one points. Thus, we have

(2.2) $$s(\hat{F}) = s(F) - (\text{rank}(x) - 2),$$
(2.3) $$t(\hat{F}) = t(F) - (\text{rank}(x) - 2).$$

Since $s(\hat{F}) < n$, we have $t(\hat{F}) < \infty$. So $t(F)$ is also finite by (2.3).

Step 2. Next we consider the case where there are no totally primitive elements in $T(F)$. Assume that $t(F) = \infty$. We take two mutually different elements $[x_1]$ and $[x_2]$ in $T(F)$. Without loss of generality, we may assume that $F_{x_1} \subset I_{x_2}^1$. Since $[x_2]$ is not totally primitive, there exists an element x_3 ($x_3 \neq x_1, x_2$) such that F_{x_3} is contained in $I_{x_2}^k$ for some $k \neq 1$. By (I2), F_{x_2} is contained in one of $(I_{x_3}^l)_{l=1,\ldots,r_{x_3}}$, here we may assume $F_{x_2} \subset I_{x_3}^1$. Then we also have $F_{x_1} \subset I_{x_3}^1$ by (I2). Since $[x_3]$ is not totally primitive, there exists an element x_4 ($x_4 \neq x_1, x_2, x_3$) such that F_{x_4} is contained in $I_{x_3}^k$ for some $k \neq 1$. Repeating this argument inductively, we can find a sequence $([x_n])_{n \in \mathbf{N}}$ such that

(2.4) $$F_{x_j} \subset I_{x_n}^1 \quad (j = 1, \ldots, n-1),$$
$$F_{x_{n+1}} \subset I_{x_n}^k \quad \text{for some } k \ (1 < k \leq r_{x_n}).$$

By Lemma 1.1 and (2.1), we have

(2.5) $$s(F) \geq N(\{[x_1], \ldots, [x_k]\}).$$

On the other hand, by (2.4), we have

(2.6) $$N(\{[x_1], \ldots, [x_k], [x_{k+1}]\}) = N(\{[x_1], \ldots, [x_k]\}) + (\text{rank}([x_{k+1}]) - 2).$$

Thus $N(\{[x_{i_1}], \ldots, [x_{i_k}]\}) \to \infty$ if $k \to \infty$. Hence $s(F) = \infty$, a contradiction. So $t(F)$ is finite. □

COROLLARY 2.6. *Suppose that $s(F) < \infty$. Then the set of all regular (resp. weakly regular) points is an open subset of S^1.*

PROOF. Since $s(F) < \infty$, $t(F) < \infty$ holds by Lemma 2.5. Thus there exists finitely many points $p_1, \ldots, p_n$ such that $S^1 \setminus (F_{p_1} \cup \cdots \cup F_{p_n})$ is the set of all regular (resp. weakly regular) points. Since each F_{p_j} $(j = 1, \ldots, n)$ is closed, the set is an open subset. □

We now prove Theorem 2.7, which is a generalization of the Bose formula.

THEOREM 2.7. *Let $F := (F_p)_{p \in S^1}$ be an intrinsic circle system. Suppose that $s(F) < \infty$ and there exists a point $p \in S^1$ such that $[p] \notin S(F)$. Then $t(F) < \infty$ and*
$$s(F) - t(F) = 2$$
holds.

PROOF. Assume that $s(F) < \infty$. If $s(F) = 0$, then this contradicts Lemma 1.1, since there exists a point such that $[p] \neq S(F)$. If $s(F) = 1$, we can conclude that $[p] \in S(F)$ for all $p \in S^1$ by Lemma 1.1. Next we suppose that $s(F) = 2$ and $t(F) \geq 1$. Then by (2.5), we have
$$2 = s(F) \geq N(\{[x_1]\}) = \text{rank}(x_1) \geq 3$$
for any $x_1 \in T(F)$, which yields a contradiction. Thus $t(F) = 0$. So we may assume $s(F) \geq 3$. Then Lemma 2.1 implies $T(F)$ is a nonempty set. Let $[x_1], \ldots, [x_{t(F)}]$ be all of the elements of $T(F)$. To we complete the proof of the theorem, we need the following lemma.

LEMMA 2.8. *Suppose that $3 \leq s(F) < \infty$ There exists an integer j ($1 \leq j \leq s(F)$) such that $[x_j]$ is totally primitive with respect to $T(F)$.*

PROOF. If $[x_1]$ is totally primitive, the proof is finished. If not, we fix a non-primitive arc $I_{x_1}^{l_1}$. Then by (I2), we may suppose that F_{x_2} lies in $I_{x_1}^{l_1}$. (If not, we can exchange $[x_2]$ for a suitable $[x_k]$ $(k > 2)$.) If $[x_2]$ is totally primitive, the proof is finished. If not, we fix a nonprimitive arc $I_{x_2}^{l_2}$ contained in $I_{x_1}^{l_1}$. Then we may assume that F_{x_3} lies in $I_{x_2}^{l_2}$. (If not, we can exchange $[x_3]$ for a suitable $[x_k]$ $(k > 3)$.) Continuing this argument, we find a totally primitive $[x_j]$ since $t(F)$ is finite. □

PROOF OF THEOREM 2.7 (continued). We will prove the formula by induction on the number $s(F)$. We have already seen that the formula is true whenever $s(F) \leq 2$. So we assume that the formula holds if $s(F) < n$ $(n \geq 3)$ and prove the assertion in the case $s(F) = n$. By Lemma 2.8, there is a totally primitive element $[x]$ in $T(F)$. Then as shown in the proof of Lemma 2.5, the induced intrinsic circle system $\hat{F}$ on $S^1/(S^1 \setminus I_x^1)$ satisfies (2.2) and (2.3). Since $s(\hat{F}) < n$, we have $s(\hat{F}) - t(\hat{F}) = 2$, which yields the formula $s(F) - t(F) = 2$. □

REMARK. Let $\gamma : S^1 \to \mathbf{R}^2$ be a C^1-regular simple closed curve with positive orientation and $C_p^\bullet$ a maximal inscribed circle of γ at $p \in \gamma$. Then $F_p := \gamma \cap C_p^\bullet$ is a typical example of an intrinsic circle system (see Section 3 and Proposition A.1 in Appendix A). We define a map $\Phi : S^1 \to \mathbf{R}^2$ by $\Phi(p) = c_p$, where c_p is the center of the circle $C_p^\bullet$. Suppose that $s(F) < \infty$. As will be seen in Appendix B, the map Φ is continuous by the C^1-regularity of the curve. Then Φ induces an injective continuous map $\varphi : S^1/F \to \mathbf{R}^2$. Since S^1/F is compact, S^1/F is homeomorphic to $\Phi(S^1)$. We denote by D the domain bounded by γ. Let $K_0 (\subset D)$

be the set of points which have more than one minimizing normal geodesic from γ. The cut locus K of γ defined in the Introduction is the closure of K_0. Then obviously $K_0 \subset \Phi(S^1)$. Since $\Phi(S^1)$ is closed, we have $K \subset \Phi(S^1)$. On the other hand, we set
$$R := \{p \in \gamma : F_p = \{p\}\}.$$
Since $s(F) < \infty$, R is a finite subset in S^1. Moreover $\Phi(S^1 \setminus R) \subset K_0$ by the definition of K_0. By the continuity of Φ, we have $\Phi(S^1) \subset K$, which implies $\Phi(S^1) = K$. Thus S^1/F is homeomorphic to K. So we can identify S^1/F with K of the cut locus of γ. We have thus seen that the concept of the intrinsic circle system characterizes the cut locus of a simple closed curve abstractly. Since S^1/F has the structure of tree by Theorem 2.7, the observation in the introduction is justified for any C^1-regular simple closed curves with $s(F) < \infty$.

3. Application to plane curves

As an application of the results of Sections 1 and 2, we give a general framework for discussing the number of vertices on a curve, which is similar to (but more elementary than) that of Och mit Grundzahl $k = 3$ (see Haupt and Künneth [**IIK1–HK3**]).

Let X be a topological space homeomorphic to S^2 with fixed orientation. We denote by $J(X)$ the set of all oriented simple closed curves. Each $\gamma \in J(X)$ separates X by two domains D_1 and D_2. We assume that D_1 is the left-hand domain bounded by γ and we set

(3.1) $$D^\bullet(\gamma) := \overline{D_1}, \qquad D^\circ(\gamma) = \overline{D_2}.$$

We call $D^\bullet(\gamma)$ the *internal domain* and $D^\circ(\gamma)$ the *external domain*.

For the sake of simplicity, we use the following notation. Let $\gamma \in J(X)$ and p, q be different points on γ. Then we denote
$$\gamma|_{[p,q]} := \{x \in \gamma \,;\, q \succeq x \succeq p\}, \quad \gamma|_{(p,q)} := \{x \in \gamma \,;\, q \succ x \succ p\}.$$

DEFINITION 3.1. Let $\gamma \in J(X)$. If a sequence $(\gamma_n)_{n \in \mathbf{N}}$ satisfies the following two properties, we write $\gamma_n \to \gamma$.
1. Let $(p_n)_{n \in \mathbf{N}}$ be a sequence in X converging to $p \in X$. If $p_n \in D^\bullet(\gamma_n)$ for all sufficiently large n, then $p \in D^\bullet(\gamma)$.
2. Let $(p_n)_{n \in \mathbf{N}}$ be a sequence in X converging to $p \in X$. If $p_n \in D^\circ(\gamma_n)$ for all sufficiently large n, then $p \in D^\circ(\gamma)$.

REMARK. This convergence properly coincides with the compact open topology on $J(X)$ or is equivalently compatible with the uniform distance on $J(X)$ induced from an arbitrary distance function $d(\,,\,)$ on X (see Greenberg and Harper [**GH**, §7]; here $d(\,,\,)$ is assumed to be compatible with the topology of X). In fact, assume $\gamma_n \to \gamma$. Let $d(\,,\,)$ be the uniform distance on $J(X)$ induced by a distance function of X. Suppose that $d(\gamma_n, \gamma) \not\to 0$. Then there is a sequence $(p_n)_{n \in \mathbf{N}}$ such that $p_n \in \gamma_n$ and $d(p_n, \gamma) > \varepsilon > 0$. Since X is compact, there is a subsequence $(p_{j_n})_{n \in \mathbf{N}}$ converging to q. Then $q \in \gamma$ since $\gamma_n \to \gamma$. But this contradicts the fact $d(p_{j_n}, \gamma) > \varepsilon > 0$.

On the other hand, assume that $(\gamma_n)_{n \in \mathbf{N}}$ converges to γ with respect to the compact open topology. Let $d(\,,\,)$ be the canonical distance function on $X = S^2(1)$. Then we have $d(\gamma, \gamma_n) \to 0$. Let $(p_n)_{n \in \mathbf{N}}$ be a sequence in X converging to $p \in X$.

Suppose that $p_n \in D^\bullet(\gamma)$ and $p \in D^\circ(\gamma) \setminus \gamma$. Let $\overline{p_n p}$ be the geodesic segment in X. Then there exists a point q_n on $\gamma \cap \overline{p_n p}$. Then we have

$$d(p_n, \gamma) \leq d(p_n, q_n) \leq d(p_n, p).$$

Since $d(p_n, p) \to 0$, we have $d(p, \gamma) = 0$, which is a contradiction. Hence $\gamma_n \to \gamma$ in the sense of the above definition.

Let $q \in X$ be a point. We interpret q as a collapse of simple closed curves. We consider two orientations of q. The point q is said to be *positively oriented* if we regard it as

(3.2) $$D^\bullet(q) = q, \qquad D^\circ(q) = X \setminus \{q\},$$

and q is said to be *negatively oriented* if we regard it as

(3.3) $$D^\circ(q) = q, \qquad D^\bullet(q) = X \setminus \{q\}.$$

In the first case, we denote q by $q^\bullet$ and in the second case q°. Then the notations $\gamma_n \to q^\bullet$ or $\gamma_n \to q^\circ$ make sense. We denote by $\partial \Gamma$ the set of all oriented points on X, that is

(3.4) $$\partial \Gamma := \{q^\bullet, q^\circ\}_{q \in X}.$$

Now we define a notion "circle system" which will produce typical examples of intrinsic circle system defined in Section 1.

DEFINITION 3.2 (Circle system). A subset Γ of $J(X)$ is called a *circle system* if the following two conditions are satisfied (we set $\hat{\Gamma} = \Gamma \cup \partial \Gamma$):
 (S1) Any distinct curves $C, C' \in \Gamma$ have at most two common points. Moreover, if $D^\bullet(C) \subset D^\bullet(C')$, then they have at most one common point.
 (S2) Let $(p_n)_{n \in \mathbf{N}}$ be a sequence in X which converges to a point $p \in X$. Let $(C_n)_{n \in \mathbf{N}}$ be a sequence in $\hat{\Gamma}$ such that $C_n \ni p_n$. Then $(C_n)_{n \in \mathbf{N}}$ has a subsequence converging to an element in $\hat{\Gamma}$.
 (S3) Let p be a point on X and A a subset of Γ such that any two elements of A have only one common point p. Then there exist $C_A^\bullet, C_A^\circ \in \hat{\Gamma}$ such that

$$D^\bullet(C_A^\bullet) \subset D^\bullet(C), \qquad D^\bullet(C) \subset D^\bullet(C_A^\circ)$$

holds for all $C \in A$.

Each element of $\hat{\Gamma}$ is called a *circle*. The third condition (S3) is needed only for the proof of Proposition A.1 in Appendix A. The followings are examples of circle systems.

EXAMPLE 1 (The Möbius plane). Let $X_1 = \mathbf{R}^2 \cup \{\infty\}$ and Γ_1 be the set of oriented circles and lines. (Since the circles are invariant under the Möbius transformations, it is natural to compactify the Euclidean plane by attaching the infinity.) Then the pair (X_1, Γ_1) satisfies the conditions of a circle system. Via the stereographic projection from the north pole of the unit sphere $S^2(1)$ in $\mathbf{R}^3$, this model is equivalent to the following one:

$X_1 := S^2(1)$,
$\hat{\Gamma}_1 :=$ the oriented intersections between $S^2(1)$ and planes.

EXAMPLE 2 (Closed strictly convex surfaces). As a canonical generalization of Example 1, the following model also satisfies the above conditions:

$X_2 :=$ a closed C^2-embedded surface in $\mathbf{R}^3$ with positive Gaussian curvature,

$\hat{\Gamma}_2 :=$ the oriented intersections between X_2 and planes.

EXAMPLE 3 (The Minkowski plane). Let $\mathcal{I}$ be a fixed C^2-regular simple closed curve with positive curvature in $\mathbf{R}^2$ enclosing the origin. We call $\mathcal{I}$ an *indicatrix*. The *Minkowski distance* $d_{\mathcal{I}}(x,y)$ associated with the indicatrix $\mathcal{I}$ is defined by

$$d_{\mathcal{I}}(x,y) := \inf\{t > 0 \,;\, \frac{1}{t}(y-x) \in D^{\bullet}(\mathcal{I})\}.$$

It satisfies the usual properties of a distance function except for the symmetry property $d_{\mathcal{I}}(x,y) = d_{\mathcal{I}}(y,x)$. The Minkowski geometry is the geometry with respect to this distance function. The indicatrix $\mathcal{I}$ is characterized as the level set

$$\mathcal{I} = \{x \in \mathbf{R}^2 \,;\, d_{\mathcal{I}}(0,x) = 1\}.$$

When $\mathcal{I}$ is the unit circle, $d_{\mathcal{I}}$ coincides with the usual Euclidean distance. A Minkowski circle C is the image of the indicatrix $\mathcal{I}$ under a translation and a homothety with a positive ratio. The point in C corresponding to the origin in $D^{\bullet}(\mathcal{I})$ is called the *center* of C, and the magnification of C with respect to $\mathcal{I}$ is called the *Minkowski radius*. We set $X_3 := \mathbf{R}^2 \cup \{\infty\}$ as a stereographic image of the unit sphere. Let Γ_3 be the set of Minkowski circles and straight lines. Then (X_3, Γ_3) satisfies condition (S1) obviously. Condition (S3) is also easily checked. (Two different lines meet only at infinity if they are parallel. So condition (S3) at $p = \infty$ is also easily checked.) Condition (S2) is verified as follows.

Case 1. First we consider the case $p \neq \infty$. Let $(p_n)_{n \in \mathbf{R}^2}$ be a sequence converging to $p \neq \infty$ and $(C_n)_{n \in \mathbf{N}}$ a sequence in Γ_3 such that $p_n \in C_n$. If $(C_n)_{n \in \mathbf{N}}$ contains either infinitely many straight lines or infinitely many oriented points, then such a subsequence of lines has a subsequence converging a line through p obviously. So we may assume that $(C_n)_{n \in \mathbf{N}}$ contains neither straight lines nor oriented points. If necessary by taking a subsequence, we may assume that $(C_n)_{n \in \mathbf{N}}$ have the same orientation. Moreover, by reversing the orientation of $(C_n)_{n \in \mathbf{N}}$ simultaneously, we may assume that $(C_n)_{n \in \mathbf{N}}$ are all positively oriented; that is, $\left(D^{\bullet}(C_n)\right)_{n \in \mathbf{N}}$ are all bounded in $\mathbf{R}^2$. Let r_n be the Minkowski radius of C_n. If $(r_n)_{n \in \mathbf{N}}$ is bounded, (S2) is easily checked. So we may assume that $r_n \to \infty$. Let L_n be the line which is tangent to C_n at p_n. Then $(L_n)_{n \in \mathbf{N}}$ contains a subsequence converging to a line L passing through p. So we may assume that $(L_n)_{n \in \mathbf{N}}$ converges to L. Let M be the maximum of the Euclidean curvature function of $\mathcal{I}$. Then any Euclidean circle of radius $1/M$ which is tangent to $\mathcal{I}$ from the left is contained in $D^{\bullet}(\mathcal{I})$. Let $E_n(r_n)$ be the positively oriented Euclidean circle with radius r_n which is tangent to C_n at p_n from the left. Then both of $E_n(r_n)$ and $E_n(r_n/M)$ converge to the line L since $r_n \to \infty$. Since

$$D^{\bullet}(E_n(r_n/M)) \subset D^{\bullet}(C_n) \subset D^{\bullet}(E_n(r_n)),$$

we can conclude $C_n \to L$.

Case 2. Next we consider the case $p = \infty$. Let $(p_n)_{n \in \mathbf{R}^2}$ be a sequence converging to ∞ and $(C_n)_{n \in \mathbf{N}}$ a sequence in Γ such that $p_n \in C_n$. Without loss of generality, we may assume that C_n is positively oriented. Suppose that $q_n \to \infty$ holds for any sequence $(q_n)_{n \in \mathbf{N}}$ such that $q_n \in C_n$. Let $x_n \in C_n$ be the point which attains the minimum of the distance function of C_n from the origin. Then we have $x_n \to \infty$, which implies $C_n \to \infty°$. Thus we may assume that there exists a sequence $(q_n)_{n \in \mathbf{N}}$ such that $q_n \in C_n$ and $q_n \to q \neq \infty$. Then it reduces to Case 1.

Hence (X_3, Γ_3) satisfies the conditions of a circle system. The vertices on curves in the Minkowski plane have been investigated by many geometers; see [**Su, Hu, Gu1**]. Here the vertex is regarded as a point where the osculating circle has the third order tangency with the curve. Later in this section, we define clean maximal (resp. minimal) vertices. Maximal (resp. minimal) vertices are defined in Appendix A; see Definition 3.6. If a closed curve in the Minkowski plane is C^3-regular, these vertices are all vertices in this sense. For the relationship between Minkowski vertices and contact geometry; see Tabachnikov [**Ta2**].

EXAMPLE 4. Let $\varphi : X_i \to X_i$ be a homeomorphism of X_i. Then $(X_i, \varphi(\Gamma_i))$, $i = 1, 2, 3$, also satisfies conditions (S1)–(S3).

DEFINITION 3.3. Let $\gamma \in J(X)$. For each $p \in \gamma$, we set

$$\mathcal{A}_p^\bullet := \{C \in \hat{\Gamma} \,;\, C \ni p,\, D^\bullet(C) \subset D^\bullet(\gamma)\},$$
$$\mathcal{A}_p^\circ := \{C \in \hat{\Gamma} \,;\, C \ni p,\, D^\circ(C) \subset D^\circ(\gamma)\}.$$

A circle $C \in \mathcal{A}_p^\bullet$ (resp. $C \in \mathcal{A}_p^\circ$) is called a $\bullet$-*maximal circle at p* (resp. $\circ$-*maximal circle at p*) if any circle $C' \in \mathcal{A}_p^\bullet$ (resp. $C' \in \mathcal{A}_p^\circ$) satisfying $D^\bullet(C) \subset D^\bullet(C')$ (resp. $D^\circ(C) \subset D^\circ(C')$) coincides with C.

DEFINITION 3.4. A point p on γ is called $\bullet$-*admissible* (resp. $\circ$-*admissible*) if there exists a unique $\bullet$-maximal (resp. $\circ$-maximal) circle at p, which is denoted by $C_p^\bullet$ (resp. C_p°). A curve $\gamma \in J(X)$ is called $\bullet$-*admissible* (resp. $\circ$-*admissible*) if all points on it are $\bullet$-admissible (resp. $\circ$-admissible).

If $(X, \Gamma) = (X_i, \Gamma_i)$ ($i = 1, 2, 3$), then every piecewise C^1-regular curve[3] in $J(X)$ whose internal angles with respect to $D^\bullet(\gamma)$ are less than or equal to π is $\bullet$-admissible (see Proposition A.1 in Appendix A). For example, the triangle as in Figure 1.1 with positive orientation is $\bullet$-admissible, but not $\circ$-admissible, because the three vertices of the triangle are not $\circ$-admissible points.

DEFINITION 3.5. Let γ be a $\bullet$-admissible (resp. $\circ$-admissible) curve. We set

$$F_p^\bullet := C_p^\bullet \cap \gamma \qquad (\text{resp. } F_p^\circ := C_p^\circ \cap \gamma).$$

PROPOSITION 3.1. *Let $\gamma \in J(X)$ be a $\bullet$-admissible (resp. $\circ$-admissible) curve. Then $F^\bullet := (F_p^\bullet)_{p \in \gamma}$ (resp. $F^\circ := (F_p^\circ)_{p \in \gamma}$) is an intrinsic circle system on $S^1 = \gamma$.*

PROOF. The condition (I1) obviously follows from the uniqueness of $C_p^\bullet$. The condition (I2) follows from (S1). Finally, we prove that $F^\bullet$ satisfies (I3). Let $(p_n)_{n \in \mathbf{N}}$ and $(q_n)_{n \in \mathbf{N}}$ be two sequences in S^1 such that $\lim_{n \to \infty} p_n = p$, $\lim_{n \to \infty} q_n = q$ and $q_n \in F_{p_n}^\bullet$. By (S2), $C_{p_n}^\bullet$ contains a convergent subsequence. So we may assume that $C_{p_n}^\bullet \to C \in \hat{\Gamma}$. If $p = q$, then $q \in F_p$ is obvious. So we may assume

---

[3] A curve $\sigma : [a, b] \to S^2$ is called a piecewise C^k-regular ($k \geq 1$) curve if there exists a finite partition $a = t_0 < t_1 \cdots < t_n = b$ of $[a, b]$ such that σ is a C^k-regular arc on each $[t_i, t_{i+1}]$ for $i = 0, \ldots, n-1$.

$p \ne q$. Since $C_{p_n}^\bullet \to C$ and $D^\bullet(C_{p_n}^\bullet) \subset D^\bullet(\gamma)$, we have $D^\bullet(C) \subset D^\bullet(\gamma)$. On the other hand, since $p_n, q_n \in D^\bullet(C_{p_n}^\bullet) \cap D^\circ(C_{p_n}^\bullet)$, we have

$$p, q \in D^\bullet(C) \cap D^\circ(C) = C.$$

Since $p \ne q$, we have $C_p^\bullet = C$ by the maximality of $C_p^\bullet$. □

Let $\gamma \in J(X)$ be a •-admissible (resp. ∘-admissible) curve. Then we set

$$\operatorname{rank}^\bullet(p) := \operatorname{rank}(F_p^\bullet) \quad (\text{resp. } \operatorname{rank}^\circ(p) := \operatorname{rank}(F_p^\circ)).$$

Namely, $\operatorname{rank}^\bullet(p)$ is the number of connected components of $C_p^\bullet \cap \gamma$.

DEFINITION 3.6. Let γ be a •-admissible (resp. ∘-admissible) curve. A point p on γ is called a *clean maximal vertex* (resp. *clean minimal vertex*) if $\operatorname{rank}^\bullet(p) = 1$ (resp. $\operatorname{rank}^\circ(p) = 1$). A point p on γ is called *•-regular* (resp. *∘-regular*) if $\operatorname{rank}^\bullet(p) = 2$ (resp. $\operatorname{rank}^\circ(p) = 2$). A point p on γ is called *weakly •-regular* (resp. *weakly ∘-regular*) if $2 \le \operatorname{rank}^\bullet(p) \le \infty$ (resp. $2 \le \operatorname{rank}^\circ(p) \le \infty$). An open arc I of γ is called *•-regular* (resp. *weakly •-regular*) if all points on I are •-regular (resp. weakly •-regular). Similarly, a ∘-*regular* (resp. *weakly ∘-regular*) arc is also defined.

By definition, I is (weakly) •-regular (resp. ∘-regular) if it is a (weakly) regular arc with respect to the intrinsic circle system $F^\bullet$ (resp. F°) (see Definition 1.2).

Let $S^\bullet(\gamma)$ (resp. $S^\circ(\gamma)$) be the set of all •-maximal (resp. ∘-maximal) circles C on γ such that $C \cap \gamma$ is connected. Moreover, let $T^\bullet(\gamma)$ (resp. $T^\circ(\gamma)$) be the set of all •-maximal (resp. ∘-maximal) circles C on γ such that $C \cap \gamma$ has at least three connected components. For an arbitrary set M, we denote its cardinality by $\#M$. We set

$$s^\bullet(\gamma) := \#\{S^\bullet(\gamma)\} \quad (\text{resp. } s^\circ(\gamma) := \#\{S^\circ(\gamma)\}),$$
$$t^\bullet(\gamma) := \#\{T^\bullet(\gamma)\} \quad (\text{resp. } t^\circ(\gamma) := \#\{T^\circ(\gamma)\}).$$

We give the following generalization of Bose's formula (0.1).

THEOREM 3.2. *Let γ be a •-admissible (resp. ∘-admissible) simple closed curve, which is not a circle. Suppose that $s^\bullet(\gamma) < \infty$ (resp. $s^\circ(\gamma) < \infty$). Then $t^\bullet(\gamma) < \infty$ (resp. $t^\circ(\gamma) < \infty$) and*

$$s^\bullet(\gamma) - t^\bullet(\gamma) = 2 \quad (\text{resp. } s^\circ(\gamma) - t^\circ(\gamma) = 2).$$

Moreover, $s^\bullet(\gamma)$ (resp. $s^\circ(\gamma)$) is equal to the the number of connected components of the set of clean maximal (resp. clean minimal) vertices and

$$t^\bullet(\gamma) = \sum_{[p] \in T^\bullet(\gamma)} (\operatorname{rank}^\bullet(p) - 2) \quad \left(\text{resp. } t^\circ(\gamma) = \sum_{[p] \in T^\circ(\gamma)} (\operatorname{rank}^\circ(p) - 2)\right).$$

PROOF. Since γ is •-admissible (resp. ∘-admissible), it induces an intrinsic circle system $F^\bullet$ (resp. F°) by Proposition 3.1. Obviously, we have that

$$s^\bullet(\gamma) = s(F^\bullet) \quad (\text{resp. } s^\circ(\gamma) = s(F^\circ)),$$
$$t^\bullet(\gamma) = t(F^\bullet) \quad (\text{resp. } t^\circ(\gamma) = t(F^\circ)).$$

Then the theorem follows from Theorem 2.7 and Definition 2.1. □

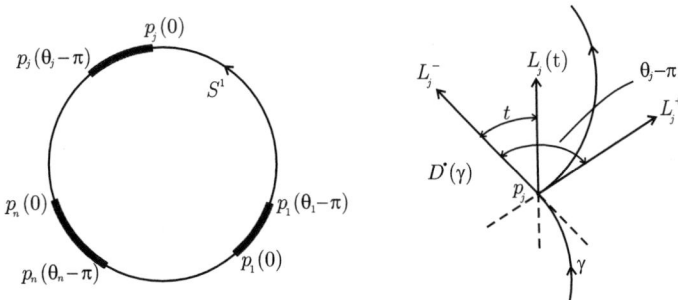

FIGURE 3.1

REMARK. If $(X, \Gamma) = (X_1, \Gamma_1)$ as in Example 1 and γ is a C^3-regular curve, then $S^{\bullet}(\gamma)$ (resp. $S^{\circ}(\gamma)$) is supported by the derivative of the curvature function (see Definition 2.2). Similarly, if $(X, \Gamma) = (X_2, \Gamma_2)$ as in Example 2 and γ is a C^3-regular curve as a space curve, then $S^{\bullet}(\gamma)$ (resp. $S^{\circ}(\gamma)$) is supported by the torsion function of γ as a space curve in $\mathbf{R}^3$. Thus in these two cases, $t^{\bullet}(\gamma) < \infty$ (resp. $t^{\circ}(\gamma) < \infty$) is equivalent to the condition $s^{\bullet}(\gamma) < \infty$ (resp. $s^{\circ}(\gamma) < \infty$).

Suppose X has C^2-differentiable structure and Γ is a circle system satisfying the additional condition (S4) in Appendix A. By Proposition A.1 in Appendix A, a piecewise C^1-regular simple closed curve γ on X is $\bullet$-admissible if its internal angle with respect to the domain $D^{\bullet}(\gamma)$ is less than or equal to π everywhere. However the above formula holds for any piecewise C^1-regular curves as follows.

THEOREM 3.3. *Let X be a C^2-differentiable sphere and Γ a C^2-differentiable circle system on X satisfying the additional condition (S4) in Appendix A. Let γ be a piecewise C^1-regular curve on X. Suppose that γ is not a circle and $s^{\bullet}(\gamma)$ (resp. $s^{\circ}(\gamma)$) is finite. Then the following identity holds*

$$s^{\bullet}(\gamma) - t^{\bullet}(\gamma) = 2 \qquad (\text{resp. } s^{\circ}(\gamma) - t^{\circ}(\gamma) = 2).$$

PROOF. If all internal angles with respect to the domain $D^{\bullet}(\gamma)$ are less than or equal to π, then γ is $\bullet$-admissible by Proposition A.1. So Theorem 3.2 can be directly applied. We consider the remaining case. Let p_j ($j = 1, \ldots, n$) be points on γ whose internal angles θ_j ($j = 1, \ldots, n$) with respect to $D^{\bullet}(\gamma)$ are greater than π. Take $2n$-points on S^1 such that

$$p_1(0) \prec p_1(\theta_1 - \pi) \prec \cdots \prec p_n(0) \prec p_n(\theta_n - \pi) \, (\prec p_1(0)).$$

Let $p_j(t_j)$ ($0 \leq t_j \leq \theta_j - \pi$, $j = 1, \ldots, n$) be a continuous parametrization of the interval $[p_j(0), p_j(\theta_j - \pi)]$. We can regard γ as an image of the continuous map $\varphi : S^1 \to \gamma$ satisfying the following two properties:
1. $\varphi([p_j(0), p_j(\theta_j - \pi)]) = \{p_j\}$ ($j = 1, \ldots, n$);
2. φ maps $(p_j(\theta_j - \pi), p_{j+1}(0))$ ($j = 1, \ldots, n$) onto $\gamma|_{(p_j, p_{j+1})}$ homeomorphically, where $p_{n+1} = p_1$.

Let L_j^- (resp. L_j^+) be the lower (resp. upper) tangent line at p_j. Then each line between L_j^- and L_j^+ can be continuously parametrized as the notation $L_j(t_j)$, where t_j ($0 \leq t_j \leq \theta_j - \pi$) is the angle from the line L_j^- (see Figure 3.1). For each

$p \in S^1$, we set
$$C_p^\bullet := \begin{cases} \text{the } \bullet\text{-maximal circle at } \varphi(p)(\varphi(p) \neq p_1, \ldots, p_n), \\ \text{the } \bullet\text{-maximal circle at } \varphi(p) \text{ with tangent line } L_j(t_j) \text{ at } \varphi(p) \\ (p = p_j(t_j), 0 \leq t_j \leq \theta_j - \pi, \ j = 1, \ldots, n). \end{cases}$$

We define a subset $F_p^\bullet$ of S^1 for each $p \in S^1$ as follows. A point $q \in S^1$ is contained in $F_p^\bullet$ if it satisfies the one of the following two conditions:
1. $\varphi(q) \neq p_1, \ldots, p_n$ and $\varphi(q) \in C_p^\bullet$;
2. there exist an integer j ($0 \leq j \leq n$) and a real number t_j ($0 \leq t_j \leq \theta_j - \pi$) such that $q = p_j(t_j)$ and $C_p^\bullet$ is tangent to $L_j(t_j)$ at p_j.

Then one can easily check that $F^\bullet := (F_p^\bullet)_{p \in S^1}$ is an intrinsic circle system. (The proof is essentially same as that of Proposition 3.1.) Similarly, we get another intrinsic circle system F°. Applying Theorem 2.7, we get the assertion. □

COROLLARY 3.4. *Let X be a C^2-differentiable sphere and Γ a C^2-differential circle system on X satisfying the additional condition* (S4) *in Appendix A. Let γ be a piecewise C^2-regular curve on X and $s(\gamma)$ the number of $\bullet$-maximal or $\circ$-maximal circles which are tangent to γ with only one connected component. Suppose that γ is not a circle and $s(\gamma)$ is finite. Then*
$$s(\gamma) - t(\gamma) = 4$$
holds, where $t(\gamma) := t^\bullet(\gamma) + t^\circ(\gamma)$.

PROOF. By Lemma A.4 in Appendix A, $\bullet$-maximal circles are not $\circ$-maximal circles unless γ is a circle. Thus the formula is obtained by summing up two formulas
$$(3.5) \qquad s^\bullet(\gamma) - t^\bullet(\gamma) = 2, \qquad s^\circ(\gamma) - t^\circ(\gamma) = 2.$$

□

REMARK 1. When the number of zeros of the curvature function is finite, the equality has been obtained in Haupt [**Hu**]. Recently, Sedykh [**Sd6**] proved the equality for generic curves on the Möbius plane using an interesting approach via singularity theory.

REMARK 2. One can formulate the following generalization of (3.5). Let A be a compact subset of X. Then the formula
$$(3.6) \qquad s - t = \chi(X \setminus A)$$
is expected under a suitable regularity of the boundary ∂A, where s (resp. t) is the number of single (resp. triple) tangent maximal inscribed circles in $X \setminus A$. (Formula (3.6) reduces to formula (3.5) when the set A is a simply connected domain.) Let K_A be the closure of the set of circles contained in $X \setminus A$ which meets A at least two points. It is naturally expected that K_A has the same homotopy type as $X \setminus A$ and has the structure of local tree. Then the formula (3.6) follows immediately under a suitable regularity assumption of ∂A. When X is the unit sphere in $\mathbf{R}^3$, K_A can be identified with the cut locus of the set A with respect to the canonical Riemannian structure on the unit sphere (see the last remark in Section 2), and (3.6) can be proved under a suitable (for example, C^2) regularity of ∂A by applying a theorem of Shiohama and Tanaka [**ST,** Theorem A]. However, if X is an arbitrary

convex surface in $\mathbf{R}^3$, the method in [**ST**] is not directly applied, because one needs a realization of the set K_A as a cut locus in X for a suitable interior metric. The author does not know the existence of such a metric.

4. Application to space curves

In this section, we apply Theorem 2.7 to convex simple closed space curves. An immersed closed C^1-curve $\gamma : S^1 \to \mathbf{R}^3$ is called *convex* if it lies on the boundary ∂H of its convex hull H. The orientation of $\mathbf{R}^3$ is given such that three vectors $(1,0,0)$, $(0,1,0)$ and $(0,0,1)$ form a positive frame. Let P be a plane in $\mathbf{R}^3$ with fixed orientation and $\{e_1, e_2\}$ an oriented frame on P. Then the positively oriented unit normal vector n is given such that the frame $\{e_1, e_2, n\}$ is an adapted to the orientation of $\mathbf{R}^3$. The upper half-space bounded by P is the side pointed by the normal vector n.

We fix a convex simple closed curve γ and assume that it is not planar. We fix an interior point o of the convex hull and consider the unit sphere S_o^2 centered at o. We denote by $\pi : \partial H \to S_o^2$ the canonical projection. Then π is a bijective continuous map. Since ∂H is compact, π is a homeomorphism. In particular, the boundary ∂H of the convex hull is homeomorphic to a sphere and γ divides ∂H into two domains. Moreover,

$$\tilde{\gamma} := \pi \circ \gamma : S^1 \to S_o^2$$

is an embedded curve. We choose the conormal vector $\nu(t)$ of $\tilde{\gamma}(t)$ such that the tangent vector $\gamma'(t)$, the conormal vector $\nu(t)$ and the inward normal vector $n(t) := -\tilde{\gamma}(t)/|\gamma(t)|$ forms an positive frame on $\mathbf{R}^3$. Then the left-hand (resp. right-hand) domain D_L (resp. D_R) of $\tilde{\gamma}$ is the open domain of S_o^2 which is the side (resp. opposite side) pointed to by the conormal vector. Let $\partial H^\bullet$ (resp. ∂H°) be the left-hand (right-hand) closed domain of γ in ∂H, which is defined by the inverse image of D_L (resp. D_R) by π. Now we fix a point p on γ arbitrarily. A plane U is called a *tangent plane* at p if it contains the tangent line L_p at p. Let $\mathcal{P}_p$ be the pencil of oriented planes which is tangent to γ at p. Then $\mathcal{P}_p$ is identified with a circle.

We denote by $V_x \in \mathcal{P}_p$ the oriented plane passing through $x \in \mathbf{R}^3 \setminus L_p$, where the orientation of V_x is chosen so that the line segment $\overline{px}$ lies in an upper half-plane bounded by the line L_p in V_x. We fix a point o in the interior of H arbitrarily. A plane $V_x(\neq V_o)$ is said to be *upper* (resp. *lower*) than V_o if $\overline{px}$ lies in the closed upper (resp. lower) half-space bounded by V_o. We give an order of $\mathcal{P}_p \setminus \{V_o\}$ such that each plane in $\mathcal{P}_p$ upper than V_o is greater than any planes in $\mathcal{P}_p$ lower than V_o. This order induces a canonical cyclic order of $\mathcal{P}_p$, which is independent of the choice of the interior point o.

An oriented plane U is called a *supporting plane* of γ at p if $p \in U$ and the curve lies entirely in the closed upper half-space bounded by U. Let $\mathcal{S}_p$ be the set of supporting plane at p whose upper half-plane does not contain any points in $\gamma \setminus L_p$. Then by definition, $\mathcal{S}_p$ is a subset of $\mathcal{P}_p$. Moreover, the set of supporting planes at p is just the closure $\overline{\mathcal{S}_p}$ of $\mathcal{S}_p$. Since γ is a convex simple closed curve, there is at least one supporting plane passing through p. Hence $\overline{\mathcal{S}_p}$ is nonempty. One can easily see that $\mathcal{S}_p$ is connected; that is, there exists $U_p^\bullet, U_p^\circ \in \mathcal{P}$ such that one of the following four possibilities occur:

(1) $\mathcal{S}_p = (U_p^\circ, U_p^\bullet)$,   (2) $\mathcal{S}_p = [U_p^\circ, U_p^\bullet)$,

(3) $\mathcal{S}_p = (U_p^\circ, U_p^\bullet]$,   (4) $\mathcal{S}_p = [U_p^\circ, U_p^\bullet]$.

The plane $U_p^\bullet$ (resp. U_p°) is called the *maximal* (resp. *minimal*) supporting plane at p. It may be possible that $U_p^\bullet$ and U_p° are the same plane. However, in this case, two planes have mutually opposite orientation and are regarded as distinct elements in $\mathcal{P}_p$. We denote by $(U_p^\bullet)^+$ (resp. $(U_p^\circ)^+$) the closed upper half-plane of $U_p^\bullet$ (resp. U_p°) bounded by L_p. If $U_p^\bullet$ and U_p° are distinct planes, H lies in the acute angular domain D bounded by $(U_p^\bullet)^+$ and $(U_p^\circ)^+$.

Later, we will need the following lemma. (Except Lemma 4.1, we do not need C^2-regularity of curves until Proposition 4.9.)

LEMMA 4.1. *Let γ be a C^2-convex simple closed space curve and $p \in \gamma$ have nonvanishing curvature. Suppose that $L_p \cap \gamma = \{p\}$. Then case (4) never occurs. Moreover, if case (2) (resp. case (3)) occurs, then U_p° (resp. $U_p^\bullet$) is the osculating plane at p.*

The lemma is well known (see Proposition 3 of [**Sd2**]) and can be proved with the standard method, so we omit the proof.

DEFINITION 4.1. We set

$$F_p^\bullet := \{q \in \gamma\,;\, \overline{pq} \subset \partial H^\bullet\}, \quad (\text{resp. } F_p^\circ := \{q \in \gamma\,;\, \overline{pq} \subset \partial H^\circ\}).$$

Now we prepare lemmas to give some sufficient conditions that $F^\bullet$ and F° are intrinsic circle systems.

LEMMA 4.2. *Let γ be a convex simple closed space curve. Then for each $p \in \gamma$, the following inclusions hold:*

$$F_p^\bullet \subset (U_p^\bullet)^+, \qquad F_p^\circ \subset (U_p^\circ)^+,$$

where $(U_p^\bullet)^+$ (resp. $(U_p^\circ)^+$) is the closed upper half-plane of $U_p^\bullet$ (resp. U_p°) bounded by L_p.

PROOF. We fix $q \in F_p^\bullet$ and will show that $q \in U_p^\bullet$. Since L_p is contained in $U_p^\bullet$, we may assume that q does not lie in L_p. Let $V_q \in \mathcal{P}_p$ be a plane passing through q. First, we show that either $V_q = U_p^\bullet$ or $V_q = U_p^\circ$ holds. In fact, we take the middle point m on the line segment $\overline{pq}$. Since $m \in \partial H^\bullet$, there exists a plane U passing through m such that H lies in the upper or lower half-space bounded by U. Then $\overline{pq} \in U$ holds, and consequently U is a supporting plane at p. Hence we have $V_q = U$, and thus $V_q = U_p^\bullet$ or $V_q = U_p^\circ$ holds.

On the other hand, H lies in the angular domain D bounded by $(U_p^\bullet)^+$ and $(U_p^\circ)^+$. We have seen that the line segment $\overline{pq}$ lies in either $U_p^\bullet$ or U_p°. Since $\pi(\overline{pq})$ lies in a left-hand side of $\tilde{\gamma}$ at p, $\overline{pq}$ lies in the upper half-space bounded by V_o. Thus we have $q \in (U_p^\bullet)^+$ and get the relation $F_p^\bullet \subset (U_p^\bullet)^+$. Similarly, $F_p^\circ \subset (U_p^\circ)^+$ is proved. □

LEMMA 4.3. *Let γ be a convex simple closed space curve and L_p the tangent line of γ at p. Suppose that there exists $q(\neq p)$ such that $q \in L_p \cap \gamma$ and the tangent line L_q at q does not coincide with L_p. Then there exists a unique supporting plane U at p. Moreover U contains the lines L_p and L_q.*

PROOF. Since γ is a convex curve, there exists at least one supporting plane U at p. Obviously U contains L_p. If U does not contain L_q, it is transversal to γ at q, which is impossible. Thus U also contains L_q. Since $L_p \neq L_q$, U is uniquely determined. □

LEMMA 4.4. *Let γ be a convex simple closed curve which has no planar open subarcs. Suppose that U is a supporting plane at $p \in \gamma$ and $p, x, y \in \gamma \cap U$ are not collinear. Then the triangle $\triangle pxy$ is contained in $\partial H^\bullet$ or ∂H°.*

PROOF. Obviously, the triangle $\triangle pxy$ on U lies in ∂H. Suppose that the triangle $\triangle pxy$ contains a point q of γ in its interior. Then $\pi(q)$ lies in the interior of $\pi(\triangle pxy)$ in S_o^2. Thus a sufficiently small open arc of $\tilde\gamma$ containing q also lies in its interior. Hence the corresponding arc of γ containing q lies in $\triangle pxy$. But this contradicts the fact that γ has no planar subarcs. Thus $\triangle xqp \subset \partial H^\bullet$ or $\triangle xqp \subset \partial H^\circ$ holds. □

PROPOSITION 4.5. *Let γ be a convex simple closed curve which has no planar open subarcs, and let p be a point on γ. Suppose that $U_p^\bullet$ satisfies the following two conditions:*
(1) the set $U_p^\bullet \cap \gamma$ does not lie in any line passing through p;
(2) $F_p^\bullet \neq \{p\}$ (resp. $F_p^\circ \neq \{p\}$).
Then it holds that $F_p^\bullet = U_p^\bullet \cap \gamma$ (resp. $F_p^\circ = U_p^\circ \cap \gamma$).

PROOF. We prove the assertion for $F_p^\bullet$. By Lemma 4.2, we have $F_p^\bullet \subset U_p^\bullet \cap \gamma$. It is sufficient to show that $U_p^\bullet \cap \gamma \subset F_p^\bullet$. By condition (1), there are points $q, q' \in U_p^\bullet \cap \gamma$ such that p, q, q' are not collinear. To prove it, we divide the proof into the the following two cases. Let $x \in U_p^\bullet \cap \gamma$ be an arbitrary point.

Case 1. Suppose that $p, q, x \in \gamma \cap U_p^\bullet$ are not collinear. Then by Lemma 4.4, either $\triangle xpq \subset \partial H^\bullet$ or $\triangle xpq \subset \partial H^\circ$ holds. But in the latter case, we have

$$\overline{pq} \subset \partial H^\bullet \cap \partial H^\circ = \gamma,$$

which contradicts the fact that γ has no planar subarcs. Thus we have $\triangle xpq \subset \partial H^\bullet$. In particular, we have $\overline{px} \subset \partial H^\bullet$, which implies $x \in F_p^\bullet$.

Case 2. Next we consider the case that $p, q, x \in \gamma \cap U_p^\bullet$ lie on a line L. Since p, q, q' are not collinear, we have $q' \notin L$. Suppose that $\overline{px} \not\subset \partial H^\bullet$. Then by Lemma 4.4, we have $\triangle pq'x \subset \partial H^\circ$. In particular $\overline{pq'} \in \partial H^\circ$. On the other hand, $\overline{pq} \subset \partial H^\bullet$ yields that $\triangle pqq' \subset \partial H^\bullet$ by Lemma 4.2. In particular,

$$\overline{pq'} \in \partial H^\circ \cap \partial H^\bullet = \gamma,$$

which is a contradiction. Hence we have $\overline{px} \subset \partial H^\bullet$. So $x \in F_p^\bullet$. □

LEMMA 4.6. *Let γ be a convex simple closed space curve. Suppose that for each $p \in \gamma$, there exists a supporting plane U satisfying $U \cap \gamma = \{p\}$. Moreover, suppose U_0 is a supporting plane of γ such that $U_0 \cap \gamma$ contains three distinct points $x, y, z \in \gamma$. Then these three points are not collinear.*

PROOF. By the assumption, we can easily see that

(4.1) $$L_z \cap \gamma = \{z\} \qquad \text{(for all } z \in \gamma\text{)}.$$

Suppose that $x, y, z \in U_0 \cap \gamma$ lie in a line L with this order. If $L = L_y$, this contradicts (4.1). So $L \neq L_y$. Then U_0 must be a unique supporting plane passing through y by Lemma 4.3. This contradicts the fact that there exists a supporting plane U such that $U \cap \gamma = \{y\}$. □

PROPOSITION 4.7. *Let γ be a convex simple closed space curve. Suppose that for each $p \in \gamma$ there exists a supporting plane U such that $U \cap \gamma = \{p\}$. Then for each $p \in \gamma$, it holds that*

(4.2) $$F_p^\bullet = U_p^\bullet \cap \gamma, \qquad F_p^\circ = U_p^\circ \cap \gamma.$$

In particular, $U_p^\bullet \neq U_p^\circ$ holds.

PROOF. We prove the first equality. (The second equality is obtained in the same manner.) If $F_p^\bullet = \{p\}$, then (4.2) is obvious. So we may assume that there exists a point $q \in F_p^\bullet$ such that $q \neq p$. By Lemma 4.2, we have $q \in U_p^\bullet$. If $\#(U_p^\bullet \cap \gamma) = 2$, (4.2) is obvious. So we may assume that $\#(U_p^\bullet \cap \gamma) > 2$. We fix a point $x \in U_p^\bullet \cap \gamma$ such that $x \neq p, q$. By Lemma 4.6, p, q, x are not collinear and thus the triangle $\triangle pqx$ is considered. Suppose that there exists a point $y \in \gamma$ in the triangle. Then the tangent line L_y separates one of three points p, q, x with the other two in the plane $U_p^\bullet$. Hence $U_p^\bullet$ must be a unique supporting plane passing through y. This contradicts the fact that there exists a supporting plane U such that $U \cap \gamma = \{y\}$. So there are no points on γ inside the triangle. In particular, $\triangle pqx \subset \partial H^\bullet$ or $\triangle pqx \subset \partial H^\circ$ holds. But if $\triangle pqx \subset \partial H^\circ$, then

$$\overline{pq} \subset \partial H^\bullet \cap \partial H^\circ = \gamma.$$

This contradicts (4.1). So $\triangle pqx \subset \partial H^\bullet$. In particular $x \in F_p^\bullet$. Thus we have $U_p^\bullet \cap \gamma \subset F_p^\bullet$. The opposite inclusion follows from Lemma 4.2. □

REMARK. The structure of generic convex space curves has been investigated in Sedykh [**Sd1**], which will be helpful in understanding the discussions of the statements of Lemmas 4.2–4.4, Proposition 4.5, Lemma 4.6 and Proposition 4.7.

THEOREM 4.8. *Let γ be a convex simple closed space curve satisfying the one of the following two conditions:*
  (a) *for each $p \in \gamma$, there exists a supporting plane U such that $U \cap \gamma = \{p\}$;*
  (b) *γ has no planar open subarcs.*
Then $(F_p^\bullet)_{p \in \gamma}$ (resp. $(F_p^\circ)_{p \in \gamma}$) is an intrinsic circle system on $S^1 = \gamma$.

PROOF. We divide the proof into three steps. (We prove the assertion for $F^\bullet$.)

Step 1. We check the property (I1). By Proposition 4.7, this is obvious for case (a). So we prove the assertion only for case (b). Let $q \in F_p^\bullet$. It is sufficient to show that $F_p^\bullet \subset F_q^\bullet$. (Opposite inclusion is obtained by interchanging the role of p and q.) If $p = q$, then the property (I1) is obvious. So we may assume that $q \neq p$.

Case 1. First we consider the case that $U_p^\bullet \cap \gamma$ does not lie in any line passing through p. If $F_p^\bullet = \{p\}$, the statement is obvious. If $F_p^\bullet \neq \{p\}$, we have the assertion by Proposition 4.5.

Case 2. So we may assume that $U_p^\bullet \cap \gamma$ lies on a line L passing through p. Let $x \in F_p^\bullet$. Then $\overline{xp}$ and $\overline{qp}$ both lie in $L \cap \partial H^\bullet$. In particular so does $\overline{qx}$, and hence $x \in F_q^\bullet$. Thus we have $F_p^\bullet \subset F_q^\bullet$.

Step 2. We show (I2). Suppose that there exist $p' \in F_p^\bullet \setminus \{p\}$ and $q' \in F_q^\bullet \setminus \{q\}$ such that $F_p^\bullet \ne F_q^\bullet$ and

(4.3) $$q \succeq p' \succeq q' \succeq p \qquad \text{on } [p, q].$$

Then $\overline{pp'}, \overline{qq'} \subset \partial H^\bullet$. Since $\overline{pp'}$ separates $\partial H^\bullet$ into two domains, $\overline{pp'} \cap \overline{qq'}$ is not empty by (4.3). Let $z \in \overline{pp'} \cap \overline{qq'}$. Then $z \ne p, p', q, q'$. (For example, if $z = p$ or $z = p'$, then $q \in F_z^\bullet = F_p^\bullet \ne F_q^\bullet$ by Step 1, which is a contradiction.) In particular, $\overline{pp'}$ and $\overline{qq'}$ cannot lie in a common line. This implies that they are transversal at a point z. By Lemma 4.2, these four points p, p', q, q' lie in $U_p^\bullet$. In particular, $U_p^\bullet \cap \gamma$ does not lie in any line passing through p. Applying Proposition 4.5 for case (b) and Proposition 4.7 for case (a), we have $F_p^\bullet = U_p^\bullet \cap \gamma \ni q$. This is a contradiction.

Step 3. Finally, we show the property (I3). Let $(p_n)_{n \in \mathbf{N}}$ and $(q_n)_{n \in \mathbf{N}}$ be two sequences in γ such that $q_n \in F_{p_n}^\bullet$, $\lim_{n \to \infty} p_n = p$ and $\lim_{n \to \infty} q_n = q$. Since $\overline{p_n q_n} \in \partial H^\bullet$, we have $\overline{pq} \in \partial H^\bullet$. Thus $F_p^\bullet \ni q$. $\square$

Let γ be a convex simple closed space curve as above. We denote by $\text{rank}^\bullet(p)$ (resp. $\text{rank}^\circ(p)$) the rank of $p \in \gamma$ with respect to $F^\bullet$ (resp. F°). By Theorem 2.7, we can get a Bose type formula for γ satisfying the assumptions of Theorem 4.8. But unfortunately in such a general setting, the points of rank one with respect to $F^\bullet$ or F° may be neither the clean nor clear vertices defined below.

DEFINITION 4.2. Let γ be a C^2-convex simple closed space curve. Then a *clear maximal* (resp. *minimal*) *vertex* is at the point with nonvanishing curvature, which is a maximum (resp. minimum) of the height function with respect to the binormal vector. Moreover, if the maximum (resp. minimum) level set of the height function is connected, it is called a *clean maximal* (resp. *minimal*) *vertex*.

We remark that $p \in \gamma$ is a clear vertex (namely, a clear maximal or clear minimal vertex) if and only if the osculating plane U at p is a supporting plane. Moreover, it is a clean vertex if and only if $U \cap \gamma$ is connected.

If γ lies in X_2 as in Section 3, Example 2, this definition of clean vertices has the same meaning as the one in Section 3. In other words, a point p of $\text{rank}^\bullet(p) = 1$ or $\text{rank}^\circ(p) = 1$ is a clean vertex in the above sense. Our next goal is to give much weaker sufficient conditions for convex simple closed space curves that $\text{rank}^\bullet(p) = 1$ (resp. $\text{rank}^\circ(p) = 1$) implies a clean or clear maximal (resp. minimal) vertex. The following assertion is essentially the same as Lemma 1 in [**Sd2**].

PROPOSITION 4.9. *Let γ be a C^2-convex simple closed space curve and $p \in \gamma$ a point with a nonvanishing curvature. Suppose that there exists a point q ($\ne p$) on γ which lies in a closed upper half-plane $(U_p^\bullet)^+$ (resp. $(U_p^\circ)^+$) of $U_p^\bullet$ (resp. U_p°) bounded by L_p. Then there exists $x \in U \cap \gamma$ ($x \ne p$) satisfying the following two properties:*
1. $x \in \overline{pq}$;
2. $x \in F_p^\bullet$ *or* $x \in F_p^\circ \cap L_p$.

PROOF. $\overline{pq} \cap \gamma$ is a closed subset of $\overline{pq}$. Suppose that there is no $x \in \overline{pq}$ such that $x \in F_p^\bullet$ or $x \in F_p^\circ$. Then we can take a sequence $(q_n)_{n \in \mathbf{N}}$ consisting of mutually different points in $\overline{pq} \cap \gamma$ such that $\lim_{n \to \infty} q_n = p$. Since the unit vectors $(q_n - p)/|q_n - p|$ converge to the unit tangent vector at p of γ, $\overline{pq}$ lies in the tangent line L_p at p. Thus $q_n \in L_p$ for all n. But this contradicts the fact that the curvature function of γ does not vanish at p. Thus there exists $x \in \overline{pq}$ such that $x \in F_p^{\bullet}$ or $x \in F_p^{\circ}$. Since $(U_p^{\bullet})^+ \cap (U_p^{\circ})^+ = L_p$, the case $x \in F_p^{\circ}$ may occur only when $x \in L_p$ by Lemma 4.2. □

DEFINITION 4.3. A convex simple closed space curve γ is called *tame* if $L_p \cap \gamma = \{p\}$ for any $p \in \gamma$.

REMARK. In Ballesteros and Romero-Fuster [**BR2**], such a curve is called strictly convex. But there is another definition of strict convexity. (The strictly convexity defined in Sedykh [**Sd5**] is stronger than that in [**BR2**].) So here we use the term "tame" to avoid confusion.

LEMMA 4.10. *Let γ be a C^2-convex simple closed space curve satisfying (a) or (b) as in Theorem 4.8. Suppose that $p \in \gamma$ has nonvanishing curvature and $L_p \cap \gamma = \{p\}$. Then $\mathrm{rank}^{\bullet}(p) = 1$ (resp. $\mathrm{rank}^{\circ}(p) = 1$) if and only if p is a clean maximal (resp. minimal) vertex.*

PROOF. If p is a clean maximal vertex, then $\mathrm{rank}^{\bullet}(p) = 1$ holds obviously. Conversely, we assume $\mathrm{rank}^{\bullet}(p) = 1$. Suppose $F_p^{\bullet} \neq \{p\}$. Then by Lemma 4.2, $F_p^{\bullet}$ is a planar arc contained in $U_p^{\bullet}$. Thus the osculating plane at p coincides with $U_p^{\bullet}$. Applying Proposition 4.5 for case (b) and Proposition 4.7 for case (a), we have $F_p^{\bullet} = U_p^{\bullet} \cap \gamma$. Thus p is a clean maximal vertex. Next we consider the case $F_p^{\bullet} = \{p\}$. Suppose that $U_p^{\bullet} \cap \gamma \neq \{p\}$. Since $L_p \cap \gamma = \{p\}$, we have $F_p^{\bullet} \neq \{p\}$ by Proposition 4.9, which is a contradiction. So we have $U_p^{\bullet} \cap \gamma = \{p\}$. Thus $U_p^{\bullet}$ is a supporting plane which does not contain any points in $\gamma \setminus L_p$. Since $L_p \cap \gamma = \{p\}$, the osculating plane at p coincides with $U_p^{\bullet}$ by Lemma 4.1. Thus p is a clean maximal vertex. Similarly, $\mathrm{rank}^{\circ}(p) = 1$ implies that p is a clean minimal vertex.□

Let γ be a convex simple closed space curve. For a plane U in $\mathbf{R}^3$, we denote by $\mathrm{rank}(U \cap \gamma)$ the number of the connected components in $U \cap \gamma$.

THEOREM 4.11. *Let γ be a C^2-convex simple closed space curve with nonvanishing curvature satisfying the one of the following conditions:*
  (1) *For each $p \in \gamma$, there exists a supporting plane U such that $U \cap \gamma = \{p\}$;*
  (2) *γ is tame and has nonplanar open subarcs.*
Suppose the number $s(\gamma)$ of connected components of clean vertices is finite. Then $t(\gamma) := t^{\bullet}(\gamma) + t^{\circ}(\gamma)$ is also finite and the following formula holds:

(4.4) $$s(\gamma) - t(\gamma) = 4.$$

Moreover, under condition (1), the following identity holds:

(4.5) $$t(\gamma) := \sum_{U \in T(\gamma)} (\mathrm{rank}(U \cap \gamma) - 2).$$

PROOF. If γ satisfies (2), then $L_p \cap \gamma = \{p\}$ holds, since γ is tame. Moreover, even when γ satisfies (1), $L_p \cap \gamma = \{p\}$ holds by (4.1). So we can apply Lemma 4.10 and conclude that $\mathrm{rank}^{\bullet}(p) = 1$ (resp. $\mathrm{rank}^{\circ}(p) = 1$) if and only if p is a clean maximal (resp. minimal) vertex. Since γ is not planar, clean minimal vertices are not clean maximal vertices. Thus equality (4.4) follows immediately from Theorem 2.7. Finally, (4.5) follows from Proposition 4.7. □

REMARK 1. The formula is an improvement of the one obtained by Romero-Fuster [**R**] for the convexly generic case and by Sedykh [**Sd5**] for the generic strictly convex case. In fact, condition (1) is equivalent to the strict convexity of curves in the sense of Sedykh [**Sd5**], and (2) is weaker than the convexity generic assumption as in [**R**].

REMARK 2. If γ is a C^2-regular curve on X_2 as in Section 3, Example 2, then γ satisfies (1) obviously. In this case, the assertion follows from Corollary 3.4 directly.

REMARK 3. If γ is C^3-differentiable, then clean vertices are zeros of the torsion function. Hence $S(F^\bullet)$ and $S(F^\circ)$ are supported by the torsion function. Thus by Theorem 2.2, $t(\gamma) < \infty$ is equivalent to $s(\gamma) < \infty$.

Next we consider convex simple closed space curves which may not satisfy the assumption of Theorem 4.11.

PROPOSITION 4.12. *Let γ be a C^3-convex simple closed space curve, which has no planar open subarcs and has at most finitely many zeros of the curvature function. Suppose that every element in the set*

$$M_\gamma := \{x \in \gamma\,;\, L_x \cap \gamma \neq \{x\},\ \ \kappa(L_x \cap \gamma) \not\ni 0\}$$

is isolated, where κ is the curvature function. Then any point p on γ satisfying $\mathrm{rank}^\bullet(p) = 1$ (*resp.* $\mathrm{rank}^\circ(p) = 1$) *is a zero of the curvature function or a clear maximal* (*resp. minimal*) *vertex.*

REMARK. If γ has nonvanishing curvature, we have a simple expression $M_\gamma = \{x \in \gamma\,;\, L_x \cap \gamma \neq \{x\}\}$. In this case, every element in M_γ is isolated if and only if M_γ is finite. In fact, if an accumulation point $p \in \gamma$ of M_γ exists, one can easily verify that $p \in M_\gamma$ using the property $\kappa(p) \neq 0$.

To prove Proposition 4.12, we prepare the following two lemmas.

LEMMA 4.13. *Let γ be a C^3-convex simple closed space curve, which has no planar open subarcs and has at most finitely many zeros of the curvature function. Let p be a point on γ with nonvanishing curvature and* $\mathrm{rank}^\bullet(p) = 1$ (*resp.* $\mathrm{rank}^\circ(p) = 1$). *Suppose that p is an isolated point in the set*

$$\{x \in \gamma\,;\, \mathrm{rank}^\bullet(x) = 1\} \qquad (\text{resp. } \{x \in \gamma\,;\, \mathrm{rank}^\circ(x) = 1\}).$$

Then p is a clear maximal (*resp. minimal*) *vertex.*

PROOF. Since γ satisfies condition (b) in Theorem 4.8, two intrinsic circle systems $F^\bullet$ and F° are induced. By assumption, there is an open arc I containing p such that all points on $I \setminus \{p\}$ are weakly regular with respect to $F^\bullet$ (resp. F°). We take a sequence $(p_n)_{n \in \mathbf{N}}$ on $I \setminus \{p\}$ such that $p_n \to p-0$. Then by Theorem 1.4, we have $\mu_+(p_n) \to p+0$. By taking a subsequence of $(p_n)_{n \in \mathbf{N}}$, we may assume that the supporting planes given by $U_n := U_{p_n}^\bullet$ (resp. $U_n := U_{p_n}^\circ$) converge to a supporting plane U at p. We set $\gamma(t_n) = p_n$ and $\gamma(s_n) = \mu_+(p_n)$. By a straightforward calculation, we have

$$U_n \ni \gamma(t_n) + \frac{\{\gamma(s_n) - \gamma(t_n)\} + (s_n - t_n)\gamma'(t_n)}{(s_n - t_n)^2}$$
$$= \gamma(t_n) + \frac{1}{2}\gamma''(t_n) + (s_n - t_n)\int_0^1 u^2 du \int_0^1 v\,dv \int_0^1 \gamma'''(t_n + (s_n - t_n)uvw)\,dw.$$

Thus U contains a vector $\gamma(p) + \gamma''(p)$. Since U contains the tangent line L_p, it is the osculating plane at p, in particular, p is a clear vertex. □

LEMMA 4.14 (Romero-Fuster and Sedykh [**RS1**, Proposition 1]). *Let $\sigma : (a,b) \to \mathbf{R}^3$ be a C^2-regular curve with nonvanishing curvature, which may not be closed. Let p be a point of σ and $q(\ne p)$ a point in $\mathbf{R}^3$. Then there is an open arc I containing p such that $q \notin L_x \cap \gamma$ for all $x \in I \setminus \{p\}$.*

As mentioned in [**RS1**], the lemma is a simple exercise.

PROOF OF PROPOSITION 4.12. Since γ satisfies condition (b) in Theorem 4.8, two intrinsic circle systems $F^\bullet$ and F° are induced. Let $p \in \gamma$ be a point satisfying $\mathrm{rank}^\bullet(p) = 1$. Assume that p has nonvanishing curvature. If $L_p \cap \gamma = \{p\}$, then p is a clean vertex by Lemma 4.10. So we may assume that $L_p \cap \gamma \ne \{p\}$. Consider the subset

$$K = \{x \in \gamma \,;\, \mathrm{rank}^\bullet(x) = 1\}.$$

If p is isolated in K, then it is a clear vertex by Lemma 4.13. So we may assume that there is a sequence $(p_n)_{n \in \mathbf{N}}$ in K which converges to p. Since $\kappa(p) \ne 0$, there exists a neighborhood I of p such that $(L_q)_{q \in I}$ are mutually distinct. Thus there exists a positive integer n_0 such that

(4.6) $$0 \notin \kappa(L_{p_n} \cap \gamma) \qquad (\text{for } n > n_0).$$

(In fact, if (4.6) fails, there is a point $q \in \gamma$ such that $q \in L_{p_n} \cap \gamma$ for infinitely many n. But this contradicts Lemma 4.14.) We fix p_n ($n > n_0$) arbitrarily. It is sufficient to show that each p_n is a clear maximal vertex. (Then the limit point p is also a clear maximal vertex.) If $L_{p_n} \cap \gamma = \{p_n\}$, then p_n is a clean maximal vertex by Lemma 4.10. So we may assume that $p_n \in M_\gamma$.

Case 1. Suppose that each p_n is isolated in K. By Lemma 4.13, p_n is a clear maximal vertex.

Case 2. Next we suppose that p_n is an accumulation point of the set $K = \{x \in \gamma \,;\, \mathrm{rank}^\bullet(x) = 1\}$. Then there is a sequence $(q_m)_{m \in \mathbf{N}}$ in K converging to p_n. Since $p_n \in M_\gamma$, q_m is not contained in M_γ for sufficiently large m. By (4.6), $0 \notin \kappa(L_{q_m} \cap \gamma)$ holds for sufficiently large m. Thus $q_m \notin M_\gamma$ implies that $L_{q_m} \cap \gamma = \{q_m\}$. By Lemma 4.10, q_m is a clean vertex. Hence the limit point p_n is a clear vertex. □

For the following applications, we recall two important facts from [**Sd2**].

LEMMA 4.15 ([**Sd2**, Proposition 4]). *Let γ be a C^3-convex simple closed space curve and $p,q \in \gamma$ be points with nonvanishing curvature and torsion. Then the straight line pq is tangent to γ at p if and only if it is tangent to the curve at q.*

LEMMA 4.16 ([**Sd2**, Proposition 7]). *Let γ be a C^2-convex simple closed space curve and let p be a point such that $0 \notin \kappa(L_p \cap \gamma)$. Then there exists an open arc I containing p such that the tangent line L_q at each $q \in I \setminus \{p\}$ is not tangent to the curve at any other point.*

REMARK. The statement of the lemma is slightly modified as in [**RS1**, Proposition 4]. As explained in [**RS1**], the proof is essentially the same as that of [**Sd2**, Proposition 7].

Lemmas 4.15 and 4.16 yield the following result.

LEMMA 4.17. *Let γ be a C^3-convex simple closed space curve whose curvature function and torsion function have only finitely many zeros. Then every element in the set M_γ is isolated.*

PROOF. Suppose that there exists a point $p \in M_\gamma$ such that a sequence $(p_n)_{n \in \mathbf{N}}$ in $M_\gamma \setminus \{p\}$ exists and converges to p. For each p_n, we can choose $q_n \in L_{p_n} \cap \gamma$ such that $q_n \neq p_n$. By Lemma 4.16, L_{p_n} is not tangent to γ at q_n. Then by Lemma 4.15, the torsion function vanishes at p_n or q_n. Since the number of zeros of the torsion function is finite, there exists a positive number $n_0 > 0$ such that $q_n = q_0$ for all $n \geq n_0$. But this contradicts Lemma 4.14. □

We get the following two corollaries.

COROLLARY 4.18 (Sedykh's four vertex theorem [**Sd2**]). *Let γ be a C^3-convex simple closed space curve with a nonvanishing curvature function. Then*

$$v(\gamma) \geq 4.$$

where $v(\gamma)$ is the number of zeros of the torsion function.

A generalization of the inequality for space curves which may not be convex will be found in Thorbergsson–Umehara [**TU**]. The inequality $v(\gamma) \geq 4$ does not hold if the curvature function of γ has zeros. (According to Barner [**Ba**, p. 210], Flohr pointed out it in the 1950s.)

COROLLARY 4.19 ([**RS1**]). *Let γ be a C^3-convex simple closed space curve. Then*

$$v(\gamma) + 2c(\gamma) \geq 4,$$

where $c(\gamma)$ is the number of zeros of the curvature function.

PROOF OF COROLLARIES 4.18 AND 4.19. We may assume that $v(\gamma)$ and $c(\gamma)$ are finite. Then γ satisfies condition (b) of Theorem 4.8. By Theorem 2.7, there exist at least four points x_1, x_2, x_3, x_4 ($x_1 \neq x_2$, $x_3 \neq x_4$) such that

$$\mathrm{rank}^\bullet(x_1) = \mathrm{rank}^\bullet(x_2) = \mathrm{rank}^\circ(x_3) = \mathrm{rank}^\circ(x_4) = 1.$$

If $c(\gamma) = 0$, then they are mutually distinct clear vertices by Lemma 4.17 and Proposition 4.12. This proves Corollary 4.18. Next, we consider the case $c(\gamma) = 1$. Since $c(\gamma) = 1$, either x_1 or x_2 (resp. x_3 or x_4) is a clear maximal (resp. minimal) vertex. Since clear minimal vertices are not clear maximal vertices, we have $v(\gamma) \geq 2$. Thus $v(\gamma) + 2c(\gamma) \geq v(\gamma) + 2 \geq 4$. □

The explicit examples of $(v, c) = (1, 1)$ or $(0, 2)$ are given in [**Sd2**] and [**RS1**]. Similar kinds of inequalities, including nonconvex space curves, are found in [**CR**] and [**RS2**].

Appendix A. Vertices on C^2-regular plane curves

As written in the Introduction, the four vertex theorem for simple closed Euclidean plane curves has been extended for various ambient spaces. On the other hand, there are many other known results for vertices on Euclidean plane curves with self-intersections; however, it is still unclear whether such a generalization works for these results. In this appendix, we give an abstract approach for the study of vertices on C^2-plane curves which may have self-intersections, and show that several known results are generalized for Minkowski plane curves and for curves on a convex surfaces with positive Gaussian curvature.

Let X be a C^r-differentiable sphere $(r \geq 1)$ and Γ a subset of C^r-regular simple closed curves satisfying the axioms of circle system (S1)–(S3) in Section 3. We call such Γ a C^r-differentiable circle system. Throughout the section, we fix a a C^2-differentiable circle system Γ on X satisfying the following additional condition.

(S4) For any $p \in X$ and a C^2-regular curve γ passing through p, there exists a unique circle $C_p \in \Gamma$ which has second order tangency with γ at p.

Such a circle C_p is called the *osculating circle* of γ at p.

Examples 1–3 in Section 3 satisfy this condition.

PROPOSITION A.1. *Let γ be a piecewise C^1-regular simple closed curve. Suppose that all internal angles of $\partial D^\bullet(\gamma)$ (resp. $\partial D^\circ(\gamma)$) are less than or equal to π. Then γ is •-admissible (resp. ∘-admissible).*

PROOF. We prove for $\partial D^\bullet(\gamma)$. (The corresponding assertion for $\partial D^\circ(\gamma)$ is obtained if one reverses the orientation of the curve.) $A_p^\bullet$ is not empty, since $p^\bullet \in A_p^\bullet$. If $A_p^\bullet = \{p^\bullet\}$, p is an admissible point by definition (see Definition 3.3). So we may assume that $A_p^\bullet \neq \{p^\bullet\}$. If p is a singular point of γ, $A_p^\bullet = \{p^\bullet\}$ holds, because the internal angle at p is less than π. Thus γ is C^1-regular at p. Then any two elements of $A_p^\bullet$ have only one common point p by Lemma A.2. Thus there exists a •-maximal circle $C_A^\bullet$ by (S3). The uniqueness of •-maximal circle at p also follows from Lemma A.2. Thus γ is •-admissible at p. □

LEMMA A.2. *Let C_1 and C_2 be two distinct circles in Γ passing through $p \in X$ with the same oriented tangent direction. Then $D^\bullet(C_1) \subset D^\bullet(C_2)$ or $D^\bullet(C_2) \subset D^\bullet(C_1)$ holds. In particular, they meet only at p.*

PROOF. By (S4), the 2-jets of C_1 and C_2 at p are mutually different. Thus there exists a sufficiently small neighborhood W of p in X such that $C_2 \cap W$ is contained in $D^\bullet(C_1)$ or $D^\circ(C_1)$. If necessary, by interchanging C_1 and C_2, we may assume that $C_2 \cap W \subset D^\bullet(C_1)$ holds. If $D^\bullet(C_2) \not\subset D^\bullet(C_1)$, C_2 must meet C_1 at least three points. By (S1), it is impossible. Thus we have $D^\bullet(C_2) \subset D^\bullet(C_1)$. Then again by (S1), we have $C_1 \cap C_2 = \{p\}$. □

LEMMA A.3. *Let γ be a piecewise C^2-regular simple closed curve. Then for each point $p \in \gamma$, the •-maximal circle $C_p^\bullet$ and the ∘-maximal circle C_p° satisfy the following relations:*

$$D^\bullet(C_p^\bullet) \subset D^\bullet(C_-) \subset D^\bullet(C_p^\circ),$$
$$D^\bullet(C_p^\bullet) \subset D^\bullet(C_+) \subset D^\bullet(C_p^\circ),$$

where C_- (resp. C_+) is the left (resp. right) limit osculating circle at p.

PROOF. Let Γ_p be the subset of Γ whose elements are tangent to γ at p. The set $\mathcal{A}_p^\bullet$ defined in Definition 3.3 can be written as

$$\mathcal{A}_p^\bullet = \{C \in \Gamma_p \cup \{p^\bullet\}\,;\, D^\bullet(C) \subset D^\bullet(\gamma)\}.$$

Let C' be a circle satisfying the relation $D^\bullet(C_-) \subsetneq D^\bullet(C')$. Then the 2-jet of C' at p is different from C_- by (S4). We denote by $\gamma|_{<p}$ the past part of γ from p. Since $\gamma|_{<p}$ has the second order tangency with C_- at p, any points on $\gamma|_{<p}$ close to p are contained in $D^\bullet(C')$. This implies $C' \notin \mathcal{A}_p^\bullet$. Thus $D^\bullet(C_p^\bullet) \subset D^\bullet(C_-)$ holds. Similarly, $D^\circ(C_p^\circ) \subset D^\circ(C_-)$ (which is equivalent to the condition $D^\bullet(C_-) \subset D^\bullet(C_p^\circ)$) can be also proved. Moreover, using the same argument, we get the corresponding assertion for C_+. □

LEMMA A.4. *Let γ be a piecewise C^2-regular curve on X and p a clean maximal (resp. minimal) vertex on γ. Suppose that C is a •-maximal (◦-maximal) circle passing through p satisfying $C \in S^\bullet(\gamma)$ (resp. $C \in S^\circ(\gamma)$). The the following assertions hold:*
  1. *C coincides with the left or light limit of osculating circle at p if $\theta \geq \pi$;*
  2. *$C = p^\bullet$ (resp. $C = p^\circ$) if $\theta < \pi$,*

where θ is the internal angle with respect to $D^\bullet(\gamma)$ (resp. $D^\circ(\gamma)$) at p. In particular •-maximal circles are not ◦-maximal circles unless γ is a circle.

PROOF. We consider the case that C is a •-maximal circle. When $\theta < \pi$, we have $C = p^\bullet$ obviously. So we may assume $\theta \geq \pi$. Let C_- (resp. C_+) be the left (resp. the right) limit of the osculating circle at p. Suppose that $C_p^\bullet \neq C_-, C_+$. Then by Lemma A.3, we have

$$D^\bullet(C_p^\bullet) \subsetneq D^\bullet(C_-) \cap D^\bullet(C_+).$$

We denote by Γ_p the subset of Γ whose elements are tangent to γ at p. Since $C_p^\bullet \neq C_-$ (resp. $C_p^\bullet \neq C_+$), the second order derivative of C_- (resp. of C_+) and $C_p^\bullet$ at p are mutually different by (S4). Moreover, by the existence of circles with given 2-jets as in (S4), there exists a sequence $(C_n)_{n \in \mathbf{N}}$ in Γ_p such that $C_n \to C_p^\bullet$ and

$$D^\bullet(C_p^\bullet) \subsetneq D^\bullet(C_n) \subsetneq D^\bullet(C_-) \cap D^\bullet(C_+) \qquad (n = 1, 2, 3, \dots).$$

Here we also used the fact that any two elements in Γ_p meet only at p by Lemma A.2. Without loss of generality, we may assume that

$$D^\bullet(C_{n+1}) \subset D^\bullet(C_n) \qquad (n = 1, 2, 3, \dots).$$

γ is approximated by C_- or C_+ at p in C^2-topology. Since C_1 and C_- (resp. C_+) have the distinct 2-jets and $\theta \geq \pi$, there exists an open subarc I containing p such that $I \setminus \{p\}$ lies in the interior of $D^\circ(C_1)$. We fix an arbitrary distance function $d(\,,\,)$ on X compatible with its topology. Since $C_p^\bullet$ and $\gamma \setminus I$ are disjoint closed subsets, the uniform distance $d(C_p^\bullet, \gamma \setminus I)$ is positive. As remarked in Section 3, the convergence $C_n \to C_p^\bullet$ is the same as that of the induced uniform distance of $J(X)$. Thus for a sufficiently large n, $d(C_n, \gamma \setminus I) > 0$. On the other hand, since $D^\bullet(C_n) \subsetneq D^\bullet(C_1)$, we have $C_n \cap I = \{p\}$. Thus $D^\bullet(C_n)$ is contained in $D^\bullet(\gamma)$. But this contradicts the maximality of $C_p^\bullet$. Thus we can conclude that $C_p^\bullet$ coincides with C_- or C_+. □

For simple closed curves, we defined clean vertices in Section 3, but for curves with self-intersections, they cannot be defined. Instead of clean vertices, we define maximal and minimal vertices on C^2-regular curves as follows:

DEFINITION A.1. A point p on γ is called a *maximal vertex* (resp. *minimal vertex*) if there exists an open subarc I of γ containing p such that $I \subset D^\circ(C_p)$ (resp. $I \subset D^\bullet(C_p)$). (In particular, all points on a circle are maximal and minimal vertices at the same time.)

In this appendix, the term "honest vertex" refers to a maximal or a minimal vertex unless otherwise stated.

REMARK. This abstract definition of an honest vertex is slightly different from the original concept in Euclidean plane curves. When γ is a Euclidean plane curve, an honest vertex should be defined as an extremal point of the curvature function. But in our general setting, we cannot define a curvature function. The honest vertices in the sense of the above definition and the extremal points of the curvature function coincide whenever the number of honest vertices is finite. On the other hand, if the number of honest vertices is infinite, honest vertices are divided into the following two cases:
1. extremal points of the curvature function;
2. an accumulate point of extremal points of the curvature function

(this observation is due to [**HK**]). The example of the graph of $t \to t^4 \sin(1/t)$ at $t = 0$ demonstrates this phenomenon, which was suggested by Dombrowski. Since we never use the curvature function in the following discussion, our definition of an honest vertex will not be confusing even when the curve has infinitely many honest vertices.

As an immediate consequence of Lemma A.4, we have the following result.

COROLLARY A.5. *Let γ be a C^2-regular simple closed curve. If p is a clean maximal (resp. minimal) vertex, then p is a maximal (resp. minimal) vertex. Furthermore, $C_p^\bullet = C_p$ (resp. $C_p^\circ = C_p$) holds.*

DEFINITION A.2. A C^2-regular curve $\sigma : [a, b] \to X$ is called a *shell* at p if $p = \sigma(a) = \sigma(b)$ and $\sigma|_{(a,b)}$ has no self-intersection. A shell is said to be *positive* (resp. *negative*) if the velocity vector $\sigma'(a)$ coincides with $\sigma'(b)$ or it points to the left (resp. right) of $\sigma'(b)$. The point p is called the *node* of the shell.

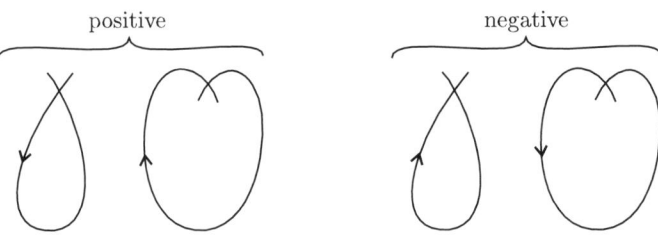

FIGURE A.1.

LEMMA A.6. *Let $\gamma : [a,b] \to X$ be a positive (resp. negative) shell. Then there exists $c \in (a,b)$, such that $C_{\gamma(c)} = C^{\bullet}_{\gamma(c)}$ (resp. $C_{\gamma(c)} = C^{\circ}_{\gamma(c)}$) and $C_{\gamma(c)} \neq C_{\gamma(a)}, C_{\gamma(b)}$.*

PROOF. By changing the orientation of the curve, we may assume that the shell is positive. Then γ is a $\bullet$-admissible curve by Proposition A.1. By Theorem 3.2, there exist two points x_1 and x_2 ($F^{\bullet}_{x_1} \neq F^{\bullet}_{x_2}$) such that $\text{rank}^{\bullet}(x_1) = \text{rank}^{\bullet}(x_2) = 1$. We set $p = \gamma(a) = \gamma(b)$. Since $F^{\bullet}_{x_1} \neq F^{\bullet}_{x_2}$, we may assume $p \notin F^{\bullet}_{x_1}$. By Corollary A.5, we have $C^{\bullet}_{x_1} = C_{x_1}$. Since $p \notin F^{\bullet}_{x_1}$, we have $C_{x_1} \neq C_{\gamma(a)}, C_{\gamma(b)}$. □

The following corollary is an abstract version of Jackson [**J1**, Lemma 4.3].

COROLLARY A.7. *A positive (resp. negative) shell $\gamma : [a,b] \to X$ has at least one maximal (resp. minimal) vertex in (a,b).*

PROPOSITION A.8. *Let $\gamma : [a,b] \to X$ be a curve which contains neither a maximal vertex nor a minimal vertex on (a,b). Then the one of the following two assertions are true*:
  (1) $\gamma|_{(a,b]}$ *lies in* $\mathcal{D}_a$;
  (2) $\gamma|_{(a,b]}$ *lies in* $D^{\bullet}(C_{\gamma(a)})^c$,
where $\mathcal{D}_a$ (resp. $D^{\bullet}(C_{\gamma(a)})^c$) is the interior (resp. the complement) of $D^{\bullet}(C_{\gamma(a)})$.

PROOF. Suppose that $\gamma|_{(a,b]}$ intersects $C_{\gamma(a)}$ first at p. Then composing γ with $C_{\gamma(a)}$ at $\gamma(a)$, we get a no-vertex shell at p. But the shell does not satisfy the conclusion of Lemma A.6, a contradiction. □

DEFINITION A.3. Let $\gamma : [a,b] \to X$ be a curve which contains neither maximal vertices nor minimal vertices on (a,b). Then γ is called a *positive scroll* (resp. *negative scroll*) if (1) (resp. (2)) of Proposition A.8 occurs.

By definition, positivity or negativity of scrolls does not depend on the choice of orientation of the scrolls. Lemma A.6 yields the following abstract version of A. Kneser's theorem [**KA**].

THEOREM A.9. *Let $\gamma : [a,b] \to X$ be a positive scroll (resp. negative scroll). Then the osculating circle $C_{\gamma(b)}$ lies in $\mathcal{D}_a$ (resp. $D^{\bullet}(C_{\gamma(a)})^c$).*

PROOF. Suppose that two osculating circles intersect. Then we can use arcs of $C_{\gamma(a)}$, γ and $C_{\gamma(b)}$ to find a shell at the one of intersection points of two circles $C_{\gamma(a)}$ and $C_{\gamma(b)}$. This contradicts Lemma A.6, since γ has no honest vertex. Thus $C_{\gamma(a)} \cap C_{\gamma(b)}$ is empty. Since $\gamma(b)$ lies in $\mathcal{D}_a$ (resp. $D^{\bullet}(C_{\gamma(a)})^c$) by Proposition A.8, we have $C_{\gamma(b)} \subset \mathcal{D}_a$ (resp. $C_{\gamma(b)} \subset D^{\bullet}(C_{\gamma(a)})^c$). □

COROLLARY A.10. *Let γ be a C^2-regular closed curve with finitely many maximal vertices. Then the number of maximal vertices is equal to the number of minimal vertices. More precisely, for any two different maximal vertices p, q on γ, there is a minimal vertex on $\gamma|_{(p,q)}$.*

PROOF. Suppose that there is no minimal vertex between p and q. Without loss of generality, we may assume that $\gamma|_{(p,q)}$ is vertex free. Since p is a maximal vertex, $\gamma|_{[p,q]}$ is a negative scroll. On the other hand, Since q is also a maximal vertex, $\gamma|_{[p,q]}$ is a positive scroll. This is a contradiction. □

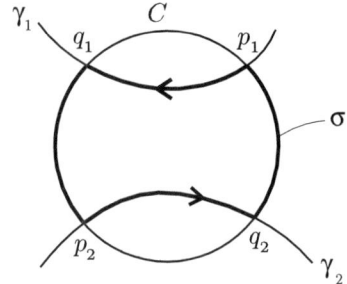

Figure A.2

As an application, we give the following $2n$-vertex theorem which is a generalization of a result of Jackson [**J1**]. (For convex curves, it was proved by Blaschke [**Bl1**]. Similar axiomatic treatments of the $2n$-vertex theorem are found in Haupt and Künneth [**HK2, HK3**].)

THEOREM A.11. *Let γ be a C^2-regular simple closed curve on (X, Γ) such that a circle $C \in \Gamma$ meets γ transversally at $p_1, q_1, \ldots, p_n, q_n \in \gamma \cap C$. Suppose that the rotational order of the crossings $p_1, q_1, \ldots, p_n, q_n$ of γ is the same as that of C. Then γ has at least $2n$ different honest vertices.*

The outline of the proof is essentially the same as in [**J1**, Theorem 7.1]. But in our general setting, we cannot apply Jackson [**J1**, Lemma 3.1]. The following lemma will replace Jackson's lemma.

LEMMA A.12. *Let C be a circle and γ_j $(j = 1, 2)$ two C^2-regular curves with finitely many honest vertices transversally intersecting C at two points p_j, q_j $(j = 1, 2)$. Suppose that $\gamma_1|_{[p_1,q_1]}$ and $\gamma_2|_{[p_2,q_2]}$ lie in $D^\bullet(C)$ and have no intersections with each other (see Figure A.2). Then there is a circle C' which lies in $D^\bullet(C)$ such that it is tangent to the three arcs $\gamma_1|_{[p_1,q_1]}$, $C|_{[q_1,p_2]}$ and $\gamma_2|_{[p_2,q_2]}$.*

PROOF. Let σ be a piecewise C^2-regular curve consisting of the four arcs $\gamma_1|_{[p_1,q_1]}$, $C|_{[q_1,p_2]}$, $\gamma_2|_{[p_2,q_2]}$ and $C|_{[q_2,p_1]}$. Since interior angles of $\partial D^\bullet(\sigma)$ at p_1, q_1, p_2, q_2 are less than π, σ is a $\bullet$-admissible curve by Proposition A.1. By Proposition 3.1, we get an intrinsic circle system $(F_p^\bullet)_{p \in \sigma}$ on σ. The four points p_1, q_1, p_2, q_2 are clean maximal vertices on σ. For each $x \in C|_{[q_1,p_2]}$, $F_x^\bullet \cap \gamma_j|_{[p_1,q_1]} = \emptyset$ $(j = 1, 2)$ never hold at the same time. (In fact, if so, we have rank$^\bullet(x) = 1$, since (S1) yields $C_x^\bullet \neq C$. Then $C_x^\bullet$ coincides with C by Lemma A.4, which is a contradiction.) Thus the arc $C|_{[q_1,p_2]}$ is a union of the following two subsets:

$$B_1 := \{x \in C|_{[q_1,p_2]} : F_x^\bullet \cap \gamma_1|_{[p_1,q_1]} \neq \emptyset\},$$
$$B_2 := \{x \in C|_{[q_1,p_2]} : F_x^\bullet \cap \gamma_2|_{[p_2,q_2]} \neq \emptyset\}.$$

By Lemma A.4, we have $F_{q_1}^\bullet = \{q_1\}$. Since $C|_{[q_1,p_2]}$ is circular and C is not contained in $D^\bullet(\sigma)$, there are no clean maximal vertices on $C|_{[q_1,p_2]}$ by Lemma A.4. Thus $C|_{[q_1,p_2]}$ is a weakly $\bullet$-regular, and we have $\lim_{x \to q_1+0} \mu_-^\bullet(x) = q_1 - 0$ (resp. $\lim_{x \to p_2-0} \mu_+^\bullet(x) = p_2 + 0$) by Theorem 1.4. So B_1 (resp. B_2) is not empty. We set

$$y_1 := \sup(B_1), \qquad y_2 := \inf(B_2),$$

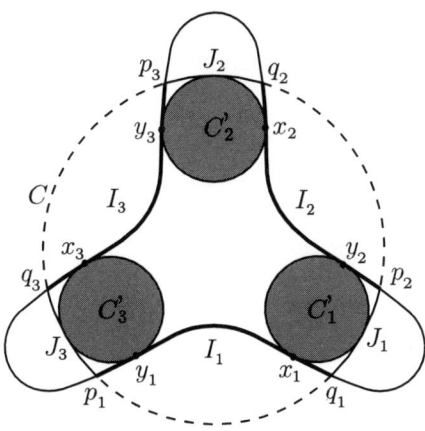

FIGURE A.3

where the lowest upper bound and the greatest lower bound are taken with respect to the canonical order of the arc $C|_{[p_1,p_2]}$. Since $C|_{[q_1,p_2]} = B_1 \cup B_2$, we have $y_1 \succeq y_2$. By the second property of intrinsic circle system (I2), we have $F^{\bullet}_{y_1} = F^{\bullet}_{y_2}$. Thus $C' := C^{\bullet}_{y_1} = C^{\bullet}_{y_2}$ is the desired circle. □

PROOF OF THEOREM A.11. We set $I_k := \gamma|_{[p_k,q_k]}$. Without loss of generality, we may assume that $\gamma \cap D^{\bullet}(C) = I_1 \cup \cdots \cup I_n$. We set

$$J_1 := C|_{[q_n,p_1]}, \quad J_2 := C|_{[q_1,p_2]}, \quad \cdots, \quad J_n := C|_{[q_{n-1},p_n]}.$$

By Lemma A.12, there exists a circle C'_k ($k = 1, \ldots, n$) which is tangent to I_k, J_k and I_{k+1}, respectively. Let x_k (resp. y_k) be a tangent point between C'_k and I_k (resp. I_{k+1}). Then there is a maximal vertex on $\gamma|_{(x_k,y_k)}$ by Lemma 1.1. (It is a clean vertex of the simple closed curve obtained by joining $\gamma|_{(x_k,y_k)}$ and C'_k, but not a clean vertex of γ in general.) Moreover, by (S1) in Section 3, we have

$$(y_1 \succ) y_n \succ x_n \succ \cdots \succ x_1 \succ y_1,$$

where $\succ$ is the rotational order of γ (see Figure A.3). Thus γ has n clean maximal vertices. By Corollary A.10, γ has n clean minimal vertices between them. □

LEMMA A.13 (The abstract version of [**KU**, Lemma 3.1]). *Let $\gamma : [a,b] \to \mathbf{R}^2$ be a positive shell at $p = \gamma(a) = \gamma(b)$.*
  1. *If γ has only one (necessary maximal) vertex, then $\gamma \setminus \{p\} \subset \mathcal{D}_a \cap \mathcal{D}_b$, where $\mathcal{D}_a$ (resp. $\mathcal{D}_b$) is the interior of the closed domain $D^{\bullet}(C_{\gamma(a)})$ (resp. $D^{\bullet}(C_{\gamma(b)})$).*
  2. *If γ has exactly two honest vertices, maximal at $t_1 \in (a,b)$ and minimal at $t_2 \in (a,b)$, then $\gamma \setminus \{p\} \subset \mathcal{D}_a$ if $t_1 < t_2$ and $\gamma \setminus \{p\} \subset \mathcal{D}_b$ if $t_2 < t_1$.*
  3. *If γ has exactly three honest vertices, two of which are maximal and the other is minimal, then either $\gamma \setminus \{p\} \subset \mathcal{D}_a$ or $\gamma \setminus \{p\} \subset \mathcal{D}_b$.*

PROOF. Lemma A.13 is an abstract version of a lemma in Kobayashi–Umehara [**KU**, Lemma 3.1]. Here we give an alternative proof. Since the shell σ is a piecewise C^2-regular curve, there exist at least two distinct ∘-maximal circles $C_1, C_2 \in S^{\circ}(\gamma)$ by Corollary 3.4. If there are no minimal vertices in σ, both of C_1 and C_2 pass through p. By Lemma A.4, C_1 and C_2 are the osculating circles at $t = a, b$, respectively. This proves (1). Next we suppose that σ has only one minimal vertex at $q = \gamma(t_2)$. Then either C_1 or C_2 does not pass through q. Without loss of generality, we may assume that $C_1 \not\ni q$. Then $C_1 \cap \gamma = \{p\}$ holds and C_1 coincides with $C_{\gamma(a)}$ or $C_{\gamma(b)}$ by Lemma A.4. This proves (3). Next we prove (2). Suppose $t_1 < t_2$ (resp. $t_2 < t_1$). Then $\mathcal{D}_b \not\ni \gamma(t_2)$ (resp. $\mathcal{D}_a \not\ni \gamma(t_2)$) by Proposition A.8. Thus we see that $C_1 \neq C_{\gamma(b)}$ (resp. $C_1 \neq C_{\gamma(a)}$), which implies that $C_1 = C_{\gamma(a)}$ (resp. $C_1 = C_{\gamma(b)}$). This proves (2). □

Using Lemma A.13, the following theorem can be proved by the same arguments as in [**KU**, Theorem 3.5].

THEOREM A.14 (The abstract version of [**KU**, Theorem 3.5]). *If a closed curve contains three positive shells or three negative shells, then it has at least six honest vertices.*

The above 6-vertex theorem is stronger than the following 6-vertex theorem.

COROLLARY A.15 (The abstract version of [**CMO**] and [**U1**]). *A closed curve has at least six honest vertices if it bounds an immersed surface other than the disc.*

PROOF. It is sufficient to show that *any closed curve γ which bounds an immersed surface with positive genus has three negative shells.* (The immersed surface is assumed to lie on the left-hand side of γ.) If γ has a positive shell, then by the proof in [**KU**, Corollary 3.7], we find three negative shells. Hence, we may assume that γ is not embedded and γ has no positive shell. Suppose that γ has at most two negative shells. Let x is a self-intersection of γ. Then γ can be expressed as a union of two distinct loops γ_1 and γ_2 at x. Each loop γ_i contains at least one shell S_i, which must be negative because γ has no positive shells. We take points $q_j \in S_j \setminus \{p_j\}$ ($j = 1, 2$), respectively, where p_j is the node of the shell S_j. Then γ can be divided into two arcs $\gamma|_{[q_1,q_2]}$ and $\gamma|_{[q_2,q_1]}$. Moreover these two closed arcs $\gamma|_{[q_1,q_2]}$ and $\gamma|_{[q_2,q_1]}$ are both embedded. (In fact, for example, if $\gamma|_{[q_1,q_2]}$ is not embedded, then we find third shell S on $\gamma|_{[q_1,q_2]}$, which must be negative. This is a contradiction.) Then by [**U1**, Theorem 3.1], γ only bounds a disc, which is a contradiction. □

Corollary A.15 for Euclidean plane curves was first proved for normal curves in [**CMO**] and extended to the general case in [**U1**]. It should be remarked that Corollary A.15 itself is obtained by Corollary A.7 using purely topological arguments. The following related result can be proved by the method in [**Pe**] using Corollary A.7.

THEOREM A.16 (The abstract version of [**Pe**, Theorem 4]). *A closed curve has at least $(4g + 2)$-vertices if it bounds an immersed surface of genus g, provided that the number of self-intersections does not exceed $2g + 2$.*

In the rest of this appendix, we consider an intersection sequence of a positive scroll and a negative scroll, which is an abstract version of [**KU**, §4]. As an application, a structure theorem for a 2-vertex curve is obtained. In [**KU**, §4], we use a corner rounding technique on curves. But this method is not valid in our general setting. So the following is the modified version of [**KU**, §4].

FIGURE A.4A   FIGURE A.4B

Let γ^- and γ^+ be positive and negative scrolls, respectively, satisfying the following two properties:
 (a) All intersections of γ^- to γ^+ are transversal.
 (b) The first crossing of γ^+ is the last crossing of γ^-.

A crossing of γ is called *positive* (resp. *negative*) if γ^+ crosses γ^- from the left (resp. right). We use small letters for positive crossings. For the sake of simplicity, we use the following notation. Let γ be an open arc and p a point on γ. Then we denote by $\gamma|_{>p}$ (resp. $\gamma|_{<p}$) the future part (resp. the past part) from p.

DEFINITION A.4 (The $*$-pairing). Let a be a positive crossing. If a crossing is the first one at which $\gamma^-|_{>a}$ meets $\gamma^+|_{<a}$, then it is expressed by a^*.

LEMMA A.17. *Let γ^+ and γ^- be positive and negative scrolls satisfying (a) and (b). If there exists a crossing a^* for a positive crossing a, then a^* is a negative crossing.*

PROOF. Suppose that a^* is a positive crossing. Let σ be a simple closed curve defined as a union of two arcs $\sigma := \gamma^-|_{[a,a^*]} \cup \gamma^+|_{[a^*,a]}$. Let $D^\bullet(\sigma)$ be the left-hand closed domain with respect to σ as in Figure A.4a or A.4b. The angle at a^* of the domain is greater than π. We consider a sufficiently small circle C, which is tangent to γ^- at a^* and lies in $D^\bullet(\sigma)$. Expand C continuously. Let $x \neq a^*$ be the first attachment of C to the heart figured domain. Then $x \neq a$, and C is tangent to γ^- or γ^+ at x. If $x \in \gamma^-$, we have $C \subset \overline{\mathcal{D}_x}$, where $\mathcal{D}_x$ is the left open domain of the osculating circle C_x. Since γ^- is a negative arc, we have $\overline{\mathcal{D}_x} \subset \mathcal{D}_{a^*}$ by Theorem A.9. Hence C cannot meet C_{a^*}, which is a contradiction because of $a^* \in C \cap C_{a^*}$. □

LEMMA A.18. *Let γ^+ and γ^- be positive and negative scrolls satisfying (a) and (b). Suppose that there exists a crossing a^* for a positive crossing a. If $\gamma^+|_{>a}$ (resp. $\gamma^-|_{<a}$) meets γ^- (resp. γ^+) at q firstly, then q lies on $\gamma^-|_{>a}$ (resp. $\gamma^+|_{<a}$).*

PROOF. Let $\sigma = \gamma^-_{[a,a^*]} \cup \gamma^+_{[a^*,a]}$ be a simple closed curve. Let C_a^- (resp. C_a^+) be the osculating circle at a with respect to γ^- (resp. γ^+). By Proposition A.8 and Definition A.3, we have $(D^\bullet(C_a^-))^c \supset \gamma^-|_{>a}$ and $(D^\bullet(C_a^+))^c \supset \gamma^+|_{<a}$, where $(D^\bullet(C_a^\pm))^c$ are the complements of $D^\bullet(C_a^\pm)$. Thus
$$(D^\bullet(C_a^-) \cap D^\bullet(C_a^+))^c = (D^\bullet(C_a^-))^c \cup (D^\bullet(C_a^+))^c \supset \sigma.$$
This implies that
(A.1) $$D^\bullet(C_a^-) \cap D^\bullet(C_a^+) \subset D^\bullet(\sigma).$$
On the other hand, $\gamma^+|_{>a} \subset D^\bullet(C_a^+)$ and $\gamma^-|_{<a} \subset D^\bullet(C_a^-)$. Suppose that $\gamma^+|_{>a}$ meets $\gamma^-|_{<a}$ at some point x. Then $x \in D^\bullet(C_a^-) \cap D^\bullet(C_a^+)$, so $x \in D^\bullet(\sigma)$ by (A.1).

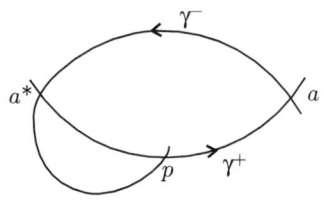
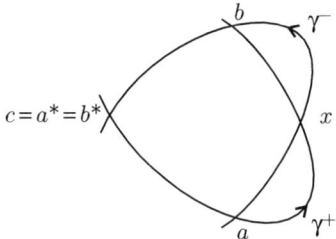

FIGURE A.5A.   FIGURE A.5B.

This means that $\gamma^+|_{>a}$ (resp. $\gamma^-|_{<a}$) meets $\gamma^-|_{[a,a^*]}$ (resp. $\gamma^+|_{[a^*,a]}$) before x (resp. after x). □

LEMMA A.19. *Let a be a positive crossing.*
(1) *a^* coincides with the first crossing at which the past part of γ^+ from a meets the future part of γ^- from a.*
(2) *If $a^* = b^*$, then $a = b$.*

PROOF. We prove the first assertion. Suppose $p \neq a^*$ is the first crossing at which $\gamma^+|_{<a}$ meets $\gamma^-|_{>a}$. Then p lies on $\gamma^+|_{[a^*,a]}$ (see Figure A.5a). Consequently, p is a positive crossing. Then $p = a^\diamond$, where $a^\diamond$ is the $*$-paring between the negative scroll γ^- and the positive scroll $\gamma^+|_{>p}$. On the other hand $p = a^\diamond$ is a negative crossing by Lemma A.17. This is a contradiction.

Next we prove (2). Suppose that $a \neq b$. Without loss of generality, we may assume that γ^- meets γ^+ first at a, next at b, and finally at $c = a^* = b^*$. Since b is a positive crossing, there is a negative crossing x on $\gamma^-|_{[a,b]}$ at which $\gamma^+|_{>a}$ meets $\gamma^-|_{[a,b]}$ first (see Figure A.5b). Now we reverse the orientation of γ^-, which is denoted by $\langle -\gamma^- \rangle$. We denote by $\#$ the $*$-pairing between the negative scroll $\langle -\gamma^- \rangle$ and the positive scroll γ^+. Then the signs of crossings are all reversed. We have $a = x^\#$. But $\gamma^+|_{>x}$ meets $\langle -\gamma^- \rangle$ at b, which contradicts Lemma A.18. □

DEFINITION A.5. *If a negative crossing does not have a $*$-pairing, then it is called a* solitary negative crossing *and is denoted by a capital letter.*

The remaining discussions in [**KU**, §4] can be easily translated to our abstract setting. In particular, the intersection sequence of γ^- consists of the following three types of words:

$$\text{Type T:} \quad A_1 A_2 \ldots A_n,$$
$$\text{Type D:} \quad [a_1 a_2 \ldots a_n] := a_1 \ldots a_n a_n^* \ldots a_1^*,$$
$$\text{Type S:} \quad [a_1 a_2 \ldots a_n : B] := a_1 \ldots a_n B a_n^* \ldots a_1^*.$$

We define the length of the each type of word by

$$|A_1 A_2 \ldots A_n| := n, \quad |[a_1 a_2 \ldots a_n]| := n, \quad |[a_1 a_2 \ldots a_n : B]| := n+1.$$

The following theorem holds by exactly the same argument as in [**KU**, §4].

THEOREM A.20. *Let γ^+ and γ^- be positive and negative scrolls satisfying (a) and (b). Then the intersection sequence W^- of γ^- is of the form $W^- = W_1 W_2 \cdots W_n$, where W_i ($i = 1, \ldots, n$) is of type T, D or S and the intersection*

sequence of γ^+ is obtained by the head picking rule as in [**KU**]. Moreover W^- satisfies the following grammar:
1. If W_i is of type D, then W_j ($j < i$) is of type T or D.
2. If W_i is of type T and W_{i+1} is of type D, then $|W_i| \leq |W_{i+1}|$. Moreover if W_{i-1} is of type D, then $|W_i| + |W_{i-1}| \leq |W_{i+1}|$ holds.
3. If W_i is of type T and W_{i-1}, W_{i+1} is of type S, then $|W_i|+|W_{i-1}| \geq |W_{i+1}|$.

An immersed curve is called *normal* if all crossings are transversal and there are only double points. The following theorem is obtained by exactly the same argument as in the proof of [**KU**, Theorems 4.8 and 4.9].

THEOREM A.21 (A structure theorem of 2-vertex curves). *Let γ be a closed normal 2-vertex curve divided by negative and positive scrolls $\gamma = \gamma^- \cup \gamma^+$. Then the intersection sequences of γ^- and γ^+ are translated mutually by the head picking rule as in [**KU**]. Moreover, the grammar of the intersection sequence of γ^- is given as follows.*
1. *The intersection sequence consists of words of type T and type S and written in the form $T_0 S_1 T_1 S_2 T_2 \cdots S_k T_k$. Each T_i ($i = 1, \ldots, k$) may possibly be empty.*
2. $|T_0| > 0$, $|T_0| \geq |S_1|$ *and* $|S_i| + |T_i| \geq |S_{i+1}|$ ($i = 1, \ldots, k$).

When $X = \mathbf{R}^2 \cup \{\infty\}$ and Γ is the set of circles in the Möbius plane (see Section 3, Example 1), the converse assertions of Theorems A.20 and A.21 are true (see [**KU**]). Moreover, in [**KU**], intersection sequences of two scrolls of the same kind are also characterized in a similar manner.

For a plane curve γ, there exists an interesting invariant $J^+(\gamma) \in \mathbf{Z}$, which is related to the linking number of the corresponding Legendrian knot in the unit sphere bundle on $\mathbf{R}^2$ (see [**A1–A3**]; Polyak [**Po**], Selwat [**Sl**] and Lin–Wang [**LW**] are also nice references). Since $J^+(\gamma)$ is not invariant under the diffeomorphism of $S^2 = \mathbf{R}^2 \cup \{\infty\}$, it is convenient to define a modified invariant

$$SJ^+(\gamma) := J^+(\gamma) + \frac{i_\gamma^2}{2},$$

where i_γ is the rotation number of γ as a plane curve. As an application of Theorem A.21, we can get the following by the same method as in [**U2**] (see also Remark in [**U2**, §1]).

THEOREM A.22 (The abstract version of [**U2**]). *Let γ be a normal closed curve in X. Suppose that $SJ^+(\gamma) > 0$, then γ has at least four honest vertices.*

Two closed normal curves $\gamma_1, \gamma_2 : S^1 \to S^2$ are called *geotopic* if there is a diffeomorphism φ on S^2 which maps the image of γ_1 onto the image of γ_2. It is an interesting problem to determine the minimum number of honest vertices that a closed normal curve with given geotopy type can have. Minimizing numbers for normal curves are determined by Heil [**He1**] for crossings (≤ 3), and in [**KU**] and Kobayashi [**Ko**] for crossings (≤ 5).

Appendix B. The continuity of the maximal circles

In this appendix we shall prove the continuity of the center of maximal circles of a simple closed curve in the Euclidean plane. This was used in the last remark in Section 2. First, we prove the following general statement.

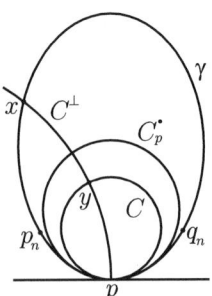

FIGURE B.1

THEOREM B.1. *Let X be a C^1-differentiable sphere and Γ a C^1-differentiable circle system. Let γ be a C^1-regular simple closed curve satisfying $s^\bullet(\gamma) < \infty$ and $c^\bullet : \gamma \to \Gamma$ a map defined by $c^\bullet(p) := C_p^\bullet$. Suppose that for each $p \in \gamma$, there exists a circle passing through p which is transversal to the tangent direction. Then $c^\bullet$ is a continuous mapping with respect to the compact open topology on Γ.*

If X is C^2 differentiable sphere and Γ is a C^2-differentiable circle system satisfying the property (S4) in Appendix A, then the assumption of Theorem B.1 is satisfied for any C^1-regular curves satisfying $s^\bullet(\gamma) < \infty$.

PROOF. If γ is a circle, the statement is obvious. So we assume γ is not a circle. It is sufficient to show that $C_{p_n}^\bullet \to C_p^\bullet$ for any $p \in \gamma$ and a sequence $(p_n)_{n \in \mathbf{N}}$ converging to p (see the Remark of Definition 3.1). For the sake of simplicity we use the notation $C_n := C_{p_n}^\bullet$. By (S2), the sequence $(C_n)_{n \in \mathbf{N}}$ has a convergent subsequence which converges to a circle C. Without loss of generality, we may assume $C_n \to C$. Obviously C is a circle contained in $D^\bullet(\gamma)$ which is tangent at p. Suppose that rank$^\bullet(p) \geq 2$. Since $s^\bullet(\gamma) < \infty$, there exists an open weakly •-regular arc I containing p by Corollary 2.6. Since $\mu_\pm^\bullet(p_n) \in C_n$ and $\lim_{p_n \to p \mp 0} \mu_\pm^\bullet(p_n) = \mu_\pm^\bullet(p) \pm 0$, the circle C contains $\mu_+^\bullet(p)$ or $\mu_-^\bullet(p)$. Since $p \neq \mu_\pm^\bullet(p)$, we have $C = C_p^\bullet$.

So we may assume rank$^\bullet(p) = 1$. Then $F_p^\bullet$ is connected. If $p_n \in F_p^\bullet$ for infinitely many p_n, then $C = C_p^\bullet$ is obvious. So we may assume that $p_n \notin F_p^\bullet$. By Corollary 2.6, there exists an open arc J containing $F_p^\bullet$ such that each point of $J \setminus F_p^\bullet$ is weakly •-regular. Since $F_p^\bullet$ is connected, $J \setminus F_p^\bullet$ consists of two disjoint weakly •-regular open arcs J_1 and J_2. By taking a subsequence, we may assume that $(p_n)_{n \in \mathbf{N}}$ is contained in J_1. (If necessary, we replace J_1 and J_2). Here, we denote by $\mu_\pm^\bullet$ the antipodal functions on J_1. Since $\lim_{p_n \to p \mp 0} \mu_\pm^\bullet(p_n) = \mu_\pm^\bullet(p) \pm 0$, either

$$q_n := \mu_+^\bullet(p_n) \text{ or } \mu_-^\bullet(p_n)$$

is contained in J_2 for sufficiently large n. We consider the case $F^\bullet \neq \{p\}$, then $F_p^\bullet$ is a closed arc, and consequently $(p_n)_{n \in \mathbf{N}}$ and $(q_n)_{n \in \mathbf{N}}$ converge to different points p, q which consist of the boundary of $F_p^\bullet$ in γ. Thus we have $p, q \in C$. Since $p \neq q$, we have $C = C_p^\bullet$.

Finally, we consider the case $F^\bullet = \{p\}$. Then, the sequences $(p_n)_{n \in \mathbf{N}}$ and $(q_n)_{n \in \mathbf{N}}$ converge to p from mutually different directions. Let $C^\perp$ be a circle passing through p which is transversal to the tangent direction of γ. Then there exists a point x such that $C^\perp|_{[p,x]} \subset D^\bullet(\gamma)$ and $C^\perp|_{(p,x)} \cap \gamma = \emptyset$ (see Figure B.1). Suppose that $C \neq C_p^\bullet$. Then $C \cap \gamma = \{p\}$ follows from the relation $D^\bullet(C) \subsetneq D^\bullet(C_p^\bullet)$. We remark that $C^\perp|_{(p,x)}$ separates $D^\bullet(\gamma)$ into two domains. Since C lies in $D^\bullet(\gamma)$ and is transversal to $C^\perp$ at p, we can conclude that C passes through a point $y\,(\neq p)$ on $C^\perp|_{(p,x)}$. By (S1), such a point y is uniquely determined, and $y \neq x$ because of $C \cap \gamma = \{p\}$. Since the sequences $(p_n)_{n\in\mathbf{N}}$ and $(q_n)_{n\in\mathbf{N}}$ converge to p from mutually different directions, there exists a point a_n (resp. b_n) such that $C^\bullet_{[p_n,q_n]}$ (resp. $C^\bullet_{[q_n,p_n]}$) meets $C^\perp|_{(p,x)}$ at a_n (resp. b_n) by the Jordan separation theorem. Since $C_n \to C$, the pair of these two points (a_n, b_n) converges to (p, y) or (y, p). Since two circles C and $C^\bullet$ have only one common point by (S1), $C^\perp|_{(p,y)}$ lies in the interior of $D^\bullet(C^\bullet_p)$. Thus a_n and b_n are interior points of $C^\bullet_p$ for any sufficiently large n. On the other hand, since p_n and q_n are exterior points of $D^\bullet(C^\bullet_p)$, two circles C_n and $C^\bullet_p$ must have at least four common points. Hence $C_n = C^\bullet_p$ and thus $C = C^\bullet_p$, which is a contradiction. $\square$

Let $\gamma : S^1 \to \mathbf{R}^2$ be a C^1-regular simple closed curve in the Euclidean plane. Assume that γ is oriented so that $D^\bullet(\gamma)$ is a bounded domain in $\mathbf{R}^2$. For each point $p \in \gamma$, let c_p be the center of the maximal circle $C^\bullet_p$. Then we have the following.

COROLLARY B.2. *Suppose that $s^\bullet(\gamma) < \infty$. Then the map $\Phi : \gamma \to \mathbf{R}^2$ defined by $\Phi(p) = c_p$ is continuous.*

PROOF. Let Γ_1 be the set of circles in the Euclidean plane X_1. It is not so hard to see that the map $\psi : \Gamma_1 \to \mathbf{R}^2 \times (0, \infty)$ defined by $\psi(C) = (z(C), r(C))$ is a homeomorphism, where $z(C)$ and $r(C)$ are the center and the radius of C, respectively. Since $\psi \circ c^\bullet$ is continuous by Theorem B.1, we have the conclusion.$\square$

Acknowledgments. I wish to thank S. Tabachnikov and G. Thorbergsson for their encouragement and fruitful discussions, and V. D. Sedykh for his many helpful remarks on an earlier version of this paper. I also wish to thank E. Heil and P. Dombrowski for helpful comments. I am deeply grateful to the Max-Planck-Institut für Mathematik in Bonn for its hospitality and wonderful working conditions.

References

[A1] V. I. Arnold, *Plane curves, their invariants, perestroikas and classifications*, Adv. Soviet Math. **21** (1994), 33-91.

[A2] ———, *The geometry of spherical curves and the algebra of quaternions*, Uspekhi Mat. Nauk **50:1** (1995), 3-68; English transl. in Russian Math. Surveys 50:1 (1995).

[A3] ———, *Topological invariants of plane curves and caustics*, University Lecture Series 5, Amer. Math. Soc., Providence, 1994, pp. 1-60.

[A4] ———, "On the number of flattening points of space curves", *Sinai's Moscow Seminar on Dynamical Systems* (L. A. Bunimovich, B. M. Gurevich and Ya. B. Pesin, eds.), Amer. Math. Soc. Transl. Ser. (2), 171, Amer. Math. Soc., Providence, RI, 1996, 11–22.

[A5] ———, *Remarks on the extatic points of plane curves*, Preprint.

[BR1] N. Ballesteros and M. C. Romero Fuster, *Global bitangency properties of generic closed space curves*, Math. Proc. Camb. Philos. Soc. **112** (1992), 519–526.

[BR2] ———, *A four vertex theorem for strictly convex space curves*, J. Geom. **46** (1993), 119–126.

[Ba] M. Barner, *Über die Mindestzahl stationärer Schmiegebenen bei geschlossenen strengkonvexen Raumkurven*, Abh. Math. Sem. Univ. Hamburg **19** (1955), 196–215.

[BF] M. Barner and F. Flohr, *Der Vierscheitelsatz und seine Verallgemeinerungen*, Mathematikunterricht **4** (1958), 43–73.

[Bi1] T. Bisztriczky, *Inflectional convex space curves*, Canad. J. Math. **36** (1984), 537–549.

[Bi2] ———, *On the four-vertex theorem for space curves*, Journal of Geometry **27** (1986), 166–174.

[Bl1]   W. Blaschke, *Kreis und Kugel*, Voget.Leipzig. (1916), 115–116.
[Bl2]   _____, *Vorlesungen über Differentialgeometrie*. II, Springer-Verlag (1923).
[Bo]   R. C. Bose, *On the number of circles of curvature perfectly enclosing or perfectly enclosed by a closed convex oval*, Math. Z **35** (1932), 16–24.
[CMO]   G. Cairns, M. McIntyre, and M. Özdemir, *A six-vertex theorem for bounding normal planar curves*, vol. 25, 1993, pp. 169–176.
[COT]   G. Cairns, M. Özdemir and E. Tjaden, *A counterexample to a conjecture of U. Pinkall*, Topology **31** (1992), 557–558.
[CS]   G. Cairns and R. W. Sharpe, *On the inversive differential geometry of plane curves*, L'Enseignement Math. **36** (1990), 175–196.
[CR]   S. R. Costa and M.C. Romero-Fuster, *Nowhere vanishing torsion closed curves always hide twice*, Geom. Dedicata **66**, 1–17.
[Gr]   W. C. Graustein, *Spherical curves that bound immersed discs*, Trans. Amer. Math. Soc. **41** (1937), 9–23.
[GH]   M. J. Greenberg and J. R. Harper, *Algebraic topology*, Mathematics Lecture Note Series, vol. 58, Benjamin, 1981.
[Gu1]   H. Guggenheimer, *On plane Minkowski geometry*, Geom. Dedicata **12** (1982), 371–381.
[Gu2]   _____, *Sign changes, extrema, and curves of minimal order*, J. Diff. Geom. **3** (1969), 511–522.
[GMO]   L. Guieu, E. Mourre and Y. Ovsienko, *Theorem on six vertices of a plane curve via the strum theory*, Preprint.
[Hu]   O. Haupt, *Verallgemeinerung eines Satzes von R.C. Bose über die Anzahl der Schmiegkreise eines Ovals, die vom Oval umschlossen werden oder das Oval umschließen*, J. Reine Angew. Math. **239/240** (1969), 339–352.
[HK1]   O. Haupt and H. Künneth, *Geometrische Ordnungen*, Springer-Verlag, Berlin-Heidelberg-New York, 1967.
[HK2]   _____, *Über einen 2n-Scheitelsatz*, Bayer. Acad. Wiss. Math.-Natur. Kl. Sitzungsber (1974), 59–72.
[HK3]   _____, *Über einen 2n-Scheitelsatz von Herren S. B. Jackson*, Bayer. Acad. Wiss. Math.-Natur. Kl. Sitzungsber (1974), 145–166.
[Hy]   T. Hayashi, *Some geometrical properties of the Fourier series*, Rend. Circ. Mat. Palermo **50** (1926), 96–102.
[He1]   E. Heil, *Some vertex theorems proved by means of Möbius transformations*, Ann. Mat. Pura Appl. **85** (1970), 301–306.
[He2]   _____, *Scheitelsätze in der Euklidishen, affinen und Minkowskischen Geometrie*, Darmstadt, 1967 (mimeographed).
[He3]   _____, *Der Vierscheitelsatz in Relativ- und Minkowski-Geometrie*, Monatshefte Math. **74** (1970), 97–107.
[He4]   _____, *Verschärfungen des Vierscheitelsatzes und ihre relativ-geometrischen Verallgemeinerungen*, Math. Nachrichten **45** (1970), 227–241.
[He5]   _____, *Sechsscheitelsätze für Ovale*, Tensor N.S. **25** (1972), 439–447.
[J1]   S. B. Jackson, *Vertices of plane curves*, Bull. Amer. Math. Soc. **50** (1944), 564–578.
[J2]   _____, *A note on arcs of finite cyclic order*, Proc. Amer. Math. Soc. **12** (1961), 364–368.
[Ka]   M. E. Kazarian, *The Chern-Euler number of circle bundle via singularity theory*, Preprint.
[KA]   A. Kneser, *Bemerkungen über die Anzahl der Extrema der Krümmung auf geschlossenen Kurven und über verwandte Fragen in einer nichteuklidischen Geometrie*, Festschrift zum 70. Geburtstag von H. Weber (1912), 170–180.
[KH]   H. Kneser, *Neuer Beweis des Vierscheitelsatzes*, Christiaan Huygens **2** (1922/23), 315–318.
[Ko]   O. Kobayashi, *Vertices of curves with complementary shell*, Preprint.
[KU]   O. Kobayashi and M. Umehara, *Geometry of Scrolls*, Osaka J. Math. **33** (1996), 441–473.
[LS]   N. D. Lane and P. Scherk, *Characteristic and order of differentiable points in the conformal plane*, Trans. Amer. Math. Soc. **81** (1956), 358–378.
[LSp]   N. D. Lane and G. Spoar, *On singular points of normal arcs of cyclic order four*, Can. Math. Bull. **17** (1974), 391–396.
[LW]   X. Lin and Z. Wang, *Integral Geometry of Plane Curves and Knot Theory*, J. Diff. Geometry **44** (1996), 74–95.

[Mo]   H. Mohrmann, *Die Minimalzahl der stationären Ebenen eines räumlichen Ovals*, Sitz Ber Kgl Bayerichen Akad Wiss. Math. Phys. K1 (1917), 1–3.

[Mu1]   S. Mukhopadhyaya, *New methods in the geometry of a plane arc I.*, Bull. Calcutta Math. Soc. **1** (1909), 31–37.

[Mu2]   _____, *Extended minimum number theorems of cyclic and sextactic points on a plane convex oval*, Coll. p. 159–174, Math. Z. **33** (1931), 648–662.

[N1]   G. Nöbeling, *Über die Anzahl der Ordnungsgeometrischen Sheitel von Kurven I*, Aequationes Math. **34** (1987), 82–88.

[N2]   _____, *Über die Anzahl der Ordnungsgeometrischen Sheitel von Kurven II*, Geom. Dedicata **31** (1989), 137–149.

[Os1]   R. Osserman, *The four or more vertex theorem*, Amer. Math. Monthly **92** (1985), 332–337.

[Os2]   _____, *Circumscribed circles*, Amer. Math. Monthly **98** (1991), 419–422.

[OT]   V. Ovsienko and S. Tabachnikov, *Sturm theory, Ghys theorem on zeros of the Schwarzian derivative and flattening of Legendrian curves*, Selecta Math., New Series **2** (1996), 297–307.

[Pe]   A. Perović, *Vertices of bounding curves with minimal self-intersections*, Preprint.

[Pi]   U. Pinkall, *On the four-vertex theorem*, Aequat. Math. **34** (1987), 221–230.

[Po]   M. Polyak, *Invariants of plane curves and Legendrian fronts via Gauss diagrams*, Preprint.

[R]   M. C. Romero-Fuster, *Convexly generic curves in $\mathbf{R}^3$*, Geom. Dedicata **28** (1988), 7–30.

[RCN]   M. C. Romero-Fuster, S. I. R. Costa and J. J. Nuño Ballesteros, *Some global properties of closed space curves*, Lecture Notes in Math., vol. 1410, Springer-Verlag, Berlin-Heidelberg-New York, 1989, pp. 286–295.

[RS1]   M. C. Romero-Fuster and V. D. Sedykh, *On the number of singularities, zero curvature points and vertices of a simple convex space curve*, Journal of Geometry **52** (1995), 168–172.

[RS2]   _____, *A lower estimate for number of zero-torsion points of space curve*, Beitrage zur Algebra und Geometrie (to appear).

[Sa]   A. Schatteman, *A four-vertex-theorem for polygons and its generalization to polytopes*, Geom. Dedicata **34** (1990), 229–242.

[Se]   P. Scherk, *Über reele geschlossene Raumkurven vierter Ordnung*, Math. Ann. **112** (1936), 743–766.

[Sd1]   V. D. Sedykh, *Structure of the convex hull of space curve*, Trudy Sem. Petrovsk. **6** (1981), 239–256; English transl., J. Sov. Math. **33** (1986), 1140–1153.

[Sd2]   _____, *The four-vertex theorem of a convex space curve*, Functional. Anal. i Prilozhen **26** (1992), no. 1, 35–41; English transl., Functional Anal. Appl. **26** (1992), no. 1, 28–32.

[Sd3]   _____, *Invariants of strictly convex manifolds*, Functsional. Anal. i Prilozhen **27** (1993), no. 3, 67–75; English transl., Functional Anal. Appl. **27** (1993), 205–210.

[Sd4]   _____, *Invariants of nonflat manifolds*, Functsional. Anal. i Prilozhen **29** (1995), 41–50; English transl., Functional Anal. Appl. **29** (1995), 180–187.

[Sd5]   _____, *The strict convexity of a convex generic manifold*, Trudy Mat. Inst. Steklov. **209** (1995), 200–219; English transl., Proc. Steklov Inst. Math. **209** (1995), 174–190.

[Sd6]   _____, *Discrete variants of the four-vertex theorem*, Preprint no. 9615, CEREMADA (URA CNRS 749), University of Paris IX Dauphine, 1996.

[Sd7]   _____, *Invariants of submanifolds in Euclidean space*, Arnold-Gelfand Mathematical Seminars: Geometry and Singularity Theory (V. I. Arnold, I. M. Gelfand, M. Smirnov, and V. I. Retakh, eds.), Birkhauser, Boston, 1996.

[Sl]   K. Selwat, *The Vassiliev invariants of curves on a plane*, Preprint.

[ST]   K. Shiohama and M. Tanaka, *Cut loci and distance sphere on Alexsandrov surface*, Actes de la Table Ronde de géométrie differentielle en l'honneur de Marcel Berger, Collection SMF Seminaires et Congrś n⁰1, CIRM, Marseille, France, 1992, pp. 533–559.

[Sp1]   G. Spoar, *Differentiability conditions and bounds on singular points*, Pacific J. Math. **61(1)** (1975), 289–294.

[Sp2]   _____, *A least upper bound for the number of singular points on normal arcs and curves of cyclic order four*, Geom. Dedicata **7** (1978), 37–43.

[Sp3]   _____, *A characterization of curves of cyclic order four*, Geom. Dedicata **24** (1987), 283–293.

[Sp4] _____, *A generalized four-vertex theorem*, J. Geom. **33** (1988), 147–154.
[Su] W. Süss, *Relativgeometrische Erweiterung eines Sechsscheitelsatzes von W. Blaschke*, Jahresber. Deutsche Math.-Ver. **37** (1928), 361–362.
[Ta1] S. Tabachnikov, *The four-vertex theorem revised-two variations on the old theme*, Amer. Math. Monthly **102** (1995), 912–916.
[Ta2] _____, *Parametrized curves, Minkowski caustics, Minkowski vertices and conservative line fields*, L'Enseignement Math. (to appear).
[Ta3] _____, *On zeros of the Schwarzian derivative*, to appear in V. Arnold's 60-th anniversary volume.
[Tm1] R. Thom, *Sur le cut-locus d'une varieté plongée*, J. Diff. Geom. **6** (1972), 577–586.
[Tm2] _____, *Problemès rencontrés dans mon parcours mathématique : un bilan*, Publ. Math. Inst. Hautes Études Sci. **70** (1989), 199–214.
[Tr] G. Thorbergsson, *Vierscheitelsatz auf Flächen nichtpositiver Krümmung*, Math. Z **149** (1976), 47–56.
[TU] G. Thorbergsson and M. Umehara, "A unified approach to the four vertex theorems. II", *Differential and Symplectic topology of knots and curves* (S. Tabachnikov, ed.), Amer. Math. Soc. Transl. Ser. (2), 190, Amer. Math. Soc., Providence, RI, 1999.
[U1] M. Umehara, *6-vertex theorem for closed planar curve which bounds an immersed surface with nonzero genus*, Nagoya Math. J. **134** (1994), 75–89.
[U2] _____, *A computation of the basic invariant J^+ for closed 2-vertex curves*, Journal of Knot Theory and Its Ramifications **6** (1997), 105–113.
[V] G. Valette, *Cubiques topologiques à 1 dimension*, Arch. Math. **16** (1965), 265–273.
[W1] B. Wegner, *A cyclographic approach to the vertices of plane curves*, Journal of Geometry **50** (1994), 186–201.
[W2] _____, *Bose's vertex theorem for simply closed polygons*, Math. Pannon. **6** (1995), 121–132.

DEPARTMENT OF MATHEMATICS, GRADUATE SCHOOL OF SCIENCE, OSAKA UNIVERSITY, MACHIKANEYAMA 1-16, TOYONAKA, OSAKA 560, JAPAN
E-mail address: umehara@math.wani.osaka-u.ac.jp

A Unified Approach to the Four Vertex Theorems. II

Gudlaugur Thorbergsson and Masaaki Umehara

Introduction

The present paper is a continuation of [**U**] by the second author. For the sake of simplicity, we will refer to that paper by I. For example, I-Theorem 2.2 means Theorem 2.2 in [**U**].

Our main goal is to study closed space curves. We will show that the following theorem can be used to improve some known results and reprove others. Our improvements will concern results of A. Kneser [**Kn**] and Segre [**Se**] on the osculating planes of spherical curves, and Sedykh's four vertex theorem [**Sd**]. We will also be able to improve Ghys' theorem on extremal points of projective line diffeomorphisms and obtain some new results about them. As was pointed out to us by E. Heil, the special case of a spherical curve in Theorem 0.1 gives a positive answer to a conjecture of Jackson about the osculating circles of plane curves in the unpublished manuscript [**J**].

THEOREM 0.1. *Let $\gamma : S^1 \to \mathbf{R}^3$ be a C^2-regular convex simple closed nonplanar space curve with nonvanishing curvature. Then γ has at least four sign changes of clear vertices, meaning that it has two clear maximal vertices p_1, p_3 and two clear minimal vertices p_2, p_4 such that $p_1 \succ p_2 \succ p_3 \succ p_4$ holds, where $\succ$ is the rotational order of the curve γ.*

A closed space curve is called *convex* if it lies on the boundary of its convex hull. Here a *clear maximal* (resp. *minimal*) *vertex* (see I-Definition 4.2) is a point which is an absolute maximum (resp. minimum) of the height function with respect to the binormal vector at that point. Moreover, if the level set of absolute maxima (resp. minima) is connected, it is called a *clean maximal* (resp. *clean minimal*) vertex.

If the curve γ in Theorem 0.1 satisfies one of some additional conditions, the four points can be chosen to be clean vertices so that the osculating planes of γ at these points are mutually different (see Corollary 1.6). To prove the theorem, we will use and further develop the method of intrinsic circle systems introduced in I-§1.

As an application of Theorem 0.1, we will improve results of A. Kneser [**Kn**] and Segre [**Se**] on spherical curves. If $\gamma : S^1 \to S^2$ is a simple closed curve, then

---

1991 *Mathematics Subject Classification.* Primary 53A04; Secondary 57M25.

©1999 American Mathematical Society

Segre proved that if p lies in the convex hull of γ without lying on γ, then the osculating planes of at least four distinct points of γ pass through p. We will show that this is still true for any convex space curve γ with nowhere vanishing curvature and a point p that lies in the interior of the domain containing the curve which is bounded by the osculating hyperplanes at the four clear vertices, whose existence is claimed in Theorem 0.1. We will also discuss how this relates to Arnold's Tennis Ball Theorem [**A2**] and the theorem of Möbius [**M**] on inflection points of curves in the projective plane.

Another easy application of Segre's theorem is the following four vertex theorem for space curves which may not be convex.

THEOREM 0.2. *Let γ be a C^2-regular simple closed curve in $\mathbf{R}^3$ with nowhere vanishing curvature. Assume there is a point p in the interior of its convex hull such that no ray starting in p intersects γ in two or more points or is tangent to γ at some point. Assume furthermore that the normal vector of γ at t makes an obtuse angle with the ray from p through $\gamma(t)$ for all t. Then the curve γ has at least four vertices.*

Here a *vertex* is a zero of the torsion function. Notice that a convex space curve γ satisfies the conditions in Theorem 0.2, and we will give an example that shows that there are also nonconvex curves that do so. Thus we have an improvement of Sedykh's four vertex theorem for convex curves in [**Sd**].

As a further application, we also use the abstract methods in Section 1 to improve a theorem of Ghys on extremal points of projective line diffeomorphisms (see [**OT**] and [**T**]) and to arrive at new results about them.

1. Compatible pairs of intrinsic circle systems

We let S^1 denote the unit circle with a fixed orientation. Let $\succ$ denote the order induced by the orientation on the complement of any interval in S^1. Any two distinct points $p, q \in S^1$ divide S^1 into two closed arcs $[p, q]$ and $[q, p]$ such that on $[p, q]$ we have $q \succ p$ and on $[q, p]$ we have $p \succ q$. We let (p, q) and (q, p) denote the corresponding open arcs. We also use the notation $p \succeq q$, which means $p = q$ or $p \succ q$.

Let A be a subset of S^1 and $p \in A$. We denote by $Z_p(A)$ the connected component of A containing p. The concept of an intrinsic circle system was introduced in I-§1 as a multivalued function on S^1 satisfying certain axioms. It was used there to prove an abstract Bose type formula. Here we recall the definition.

DEFINITION 1.1. A family of nonempty closed subsets $F := (F_p)_{p \in S^1}$ of S^1 is called an *intrinsic circle system on S^1* if it satisfies the following three conditions for any $p \in S^1$:
 (I1) We have $p \in F_p$ for each $p \in S^1$. If $q \in F_p$, then $F_p = F_q$.
 (I2) If $q \in S^1 \setminus F_p$, then $F_q \subset Z_q(S^1 \setminus F_p)$. Or equivalently, if $p' \in F_p$, $q' \in F_q$ and $q \succeq p' \succeq q' \succeq p(\succeq q)$, then $F_p = F_q$ holds.
 (I3) Let $(p_n)_{n \in \mathbf{N}}$ and $(q_n)_{n \in \mathbf{N}}$ be two sequences in S^1 such that $\lim_{n \to \infty} p_n = p$ and $\lim_{n \to \infty} q_n = q$, respectively. Suppose that $q_n \in F_{p_n}$ $(n = 1, 2, 3, ...)$. Then $q \in F_p$ holds.

We will now give an application of I-§§1–2 by discussing pairs of intrinsic circle systems $(F^\bullet, F^\circ)$ satisfying the following compatibility condition. We will let rank$^\bullet(p)$ (resp. rank$^\circ(p)$) denote the number of connected components of $F_p^\bullet$ (resp. F_p°).

DEFINITION 1.2. A pair of intrinsic circle systems $(F^\bullet, F^\circ)$ is said to be *compatible* if it satisfies the following two conditions:
(C1) $F_p^\bullet \cap F_p^\circ = \{p\}$ for all $p \in S^1$;
(C2) Suppose that rank$^\bullet(p) = 1$ (resp. rank$^\circ(p) = 1$). Then there are no points of rank$^\circ = 1$ (resp. rank$^\bullet = 1$) in a sufficiently small neighborhood of p.

The following are examples of compatible pairs of intrinsic circle systems.

EXAMPLE 1. Let $\gamma : S^1 \to \mathbf{R}^2$ be a C^2-regular simple closed curve which is not a circle. Let Γ be the set of all oriented circles and lines in $\mathbf{R}^2$. The curve γ separates the plane into two closed domains. We denoted by $D^\bullet(\gamma)$ the compact domain bounded by γ and by $D^\circ(\gamma)$ the noncompact one. We assume that γ is positively oriented, meaning that the compact domain $D^\bullet(\gamma)$ is on the left of γ. For each $p \in \gamma$, there is an element $C_p^\bullet \in \Gamma$ (resp. $C_p^\circ \in \Gamma$) which has the smallest (resp. largest) curvature among $C \in \Gamma$ that are tangent to γ in p and satisfy $C \subset D^\bullet(\gamma)$ (resp. $C \subset D^\circ(\gamma)$). We call $C_p^\bullet$ (resp. C_p°) the maximal (resp. minimal) circle at p. Now we set

$$(1.1) \qquad F_p^\bullet := \gamma \cap C_p^\bullet, \qquad F_p^\circ := \gamma \cap C_p^\circ.$$

Then $(F^\bullet, F^\circ)$ is a pair of intrinsic circle systems (see I-Proposition 3.1). Condition (C1) of Definition 1.2 trivially holds. For a point p of rank$^\bullet(p) = 1$ (resp. rank$^\circ(p) = 1$), the osculating circle C_p at p coincides with $C_p^\bullet$ (resp. C_p°) (see I-Corollary A.5). In particular, condition (C2) of Definition 1.2 is also satisfied. Thus $(F^\bullet, F^\circ)$ is a compatible pair. Instead of Γ, we can use a system of Minkowski circles in the plane (see I-§3 for details).

EXAMPLE 2. An immersed closed space curve $\gamma : S^1 \to \mathbf{R}^3$ is called convex if it lies on the boundary ∂H of its convex hull H. We fix a nonplanar C^2-convex simple closed curve γ and assume that its curvature function is positive. The boundary ∂H of the convex hull is homeomorphic to a sphere and γ divides ∂H into two domains. Let $\partial H^\bullet$ (resp. ∂H°) be the left-hand (right-hand) closed domain bounded by γ. We set

$$(1.2) \qquad F_p^\bullet := \{q \in \gamma\,;\, \overline{pq} \subset \partial H^\bullet\}, \quad (\text{resp. } F_p^\circ := \{q \in \gamma\,;\, \overline{pq} \subset \partial H^\circ\}).$$

By I-Theorem 4.8, $F^\bullet$ and F° are intrinsic circle systems if γ satisfies one of the following two conditions:
1. For each point $p \in \gamma$, there exists a supporting plane U_p such that $U_p \cap \gamma = \{p\}$; or
2. γ has no planar open subarcs.

Moreover, $(F^\bullet, F^\circ)$ is a compatible pair if γ satisfies one of the following three conditions:
   (a) γ satisfies (1);
   (b) γ satisfies (2) and any tangent line of γ meets γ in only one point; or
   (c) γ is a C^3-convex space curve whose torsion function has only finitely many zeros.

In fact, (C1) of Definition 1.2 is satisfied by definition. (See the Introduction for definitions of concepts we are now going to use.) When γ satisfies (a) or (b), a point p on γ is of rank$^\bullet = 1$ (resp. rank$^\circ = 1$) if and only if it is a clean maximal (resp.

minimal) vertex by I-Lemma 4.10. When γ satisfies (c), a point p of rank$^\bullet = 1$ (resp. rank$^\circ = 1$) is a clear maximal (resp. minimal) vertex by I-Proposition 4.12 and I-Lemma 4.17. A clear maximal vertex cannot be a clear minimal vertex because γ is not planar. Thus (C2) of Definition 1.2 also follows in any of these three cases.

We will give a further example in Section 3.

From now on, we fix a compatible pair $(F^\bullet, F^\circ)$ of intrinsic circle systems.

DEFINITION 1.3. If rank$^\bullet(p) = 2$ (resp. rank$^\circ(p) = 2$), p is called a $\bullet$-*regular* (*resp.* $\circ$-*regular*) *point*. If rank$^\bullet(p) \geq 2$ (resp. rank$^\circ(p) \geq 2$), p is called *weakly* $\bullet$-*regular* (*weakly* $\circ$-*regular*). An open arc I of S^1 is called $\bullet$-regular (resp. $\circ$-regular) if all of its points are $\bullet$-regular (resp. $\circ$-regular). More generally, an open arc I of S^1 is called weakly $\bullet$-regular (resp. weakly $\circ$-regular) if all of its points are weakly $\bullet$-regular (resp. weakly $\circ$-regular).

We remark that a point p is (weakly) $\bullet$-regular (resp. $\circ$-regular) if and only if it is a (weakly) regular point with respect to the the intrinsic circle system $F^\bullet$ (resp. F°) (see I-Definition 1.2). Let $I = (x_1, x_2)$ be a weakly $\bullet$-regular (resp. weakly $\circ$-regular) arc. Then for each point on $p \in [x_1, x_2]$, the antipodal functions $\mu^\bullet_\pm$ (resp $\mu^\circ_\pm$) are induced (see I-Definition 1.4).

We now come to the main result of this section.

THEOREM 1.4. *Let $(F^\bullet, F^\circ)$ be a compatible pair of intrinsic circle systems. Then there are four points $p_1, p_2, p_3, p_4 \in \gamma$ satisfying $p_1 \succ p_2 \succ p_3 \succ p_4 \,(\succ p_1)$ such that*

$$\mathrm{rank}^\bullet(p_1) = \mathrm{rank}^\circ(p_2) = \mathrm{rank}^\bullet(p_3) = \mathrm{rank}^\circ(p_4) = 1.$$

REMARK. E. Heil has pointed out to us that Jackson explicitly conjectured in the unpublished manuscript [**J**] that " If A is a simple closed curve which is strongly conformally differentiable, then it is not possible to divide A into two arcs such that all positive global vertices are on one arc and all negative global vertices are on the other." He then goes on to say that "So far, however, the writer has been unable either to prove or disprove the statement." In Jackson's terminology a curve is said to be *conformally differentiable* if there is a unique osculating circle at each of its points. Theorem 1.4 applies to such curves because conformal differentiability implies condition (I3) in the definition of an intrinsic circle system. A *positive (negative) global vertex* is a point p on the curve with rank$^\bullet(p) = 1$ (rank$^\circ(p) = 1$) with respect to the intrinsic circle system induced by the usual circle system of the plane.

PROOF. Suppose there are less than four sign changes of rank one points. Since the number of sign changes is even, it must be exactly two. We set

$$V^\bullet := \{x \in \gamma \,;\, \mathrm{rank}^\bullet(x) = 1\},$$
$$V^\circ := \{x \in \gamma \,;\, \mathrm{rank}^\circ(x) = 1\},$$

and denote by $\overline{V^\bullet}$ and $\overline{V^\circ}$ their closures. Let I be the connected component of $S^1 \setminus \overline{V^\circ}$ containing $\overline{V^\bullet}$. We set

$$x_1 := \sup_{\overline{I}}(\overline{V^\bullet}), \qquad x_2 := \inf_{\overline{I}}(\overline{V^\bullet}).$$

By condition (C2) of Definition 1.2, it holds that $x_1, x_2 \in I$. Then the open interval $J := (x_1, x_2)$ is a weakly •-regular arc, and so the antipodal functions $\mu_\pm^\bullet$ are defined on J. On the other hand, I is a weakly ○-regular arc and so $\mu_\pm^\circ$ are defined on it. By I-Corollary 1.2 and I-Definition 1.4, we have

$$\mu_\pm^\bullet(\overline{J}) \subset \overline{I}, \qquad \mu_\pm^\circ(\overline{I}) \subset \overline{J}.$$

We set

$$A = \{p \in \overline{J}\,;\, \mu_-^\circ(\mu_-^\bullet(p)) \succ p \text{ on } \overline{J}\},$$
$$B = \{p \in \overline{J}\,;\, \mu_+^\circ(\mu_+^\bullet(p)) \prec p \text{ on } \overline{J}\}.$$

We suppose that $p \in \overline{J} \setminus A$. Since $F_{\mu_-^\bullet(p)}^\bullet \cap F_{\mu_-^\bullet(p)}^\circ = \mu_-^\bullet(p)$, we have $p \neq \mu_-^\circ(\mu_-^\bullet(p))$. Then we have

$$p \succ \mu_-^\circ(\mu_-^\bullet(p)) \qquad \text{on } \overline{J}.$$

Hence (I2) of Definition 1.1 for F° yields that

$$(p \succ)\mu_-^\circ(\mu_-^\bullet(p)) \succeq \mu_+^\circ(\mu_+^\bullet(p)) \succ \mu_+^\bullet(p) \succeq \mu_-^\bullet(p) \qquad \text{on } [\mu_-^\bullet(p), p].$$

This implies $p \in B$. Thus we have

(1.3) $$\overline{J} = A \cup B.$$

We will now use Lemma 1.5 below. It says there that A is nonempty. We set

$$q := \sup_{\overline{J}}(A).$$

If $q \in A$, then Lemma 1.5 also yields that there is $y \in A$ such that $y \succ q$. Thus $q \notin A$, that is $q \in B$ by (1.3). Then by Lemma 1.5, there exists $z \in B$, $q \succ z$, such that $(z,q) \cap A = \emptyset$, contradicting that $q := \sup_{\overline{J}}(A)$. □

LEMMA 1.5. *The sets A and B are nonempty subsets of the arc $\overline{J}$. Moreover, for each $x \in A$ (resp. $x \in B$), $[x,y] \subset A$ and $(x,y) \cap B = \emptyset$ (resp. $[y,x] \subset B$ and $(y,x) \cap A = \emptyset$) hold, where $y := \mu_-^\circ(\mu_-^\bullet(x))$.*

PROOF. We prove the assertion for A. (The corresponding assertion for B follows if one reverses the orientation of S^1.) First we prove $x_1 \in A$. In fact, if $x_1 \notin A$, then $x_1 \in B$. Then

$$x_1 \succ \mu_+^\circ(\mu_+^\bullet(x_1)) \qquad \text{on } \overline{J}.$$

But this contradicts the fact that x_1 is the smallest point contained in $\overline{J}$. This implies $x_1 \in A$. In particular, A is not empty. Now we fix an element $x \in A$ arbitrarily. Then by definition, we have

$$y = \mu_-^\circ(\mu_-^\bullet(x)) \succ x \qquad \text{on } \overline{J}.$$

We fix a point z on the interval (x,y) arbitrarily. Since $z \succ x$ on J, applying I-Lemma 1.3, we have

$$\mu_-^\bullet(x) \succeq \mu_\pm^\bullet(z) \qquad \text{on } \overline{I}.$$

Then applying I-Lemma 1.3 again, we have

$$\mu_\pm^\circ(\mu_\pm^\bullet(z)) \succeq \mu_-^\circ(\mu_-^\bullet(x))(\succ z).$$

This implies $z \in A$ and $z \notin B$. □

COROLLARY 1.6. *Let $\gamma : S^1 \to \mathbf{R}^3$ be a C^2-regular convex simple closed nonplanar space curve with nonvanishing curvature satisfying one of the following two conditions:*
  (a) *For each point $p \in \gamma$, there exists a supporting plane U_p such that $U_p \cap \gamma = \{p\}$; or*
  (b) *γ contains no planar open subarcs and no tangent line of γ meets γ in more than one point.*
Then γ has at least four sign changes of clean vertices. Moreover, at these four clean vertices the osculating planes of γ are mutually different.

PROOF. As mentioned in Example 2, the pair of intrinsic circle systems $(F^\bullet, F^\circ)$ is compatible if γ satisfies either one of the conditions in (a) and (b). We now apply Theorem 1.4. We know that each rank one point on γ is a clean vertex by I-Lemma 4.10. From this it follows immediately that the four osculating planes are mutually distinct. □

We now can prove Theorem 0.1 in the Introduction. Let $\mathfrak{C}$ be the set of nonplanar C^2-convex simple closed space curves and $\mathfrak{C}_0$ be the subset of $\mathfrak{C}$ consisting of convex space curves which satisfy condition (a) of Corollary 1.6. It is sufficient to prove the following lemma.

LEMMA 1.7. *For each $\gamma \in \mathfrak{C}$, there exists a sequence $(\gamma_n)_{n \in \mathbf{N}}$ in $\mathfrak{C}_0$ converging to γ with respect to the uniform C^2-topology.*

In fact, assuming the lemma, we can prove the theorem as follows. Let γ be a C^2-regular convex nonplanar space curve with nonvanishing curvature. By Lemma 1.7, there exists a sequence $(\gamma_n)_{n \in \mathbf{N}}$ in $\mathfrak{C}_0$ converging to γ with respect to the uniform C^2-topology. Since the curvature function of γ never vanishes, the same is true for γ_n if n is sufficiently large. By Corollary 1.6, each γ_n has two clean maximal vertices $p_1^{(n)}, p_3^{(n)}$ and two clean minimal vertices $p_2^{(n)}, p_4^{(n)}$ such that $p_1^{(n)} \succ p_2^{(n)} \succ p_3^{(n)} \succ p_4^{(n)}$ holds. If necessary, by going to a subsequence, we may assume that there exist points (p_1, p_2, p_3, p_3) such that

$$\lim_{n \to \infty} (p_1^{(n)}, p_2^{(n)}, p_3^{(n)}, p_4^{(n)}) = (p_1, p_2, p_3, p_3).$$

Since $p_1^{(n)}$ and $p_3^{(n)}$ (resp. $p_2^{(n)}$ and $p_4^{(n)}$) are clean maximal (resp. minimal) vertices, the limit points $p_1^{(n)}$ and $p_3^{(n)}$ (resp. $p_2^{(n)}$ and $p_4^{(n)}$) are clear maximal (resp. minimal) vertices. Since γ is not a plane curve, a clear maximal vertex is not a clear minimal vertex. Thus the four points are mutually distinct and satisfy $p_1 \succ p_2 \succ p_3 \succ p_4$. □

PROOF OF LEMMA 1.7. To prove the lemma, we recall some fundamental properties of convex bodies. A convex bounded open region Ω in $\mathbf{R}^3$ is called a *convex body*. We fix an interior point o of Ω. Without loss of generality we may assume that o is the origin of $\mathbf{R}^3$. We set

$$\rho(x) := \inf\{t > 0 \,;\, x \in t\Omega\} \qquad (x \in \mathbf{R}^3),$$

where $t\Omega := \{tx : x \in \Omega\}$. Notice that $\pi : \partial\Omega \to S^2$ defined by $\pi(x) := x/|x|$ is a homeomorphism. We have the expression

(1.4) $$\rho(x) = |x|\rho(x/|x|) = |x|\pi^{-1}(x/|x|).$$

Thus $\rho : \mathbf{R}^3 \to \mathbf{R}$ is a continuous function. Moreover the function ρ satisfies the following properties (see Proposition 1.1.5 in [**KR**]):
1. For each $x \in \mathbf{R}^3$, $\rho(x) \geq 0$ and $\rho(x) = 0$ if and only if $x = o$;
2. $\rho(ax) = a\rho(x)$ for any real number $a \geq 0$;
3. $\rho(x+y) \geq \rho(x) + \rho(y)$ for all $x, y \in \mathbf{R}^3$.

Furthermore Ω can be expressed as

(1.5) $$\Omega = \{x \in \mathbf{R}^3 \,;\, \rho(x) < 1\}.$$

The convex body Ω is called *strictly convex*, if for each boundary point $p \in \Omega$, there is a plane U passing through p such that $\Omega \cap U = \{p\}$. One can easily show that Ω is strictly convex if ρ satisfies the condition that

(1.6) $$\rho(x+y) < \rho(x) + \rho(y)$$

for any two linearly independent $x, y \in \mathbf{R}^3$. Conversely, if there exists a continuous function $\rho : \mathbf{R}^3 \to \mathbf{R}$ satisfying the three properties (1)–(3), the open subset Ω defined in (1.5) is a convex body.

Now we fix a C^2-convex space curve $\gamma : [0,1] \to \mathbf{R}^3$. Let Ω_γ be its convex hull. Without loss of generality, we may assume that Ω_γ contains the origin o in its interior. Let ρ be the continuous function satisfying (1)–(3) associated to the convex body Ω_γ. We set

(1.7) $$\tilde{\gamma} := \pi(\gamma),$$

where $\pi : \partial\Omega \to S^2$ is the projection defined above. Then $\tilde{\gamma} : [0,1] \to S^2$ is a C^2-regular embedding. By (1.4), we have $\gamma = \{\rho(\tilde{\gamma})\}\tilde{\gamma}$. We set

$$\rho_n(x) := \rho(x) + \frac{|x|}{n}.$$

Then ρ_n satisfies (1)–(3) and also (1.6). Thus the associated convex body Ω_n is strictly convex. We set

(1.8) $$\gamma_n(t) = \rho_n(\tilde{\gamma}(t)) \cdot \tilde{\gamma}(t), \qquad t \in [0,1].$$

Notice that $\rho_n(\tilde{\gamma}(t))$ is clearly C^2 in t although ρ_n is only continuous. Thus each curve γ_n is a C^2-regular simple closed curve that lies in the boundary of Ω_n. Since Ω_n is strictly convex, γ_n satisfies condition (a) of Corollary 1.6. Moreover, it is obvious that γ_n converges to γ in the uniform C^2-topology. □

Next we give an example of a plane curve γ with the following two properties:
1. There are only four sign changes of clear vertices, although γ has more than four clean vertices;
2. The number of clean or clear maximal vertices is not equal to the number of clean or clear minimal vertices.

In particular, this example shows that we cannot improve the number of sign changes to $2n$ in Theorem 1.4 when the curve γ meets a circle in $2n$ points such that the rotational order on γ and the circle coincide. In fact, since the number of clean vertices exceeds four, we may assume that the number of maximal clean vertices is at least three. By I-Theorem 2.7, we have $t(F^\bullet) \geq 1$, i.e., there is a triple tangent enclosed circle C. Expanding C slightly by a homothety with the same center as C, we get a circle C_ε which meets γ in six points whose order on C_ε coincides with that on γ.

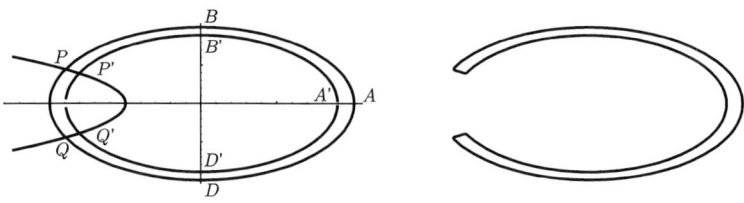

FIGURE 1-A FIGURE 1-B

The curve γ is constructed as follows. Consider the ellipse

$$\gamma_1 : \quad \frac{x^2}{4} + y^2 = 1$$

that we assume to be positively oriented (i.e., with an inward pointing normal vector). We shrink it by a homothety with the dilation factor $\sqrt{1-\varepsilon}$, where $\varepsilon > 0$ is a sufficiently small number. Then we have another ellipse

$$\gamma_2 : \quad \frac{x^2}{4} + y^2 = 1 - \varepsilon$$

that we assume to be negatively oriented (i.e., with on outward pointing normal vector). We also consider the parabola

$$x = -4y^2 - 1$$

oriented in the negative direction of the y-axis.

Then γ_1 (resp. γ_2) meets the parabola in two points P, Q (resp. P', Q'), as in Figure 1-a. Note that with the orientation chosen above and notation as in Figure 1-a the arcs QD, $D'Q'$ and PP' are curvature decreasing and BP, $P'B'$ and $Q'Q$ are curvature increasing. We can round the corners at P, Q (resp. P', Q') introducing exactly one (resp. three) vertices using the method of Proposition 2.3 in [**KU**], and get a simple closed curve γ as in Figure 1-b with the following vertices:

$$A, B, P, P'_{-1}, P'_0, P'_1, B', A', D', Q'_{-1}, Q'_0, Q'_1, Q, D.$$

Here P'_{-1}, P'_0, P'_1 (resp. Q'_{-1}, Q'_0, Q'_1) are the vertices which appear after rounding the corner at P' (resp. Q'). More precisely, P'_0, Q'_0 (resp. $P'_{-1}, P'_1, Q'_{-1}, Q'_1$) are local maxima (resp. minima) of the curvature function. By construction, γ has 14 vertices. The clean (clear) maximal vertices are P, P'_0, Q, Q'_0. The clean (clear) minimal vertices are B, D, A'. (In fact, if ε is sufficiently small, other vertices cannot be clear vertices.) Thus the number of clean maximal vertices is greater than the number of clean minimal vertices. Moreover, the clean vertices lie on the curve γ in the order

$$B, P, P'_0, A', Q'_0, Q, D,$$

so the number of sign changes of clean vertices is only four. After finishing the paper we noticed that Jackson gives an example in [**J**] with similar properties.

We end this section by giving a further corollary of Theorem 1.4. It is a refinement of the four vertex theorem for convex curves.

COROLLARY 1.8. *Let $\gamma : S^1 \to \mathbf{R}^2$ be a C^2-regular convex plane curve with length 2π, which is not a circle. Then the function $\kappa^2 - 1$ changes sign at least four times, where $\kappa : S^1 \to \mathbf{R}$ is the curvature of γ.*

PROOF. We may assume that $\kappa \geq 0$. We use the notation in Example 1. Let $p_1, ..., p_4$ be points as in Theorem 1.4 and C_j ($j = 1, ..., 4$) the osculating circle at p_j. Then as mentioned in Example 1, we have

(1.9) $$C_i = C_{p_i}^{\bullet} \subset D^{\bullet}(\gamma) \qquad (i = 1, 3)$$

and

(1.10) $$C_j = C_{p_j}^{\circ} \subset D^{\circ}(\gamma) \qquad (j = 2, 4).$$

Since γ is not a circle, the lengths of C_1, C_3 (resp. C_2, C_4) are less (resp. greater) than 2π by (1.9) and (1.10). In particular, $\kappa > 1$ at p_1, p_3 and $\kappa < 1$ at p_2, p_4. $\square$

It should be remarked that there is an elementary proof of Corollary 1.8 using integration that we now want to explain.

AN ALTERNATIVE PROOF OF COROLLARY 1.8. We parametrize the curve according to arclength t ($0 \leq t \leq 2\pi$) and write $\gamma(t) = (x(t), y(t))$. We assume that $\kappa \geq 0$. Since the length and total curvature of γ are both 2π, we have the following identity

(1.11) $$\int_0^{2\pi} (\kappa(t) - 1)(ax'(t) + by'(t) + c)dt = 0,$$

where a, b, c are arbitrary real numbers. If we set $a = b = 0$ and $c = 1$, this implies that $\kappa - 1$ changes sign an even number of times. Suppose that the sign changes only twice. Then we may assume that there exists $t_0 \in [0, 2\pi]$ such that $\kappa - 1 \geq 0$ on $[0, t_0]$ and $\kappa - 1 \leq 0$ on $[t_0, 2\pi]$. We can choose the numbers a, b, c so that the line $ax + by + c = 0$ passes through the two points $\gamma'(0)$ and $\gamma'(t_0)$. Then $(\kappa(t) - 1)(ax'(t) + by'(t) + c)$ is a nonpositive or nonnegative function. By (1.11), this implies κ is identically 1, which is a contradiction. $\square$

2. Applications to space curves

With the methods of Section 1 at hand it is remarkably easy to improve some well-known theorems and show how others follow as corollaries. In this section we will use Theorem 0.1 (which is a corollary of Theorem 1.4) to improve a result of A. Kneser [**Kn**] and Segre [**Se**] on spherical curves. Notice that a new proof of Segre's theorem was given by Weiner in [**We**]. We will also discuss how this improvement relates to the Tennis Ball Theorem of Arnold [**A2**] and the Theorem of Möbius [**M**] that a simple closed curve in the projective plane that is not nullhomotopic has at least three inflection points. We will also prove a four vertex theorem for closed simple space curves that improves a result of Sedykh [**Sd**]. Instead of convexity we assume that the curve satisfies a condition of starshapedness. We do assume that the curvature never does vanish and our theorem seems to improve existing results on this class of curves.

Assume that $\gamma : S^1 \to S^2$ is simple, closed and regular. Then Segre [**Se**] proved the following (p. 243):
(1) If $p \in \mathbf{R}^3$ lies in the convex hull of γ without lying on γ, then the osculating planes of at least four distinct points of γ pass through p.
(2) If p lies on γ and p is not a vertex, then the osculating planes of at least three distinct points of γ, all of which differ from p, pass through p.

(3) If p is a vertex of γ, then the osculating planes of at least two distinct points of γ, both of which differ from p, pass through p.

A. Kneser proved part (2) of the above as an application of the theorem of Möbius. This was a step in his proof of the four vertex theorem for simple closed curves in the Euclidean plane. Segre does not say explicitly in the statement of his theorem that the points he finds in (2) and (3) all differ from p, but this seems to be what he proves.

Segre pointed out on p. 258 of his paper that the claim in (1) also holds for a point p in the convex hull of a space curve $\gamma : S^1 \to \mathbf{R}^3$, if $p \notin \gamma$ and no ray starting in p is tangent to γ or meets γ in two or more points. This is proved by simply projecting γ radially onto a sphere with center in p. We will also use this argument in the proof of Theorem 2.2 below.

Our improvements of the results of A. Kneser and Segre will consist in the following.
1. We will consider more general space curves.
2. Instead of the convex hull of the curve, we can prove (1) above for points p in the interior of a certain polyhedron Δ that contains the curve and only has vertices of the curve in its boundary.
3. We will show that the osculating planes found in the theorem of A. Kneser and Segre can under certain conditions been chosen to be at points different from the vertices of the curve. This will be needed to prove the Tennis Ball Theorem and the Möbius Theorem as corollaries. Segre also discusses such things in his proof without summarizing them in the statement of his theorem. We will give an example that shows that this is not true for all choices of p in the boundary of the convex hull of the curve. This example will also show that some of the osculating planes at the different points A. Kneser and Segre find might coincide.

We will be dealing with a C^2-regular simple closed convex nonplanar space curve $\gamma : S^1 \to \mathbf{R}^3$ with nowhere vanishing curvature. Notice that this situation is more general than the one in the theorem of A. Kneser and Segre.

We know from Theorem 0.1 that γ has at least four clear vertices $t_1 \succ t_2 \succ t_3 \succ t_4$ such that t_1 and t_3 are maxima of the height functions in the direction of the binormal vectors at the respective points, and t_2 and t_4 are minima of the corresponding height functions. Let Π_i be the osculating plane of γ at t_i. Then γ is contained in one of the closed halfspaces bounded by Π_i. We denote this halfspace by S_i. The binormal of γ at t_i points out of S_i if i is 1 or 3 and in to S_i if i is 2 or 4. We cannot exclude that $\Pi_1 = \Pi_3$ and $\Pi_2 = \Pi_4$, but it follows that $\Pi_i \neq \Pi_{i+1}$ since γ is not planar. Let Δ denote the intersection of these halfspaces, or the closure of the connected component of $\mathbf{R}^3 \setminus (\Pi_1 \cup \Pi_2 \cup \Pi_3 \cup \Pi_4)$ containing γ, which is the same thing. Then Δ is a simplex if no two of the osculating planes Π_i coincide or are parallel. Otherwise, Δ is a polyhedron that might be unbounded.

We denote the closure of the connected component of $\mathbf{R}^3 \setminus (\Pi_1 \cup \Pi_2)$ containing γ by S_{12}. Let Π_t denote the osculating plane of γ at t. We want to show that for every point $p \in S_{12}$ there is a $t \in [t_1, t_2]$ such that $p \in \Pi_t$. Notice that the conditions on γ as well as the conclusions we are aiming at can all be phrased in terms of projective geometry. We can therefore, if necessary, apply a projective transformation that sends the line in which Π_1 and Π_2 intersect to infinity. Hence we can assume that the planes Π_1 and Π_2 are parallel and the curve γ lies in between the planes. The set S_{12} is thus a slab. The binormal vector of γ at t_1 points out of the slab and at t_2 it points into it.

Now assume that $p \in S_{12}$ is not contained in any osculating plane Π_t for $t \in (t_1, t_2)$. Let q_t be the point on Π_t closest to p and let v_t be the unit vector at q_t pointing in the direction from p to q_t. The vector v_t is perpendicular to Π_t for all $t \in [t_1, t_2]$. Notice that v_{t_1} points out of the slab S_{12} since p lies in its interior, i.e., v_{t_1} points in the same direction as the binormal vector at $\gamma(t_1)$. By continuity, v_t points in the same direction as the binormal vector at $\gamma(t)$ for all $t \in [t_1, t_2]$. It follows that the binormal vector at $\gamma(t_2)$ points out of the slab, which is a contradiction. We have thus proved that for every point $p \in S_{12}$ there is a $t \in [t_1, t_2]$ such that $p \in \Pi_t$.

We can repeat the argument above for the pairs (t_2, t_3), (t_3, t_4) and (t_4, t_1) and prove that for every point p in the interior of $\Delta = S_{12} \cap S_{23} \cap S_{34} \cap S_{41}$, there is an $s_i \in (t_i, t_{i+1})$ such that $p \in \Pi_{s_i}$. It follows that every point in the interior of Δ lies in the osculating planes of four different points of γ.

This already improves part (1) of the theorem of Segre for the following reason. A point p in the convex hull of γ which neither lies in the interior of Δ nor on γ is in the convex hull of the vertices in one of the osculating planes Π_i. Since γ is a spherical curve in Segre's theorem we can choose the points $t_1, \ldots, t_4$ to be clean vertices. It follows that Π_i contains an open arc of vertices of γ and that Π_i is the osculating plane at each point of this arc. There are therefore infinitely many points of γ whose osculating planes contain p. Notice that we do not claim that these osculating planes are different. We will see in an example below that the number of osculating planes containing such a point p can be three. One does of course not need to assume that γ is spherical to prove the claim in part (1) of Segre's theorem. It is enough to have a condition that guarantees that the vertices $t_1, \ldots, t_4$ are clean.

Part (2) of the theorem of A. Kneser and Segre also follows since any point on γ which is not a vertex lies in the interior of Δ.

Assume that the osculating planes $\Pi_1, \ldots, \Pi_4$ are all different, and let p be in the interior of the face $\Delta \cap \Pi_1$. Then the above argument shows that there are $s_2 \in (t_2, t_3)$ and $s_3 \in (t_3, t_4)$ such that $p \in \Pi_{s_2}$ and $p \in \Pi_{s_3}$. Since p is also in Π_1 we see that every point in the open face in $\Delta \cap \Pi_1$ lies in the osculating planes of at least three different points of γ. This improves part (3) of the theorem of Segre since the planes $\Pi_1, \ldots, \Pi_4$ are all different if γ is spherical. The argument cannot be applied to points in the open 1-simplices of Δ in $\Pi_1 \cap \Pi_3$ and $\Pi_2 \cap \Pi_4$, but they contain no points on γ.

We summarize what we have proved in the next theorem.

THEOREM 2.1. *Let γ be a C^2-regular convex space curve with nowhere vanishing curvature. Let Δ be the closed polyhedron containing γ that is bounded by osculating supporting planes Π_1, Π_2, Π_3 and Π_4 at $t_1 \succ t_2 \succ t_3 \succ t_4$, respectively, such that Π_1 and Π_3 are maximal and Π_2 and Π_4 are minimal. Then every point in the interior of Δ is contained in osculating planes of at least four different points of γ. If the planes Π_1, Π_2, Π_3 and Π_4 are all different, then every point in the open faces of Δ lie in the osculating planes of three different points of γ.*

As we will see below, it is quite important in applications to know that the osculating hyperplanes that we find do not belong to honest vertices. Recall that t is an *honest vertex* if the following holds. Let Π_t be the osculating plane of γ at

$\gamma(t)$ and s_1 and s_2 are such that $\gamma([s_1, s_2])$ is the connected component of $\gamma \cap \Pi_t$ containing $\gamma(t)$. Then there is an $\varepsilon > 0$ such that $\gamma(s_1 - \varepsilon, s_1)$ and $\gamma(s_2, s_2 + \varepsilon)$ lie on the same side of Π_t. In loose terms, t is an honest vertex if γ does not cross Π_t in t.

The question is therefore whether one can improve Theorem 2.1 and show that the osculating planes whose existence is claimed do not belong to honest vertices. We will see that the answer is yes and no. Notice that Segre discusses this question in his proof without summarizing the results in the statement of his theorem.

In part (1) of the next theorem we give an easy positive answer to this question if p is contained in the interior of the convex hull of γ. The method of proof is based on Segre's observation on p. 258 in his paper on how to use the spherical curve case to get similar information on more general curves. In part (2) we generalize Sedykh's four vertex theorem [Sd] that says that a convex simple closed space curve with nowhere vanishing curvature has at least four honest vertices.

THEOREM 2.2. *Let $\gamma : S^1 \to \mathbf{R}^3$ be a C^2-regular simple closed curve in $\mathbf{R}^3$ with nowhere vanishing curvature. Assume there is a point p in the interior of the convex hull of γ such that no ray starting in p intersects γ in two or more points or is tangent to γ in some point.*

(1) *Then there are four distinct points on γ, none of which is an honest vertex, whose osculating planes contain p.*
(2) *Assume furthermore that the normal vector of γ at t makes an obtuse angle with the ray from p through $\gamma(t)$ for all t. Then γ has at least four vertices.*

Here *vertices* are zeros of the torsion function.

PROOF. We project γ radially from p onto a sphere $S^2(p)$ with center in p. We denote the new curve by $\tilde{\gamma}$. It follows that p lies in the interior of the convex hull of $\tilde{\gamma}$ and that an osculating plane of $\tilde{\gamma}$ that passes through p and does or does not correspond to an honest vertex of $\tilde{\gamma}$ has the same properties with respect to γ. It is therefore enough to prove the claim in (1) for $\tilde{\gamma}$. Notice that $\tilde{\gamma}$ satisfies the conditions that we assumed in Theorem 2.1, whereas γ might not. Using the same notation as above, let t_1 and t_2 be points such that t_1 is a maximum of the height function in the direction of the binormal vector at t_1, and t_2 is a minimum of the height function in direction of the binormal vector at t_2. Let Π_i be the osculating plane of $\tilde{\gamma}$ at t_i. Then $\tilde{\gamma}$ lies between the planes Π_1 and Π_2. Denote the osculating plane at t by Π_t and let D_t^+ and D_t^- denote the disjoint open halfspaces bounded by Π_t such that the binormal of $\tilde{\gamma}$ at t points into D_t^+. We define the following sets

$$A = \{t \in [t_1, t_2]; p \notin D_t^+\}$$

and

$$B = \{t \in [t_1, t_2]; p \notin D_t^-\}.$$

Since p is an interior point of the convex hull, we have $t_1 \in A$ and $t_2 \in B$. Moreover, we set

$$s_1 = \sup(A),$$

and

$$s_1' = \inf(B \cap [s_1, t_2]).$$

If $s_1 = s_1'$, then the osculating plane Π_{s_1} of $\tilde{\gamma}$ at s_1 passes through p and s_1 is clearly not an honest vertex since no neighborhood of s_1 maps onto one side of Π_{s_1}. If $s_1 < s_1'$, then for every $t \in [s_1, s_1']$, the osculating plane at t passes through p.

Hence the osculating circles of $\tilde{\gamma}|_{[s_1,s_1']}$ on $S^2(p)$ are all great circles and therefore intersect. It follows that the torsion of $\tilde{\gamma}$ vanishes for all $t \in [s_1, s_1']$ since the osculating circles of a spherical curve segment with nonvanishing torsion do not intersect. As a consequence $\tilde{\gamma}|_{[s_1,s_1']}$ is planar. It follows that $\tilde{\gamma}|_{[s_1,s_1']}$ is a great circle arc since it is contained in the osculating plane Π_{s_1}. No neighborhood of $[s_1, s_1']$ maps onto one side of Π_{s_1}. Hence we have proved that s_1 is not an honest vertex. We repeat this argument for the segments (t_2, t_3), (t_3, t_4), (t_4, t_1) and get the four points s_1, s_2, s_3, s_4 as claimed in (1).

To prove (2) notice that the spherical curve $\tilde{\gamma}$ on $S^2(p)$ changes its convexity in s_1 and s_2. We can assume that $\tilde{\gamma}|_{(s_1,s_2)}$ is a locally convex spherical curve. Then the curve $\tilde{\gamma}$ crosses its osculating plane at s_1 in the direction of its binormal vector in s_1 and it crosses its osculating plane at s_2 in the direction opposite of the binormal vector at s_2 (or the other way around).

We would like to prove that we have similarly for the curve γ that it crosses its osculating plane at s_1 in the direction of its binormal vector in s_1 and it crosses its osculating plane at s_2 in the direction opposite of the binormal vector at s_2 (or the other way around). To see this, notice that the osculating planes of γ and $\tilde{\gamma}$ coincide at the points s_1 and s_2. It is therefore only left to prove that the orientations of the planes also coincide. The assumption that the normal vectors of both γ and $\tilde{\gamma}$ at s_1 and s_2 make an obtuse angle with the ray from p implies that the orientations of the osculating hyperplanes do not change. Suppose that there are no vertices on the interval (s_1, s_2). Then the torsion function τ has no zeros on the interval. Hence one of the following two cases occurs:

(1) $\tau > 0$ on (s_1, s_2);
(2) $\tau < 0$ on (s_1, s_2).

Suppose the case (1) (resp. the case (2)) occurs. Without loss of generality, we may assume that the curvature function is positive. Then at both points s_1 and s_2, the curve γ crosses the osculating planes in the direction (resp. opposite direction) of the binormal vectors, which is a contradiction. So the torsion function has a zero in the interval (s_1, s_2). By repeating this argument for $(s_2, s_3), (s_3, s_4)$ and (s_4, s_1), we find four different vertices. □

REMARK. Notice that we only use in the proof of Theorem 2.2(2) that the angle between the ray from p to $\gamma(t)$ and the normal vector at $\gamma(t)$ is obtuse if the osculating hyperplane at $\gamma(t)$ passes through p.

EXAMPLES. (i) First we give an example of a space curve γ that satisfies the conditions of Theorem 2.2(2) without being convex. Since convex space curves satisfy these conditions, we have a generalization of Sedykh's theorem.

Consider a starshaped open domain Ω with respect to the origin in $\mathbf{R}^3$. Suppose that the boudary of Ω is an embedded surface. We set

$$\rho(x) := \inf\{t > 0; x \in t\Omega\},$$

where $t\Omega := \{tx; x \in \Omega\}$. Similarly as discussed in the proof of Theorem 1.7, Ω can be expressed as

$$\Omega = \{x \in \mathbf{R}^3; \rho(x) < 1\}.$$

For each positive integer n, we set

$$\rho_n(x) = |x| + (1/n)\left(\rho(x) - |x|\right).$$

Then $\rho = \rho_1$ holds and the set
$$\Omega_n = \{x \in R^3; \rho_n(x) < 1\}$$
is star-shaped with respect to the origin. Moreover, $(\Omega_n)_{n \in \mathbf{N}}$ converges to the unit ball when n goes to infinity. Consider a simple closed C^2-regular curve γ on the unit sphere which has only finitely many inflection points. We set
$$\gamma_n(t) = \rho_n(\gamma(t)) \cdot \gamma(t).$$
Then γ_n is a curve on the starshaped surface $\partial \Omega_n$. We would like to show that γ_n satisfies the conditions of Theorem 2.2(2) for n sufficiently large. We have that
$$\left\langle \frac{\gamma(t)}{\|\gamma(t)\|}, N(t) \right\rangle \leq \alpha^2 < 0$$
for all t, where N is the normal vector of γ and $\langle , \rangle$ denotes the inner product in $\mathbf{R}^3$. In other words, the normal vector at $\gamma(t)$ makes an obtuse angle with the ray from p to $\gamma(t)$. Since γ_n converges to γ in the C^2-topology when n goes to infinity, the curve γ_n also satisfies the property
$$\left\langle \frac{\gamma_n(t)}{\|\gamma_n(t)\|}, N_n(t) \right\rangle < 0$$
for all sufficiently large n, where N_n denotes the normal vector of γ_n. Since Ω is not convex and can be chosen arbitrarily, γ_n is not convex for a suitable choice of Ω.

(ii) The condition on the angle between the ray from p through $\gamma(t)$ and the normal vector is necessary in Theorem 2.2(2) as the following example constructed by Costa [C] shows, which Sedykh pointed to us. On a circular torus of revolution there is a torus knot without vertices that intersects every meridian exactly once. This curve satisfies the starshapedness condition in the theorem with p being the point on the rotation axis closest to the torus, but not the condition on the angle between the normal vector and the ray starting in p. When applying the radial projection in the proof of Theorem 2.2(2) to this curve, one changes the orientation of some of the osculating planes passing through p.

REMARK. Theorem 2.1 has some well-known results as immediate applications that we now would like to explain.

Arnold's Tennis Ball Theorem [A2] says that a C^2-regular simple closed curve γ on the unit sphere S^2 that divides the area of the sphere into two equal parts has at least four inflection points. Here an inflection point is a point p on the curve with the following property. Near the connected component of the intersection of γ with the tangent great circle C that contains p, the curve does not lie on one side of C. Notice that the origin must lie in the interior of the convex hull of γ since we may assume that the curve is not a great circle. In fact, one can slightly generalize the Tennis Ball Theorem by assuming that the origin lies in the interior of the convex hull of γ instead of assuming that γ divides the area of S^2 into two equal parts. There are therefore at least four different points on γ that are not honest vertices whose osculating planes pass through the origin. It follows that the osculating circles on S^2 of γ at these points are great circles. Thus they are inflection points since they are not honest vertices. This application was pointed out to us by S. Tabachnikov.

As a matter of fact Arnold deduces his Tennis Ball Theorem in [**A2**] as an immediate corollary from the following result which also clearly follows from Theorem 2.1.

($*$) Let γ be a regular simple closed curve on S^2 with at most two inflection points. Then γ is contained in a hemisphere.

This he proves by using the result from [**A1**] saying that

($**$) a regular curve segment $\hat{\gamma}$ in S^2 without self-intersections and inflection points must be contained in a hemisphere.

Conversely, ($*$) implies ($**$). This can be seen by taking another curve segment $\tilde{\gamma}$ in constant distance from $\hat{\gamma}$ and joining the endpoints only introducing one inflection point at each end, thus giving us a curve γ as in ($*$).

It is also interesting to notice that ($*$) implies the part of Segre's theorem saying that if γ is a simple closed spherical curve and p a point in the interior of the convex hull of γ, then there are at least four points on γ none of which is an honest vertex whose osculating planes contain p. To see this we use the idea of Segre that we already used in the proof of Theorem 2.2 to project γ centrally from p onto a curve $\tilde{\gamma}$ on $S^2(p)$ using the property that the osculating planes of γ and $\tilde{\gamma}$ passing through p are the same.

E. Heil has pointed out to us that a claim very similar to ($*$) can be found on p. 32 in Blaschke's textbook [**Bl**] as item 12 in §21, *Aufgaben und Lehrsätze* (Exercises and Theorems). In fact, Blaschke claims that a regular simple closed curve on the sphere with only two osculating planes passing through the center of the sphere is contained in a hemisphere. It seems that Blaschke meant this rather as a theorem than an exercise since it does not seem to follow easily from the material presented in the book. We do not know whether Blaschke or some of his contemporaries ever published a proof of this result.

On the same page, in item 11, Blaschke says that a regular, simple closed space curve has at least four points with a stationary osculating plane if there is for every point on the curve a plane meeting the curve only in that point. Blaschke owes this result to Carathéodory whose proof we do not know. In other sources this statement is attributed to Mohrmann (see e.g., [**Bo**], p. 232), where one also finds a sketch of a proof of this claim that the author has from his correspondence with Blaschke. In some recent papers this claim in a somewhat more general form has been called Scherk's conjecture.[1] In any case, one needs to add here that the curve has nonvanishing curvature since otherwise there is a counterexample due to Flohr from the 1950's (unpublished) and also a recent one due to Sedykh [**Sd**]. (See also the Introduction of part I [**U**] for related references.) Theorem 2.2(2) is the most general result known to us in this direction.

REMARK. A closed curve in the projective plane which is not nullhomotopic must have at least one inflection point. Here an inflection point is defined as for spherical curves by replacing "tangent great circles" by "tangent projective lines". One sees this as follows. If γ does not have an inflection point, then we have a continuous normal vector field along it pointing to the side γ is curving. This is a contradiction since a closed curve that is not nullhomotopic does not admit any

---

[1]This does not seem to be appropriate for chronological reasons (the paper of Mohrmann and Blaschke's book were published long before Scherk's first paper). We have also not been able to find any such conjecture in Scherk's work.

such normal vector field. This argument can be used to prove that the number of inflection points of such a curve must be odd, if it is finite.

Möbius proved the following theorem in [**M**]. If γ is a C^2-regular simple closed curve in the projective plane, then it has at least three inflection points. This follows from Theorem 2.2 by the following argument. Let us think of the projective plane as the plane $z = -1$ in $\mathbf{R}^3$ with a line added at infinity. Let $\tilde{\gamma}$ be the curve on S^2 whose points are the intersections of the lines with S^2 that connect the origin with the points on γ. There are two points on $\tilde{\gamma}$ corresponding to a point on γ. Notice that $\tilde{\gamma}$ is connected since γ is not nullhomotopic. It is clear that the origin lies in the interior of the convex hull of $\tilde{\gamma}$, since we may assume that γ is not a line. It follows that there are at least four different points on $\tilde{\gamma}$, none of which is an honest vertex, whose osculating planes pass through the origin. These four different points correspond to at least two different points in the projective plane that are inflection points. Since γ has an odd number of inflection points, there are at least three inflection points. Of course one can also deduce the Möbius theorem from the Tennis Ball Theorem.

The next theorem is a consequence of the theorem of Möbius. The proof is similar to an argument by A. Kneser [**Kn**].

THEOREM 2.3. *Let γ be a C^3-regular convex simple closed curve in $\mathbf{R}^3$ with nowhere vanishing curvature, and let p be a point on γ. Assume that no ray starting in p meets γ in two or more points or is tangent to γ in some point. If p is not an honest vertex of γ, then there are at least three different points on γ, all of which differ from p and none of which is an honest vertex, such that the osculating planes at these points contain p. If p is an honest vertex, there are at least two such points.*

PROOF. Let P be a plane supporting γ in p and let Π be the osculating plane of γ in p. The planes P and Π can only coincide if p is an honest vertex. Let L be the tangent line of γ in p. We project γ radially from p onto a sphere $S^2(p)$ with center in p. Notice that the image curve which we denote by $\tilde{\gamma}$ lies in a hemisphere of $S^2(p)$ that is bounded by the great circle $P \cap S^2(p)$ and is not closed since p corresponds to the two antipodal points in $L \cap S^2(p)$. Let $\hat{\gamma}$ be the curve in the projective plane $\mathbf{P}^2\mathbf{R}$ that we get by composing $\tilde{\gamma}$ with $\pi : S^2(p) \to \mathbf{P}^2\mathbf{R}$ where π is the identification of antipodal points.

Without loss of generality, we may assume that the point p is the origin. Then the curve $\tilde{\gamma}$ is given by
$$\tilde{\gamma}(t) = \gamma(t)/|\gamma(t)|.$$
We assume that $\gamma(0) = p$. By the Bouquet formula, it holds that
$$\gamma(t) = \mathbf{e}t + \kappa(t)\mathbf{n}\frac{t^2}{2} + o(t^2),$$
where $\mathbf{e}$ is the unit tangent vector at p, $\mathbf{n}$ is the unit normal vector at p, $\kappa(t)$ the curvature and t is the arc length parameter. So it can be easily shown that $\hat{\gamma} : S^1 \to \mathbf{P}^2\mathbf{R}$ is an immersion (at $t = 0$) if and only if $\kappa(0) \neq 0$.

Notice that $\hat{\gamma}$ is still C^3 except in the point $\hat{p}$ corresponding to p where it might only be C^2. Let λ be the line in $\mathbf{P}^2\mathbf{R}$ that is the image of $\Pi \cap S^2(p)$. Then λ is tangent to $\hat{\gamma}$ in $\hat{p}$. If p is an honest vertex, then γ does not cross Π in p. Consequently, $\hat{\gamma}$ *crosses* λ in $\hat{p}$, and we see that $\hat{p}$ is an inflection point of $\hat{\gamma}$. Similarly we see that if p is not an honest vertex, then $\hat{p}$ is not an inflection point of

$\hat{\gamma}$. Let l be the line in $\mathbf{P}^2\mathbf{R}$ that is the image of $P \cap S^2(p)$. The mod 2 intersection number between $\hat{\gamma}$ and l is equal to one, since $\hat{\gamma}$ changes sides of l locally around $\hat{p}$ and only there. It follows that $\hat{\gamma}$ is not nullhomotopic. By the theorem of Möbius, $\hat{\gamma}$ has at least three inflection points. Notice that an inflection point of $\hat{\gamma}$ different from $\hat{p}$ corresponds to a point on $\tilde{\gamma}$ with osculating circle on $S^2(p)$ being a great circle. Hence the corresponding osculating plane of γ passes through p. It follows that if p is not an honest vertex, there are at least three points, all of which are different from p whose osculating planes pass through p. It is also clear that none of these points is an honest vertex. If p is an honest vertex there are at least two such points. □

We end this section by giving an example. We will need the following lemma in which we denote the closed unit ball by B_1.

LEMMA 2.4. *Let $\gamma : S^1 \to S^2$ be a smooth regular spherical curve which may have self-intersections, and let x be a point in B_1. Assume there are points $p, q \in \gamma$ such that the osculating planes at p and q pass through x. Then there is at least one honest vertex on the open arc $\gamma|_{(p,q)}$.*

PROOF. We divide the proof into two parts.

Case 1. If x lies on the sphere S^2, then x is an intersection point between the osculating circle C_p at p and the osculating circle C_q at q. Then the assertion follows immediately from the result of A. Kneser that the osculating circles of an arc without vertices do not intersect.

Case 2. Assume that x lies in the interior of B_1. Let Π_p and Π_q be the osculating planes of the curve γ at p and q, respectively. We set

$$L := \Pi_p \cap \Pi_q.$$

Since $x \in \Pi_p \cap \Pi_q$, L is not empty. Thus L is a line passing through the point x. Since x is an interior point of B_1, the line L must meet S^2 in two points. Let y be one of them. Then we have $y \in \Pi_p \cap \Pi_q$, and the assertion follows from the first case. □

REMARK 1. Notice that Lemma 2.4 together with the theorem of A. Kneser and Segre immediately implies the four vertex theorem. This is very similar to A. Kneser's original proof.

REMARK 2. Using the same argument, one can easily generalize the assertion of the lemma to a simple closed C^2-curve γ on a smooth convex surface with positive Gaussian curvature and a point x in the compact region bounded by the surface. (Instead of the result of Kneser, apply I-Theorem A.9 to Example 2 in I-§3.)

We give an example of a simple closed spherical curve γ with the following property. *There is a point q in the boundary of the convex hull of γ such that there are four distinct points whose osculating planes contain q, but there are only two such points which are not honest vertices.* This example seems to contradict section 9 in Segre's paper [**Se**], where he claims that if p is in the convex hull of γ, but not in its image, then there must be more than two points which are not honest vertices and whose osculating planes contain p. Notice that we have proved in Theorem 2.2 that if p lies in the *interior* of the convex hull of γ, then there are at least four such points.

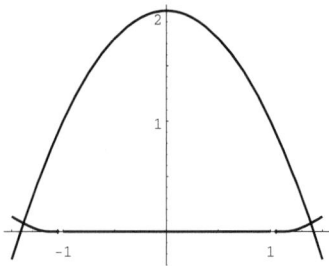

FIGURE 2

We first prove the following result.

There is a smooth simple closed curve $\gamma : S^1 \to \mathbf{R}^2$ satisfying the following two properties:

1. γ has three isolated clean vertices p'_1, p'_2, p'_3 and a closed arc $I := \gamma|_{[x',y']}$ consisting of clean maximal vertices;
2. The curvature function of γ never vanishes except on p'_1, p'_2, p'_3 and I.

To see this, consider the functions

$$f(x) := \begin{cases} e^{-1/(|x|-1)} & (|x| \geq 1), \\ 0 & (|x| < 1), \end{cases}$$

and

$$g(x) = -x^2 + 2.$$

The graphs of these functions meet in two points that we denote by p and q, see Figure 2. Denote by $\tilde{\gamma}$ the closed curve that we get by joining the graphs of f and g between p and q. Using the method of rounding corners in Proposition 2.3 in [**KU**], we find a curve γ' that agrees with $\tilde{\gamma}$ except close to p and q and has exactly the following vertices: one vertex close to p, another one close to q, one at the top of the graph of g and finally the whole interval $[0, 1]$ on the x-axis.

Set $\gamma = \pi \circ \gamma' : S^1 \to S^2$, where $\pi : \mathbf{R}^2 \to S^2$ is the inverse of the stereographic projection and γ' is a curve with the two properties (1) and (2). Set

$$x := \pi(x'), \quad y := \pi(y'), \quad p_j := \pi(p'_j) \qquad (j = 1, 2, 3).$$

Without loss of generality, we may assume that $(x \succ) \; p_3 \succ p_2 \succ p_1 \succ y \succ x$ holds. Then $\gamma|_{[x,y]}$ lies in a common osculating supporting plane Π_0. Let q be the midpoint of the line segment $\overline{xy}$. Then q lies in the boundary of the convex hull of γ. On $\gamma|_{(y,p_1]}$ and $\gamma|_{[p_3,x)}$ there is no point whose osculating plane passes through q. In fact if such a point would exist, then there would be an honest vertex on $\gamma|_{(y,p_1)}$ (resp. $\gamma|_{(p_3,x)}$) by Lemma 2.4, which is a contradiction. By an intermediate argument as in the proof of Theorem 2.1, there is at least one point on $\gamma|_{(p_1,p_2)}$ (resp. $\gamma|_{(p_2,p_3)}$) whose osculating plane passes through q. We denoted it by z_1 (resp. z_2). Since $\gamma|_{(p_1,p_2)}$ and $\gamma|_{(p_2,p_3)}$ have no vertices, there is no other such point on $\gamma|_{(p_1,p_2)}$ (resp. $\gamma|_{(p_2,p_3)}$) by Lemma 2.4. Thus the points whose osculating planes pass through q are exactly z_1, z_2 and all points on $\gamma|_{[x,y]}$. Among them only z_1 and z_2 are not honest vertices.

Thus the curve γ and the point q in the boundary of its convex hull are the example we have been looking for.

3. Extremal points of projective line diffeomorphisms

Let $\mathbf{P}^1\mathbf{R}$ denote the real projective line. Let $f : \mathbf{P}^1\mathbf{R} \to \mathbf{P}^1\mathbf{R}$ be a diffeomorphism and $x \in \mathbf{P}^1\mathbf{R}$. Then there is a unique projective transformation

$$A_x : \mathbf{P}^1\mathbf{R} \to \mathbf{P}^1\mathbf{R}$$

whose 2-jet at x coincides with that of f. Let us call A_x the *osculating map of f at x*. If the osculating map A_x has the same 3-jet as f in x, we call x a *projective point of f*. One owes to Ghys the beautiful theorem that any diffeomorphism $f : \mathbf{P}^1\mathbf{R} \to \mathbf{P}^1\mathbf{R}$ has at least four distinct projective points, see [**OT**] and [**T**]. In fact he proved somewhat more: f has at least four extremal points (see the definition below).

Ghys' theorem is of course reminiscent of the four vertex theorem. We will reprove it here as an application of the theory of compatible intrinsic circle systems in Section 1. As a consequence we can prove a Bose type formula as well as an analogue of part (2) of the theorem of A. Kneser and Segre that was discussed in the last section.

Ghys proved his theorem by translating it into an equivalent statement about curves in the Lorentz plane. This will also be our approach. For a proof using Sturm theory see [**OT**] and [**T**]. It is interesting to notice that in one of the first papers on vertices [**Kn**], A. Kneser already considers curves in the Lorentz plane and arrives at a result that we will use below. More precisely, he proves that two osculating hyperbolas of a curve segment γ with no timelike tangent vectors in the Lorentz plane do not meet if the curvature of γ does not have local extrema. The corresponding statement for curves in the Euclidean plane is well known and can be found in many elementary textbooks on differential geometry. It was also first proved by A. Kneser in the same paper.

We will always assume for simplicity that the diffeomorphism f is orientation preserving. The orientation reversing case then follows by composing f with an orientation reversing projective transformation.

As will become clear when we set up the correspondence with Lorentz geometry, the result of A. Kneser that we mentioned above implies the following fact about a diffeomorphism f: If $x \in \mathbf{P}^1\mathbf{R}$ is not a projective point of f, then the fixed point x of $f \circ A_x^{-1}$ is either an attractor (sink) or an expellor (source). Consequently, if x is neither an attractor nor an expellor of $f \circ A_x^{-1}$, then x is a projective point of f. We call a point x an *extremal point of f*, if there is a neighborhood I of the connected component F_x of the fixed point set of $f \circ A_x^{-1}$ containing x such that $f \circ A_x^{-1}$ attracts the points in I on one side of F_x toward F_x and expels those on the other side. If $f \circ A_x^{-1}$ attracts points in I on the left of F_x and expels points on the right, then we call the points in F_x *minimal points of f*; otherwise they are called *maximal points*.

Another consequence is that if x is not an extremal point, then f and A_x agree in a disconnected set. Hence a point x with the property that f and A_x only agree on an interval containing x must be an extremal point. We call such an x a *clean extremal point of x*. It is now clear what we mean by a *clean minimal* and a *clean maximal point of f*. Notice that the notions of maximal and minimal points depend on the orientation of $\mathbf{P}^1\mathbf{R}$ given by the parametrization. Changing the orientation, a minimal point becomes maximal and vice versa.

We are now in a position to state a more precise version of Ghys' theorem.

THEOREM 3.1. *Let $f : \mathbf{P}^1\mathbf{R} \to \mathbf{P}^1\mathbf{R}$ be a diffeomorphism. Then:*
(1) *The number of connected components of extremal points of f is even if it is finite; more precisely, between any two minimal points there is a maximal point and vice versa;*
(2) *f has at least two distinct connected components of clean minimal points and at least two distinct connected components of clean maximal points.*

The proof of the theorem follows immediately when we have associated to a diffeomorphism f a pair of compatible intrinsic circle systems.

REMARK. Notice that we do not claim that there are clean maximal points between two connected components of clean minimal points. This is not expected to be true (see a counterexample in the similar case of planar curves in Figure 1).

Before we associate a pair of compatible intrinsic circle systems to a diffeomorphism f of $\mathbf{P}^1\mathbf{R}$, we explain how it relates to planar Lorentz geometry.

By composing f with a rotation if necessary, we can assume that f has a fixed point. We choose the fixed point as the point at infinity and restrict f to the reals, $\mathbf{R} = \mathbf{P}^1\mathbf{R} - \{\infty\}$, keeping the notation f for the restriction. We are therefore in the situation of a surjective strictly increasing function $f : \mathbf{R} \to \mathbf{R}$ with a nowhere vanishing positive derivative. The orientation preserving projective transformations of $\mathbf{P}^1\mathbf{R}$ correspond to the linear fractional transformations

$$P(x) = \frac{\alpha x + \beta}{\gamma x + \delta}$$

with $\alpha\delta - \beta\gamma = 1$. Their graphs are the nonvertical lines with positive inclination if they have ∞ for a fixed point (i.e., $\gamma = 0$); otherwise their graphs are hyperbolas of the type

$$(x - a)(y - b) = -c^2$$

with $c \neq 0$. Notice that a hyperbola satisfying the above equation is asymptotic to the perpendicular lines $x = a$ and $y = b$ and corresponds to a linear fractional transformation sending a to ∞. The branches of the hyperbola lie in the second and the fourth quadrant of $\mathbf{R}^2$ with $x = a$ and $x = b$ as axes. The osculating map of f at a point x_0 corresponds to such a line or hyperbola that has second order contact with the graph of f in (x_0, y_0) where $y_0 = f(x_0)$. To simplify the exposition, we call the lines and hyperbolas of the above type *admissible hyperbolas* or simply *hyperbolas*. Notice that a point x is an extremal point of f if and only if the osculating hyperbola at x is locally around the connected component containing x of the set where it coincides with the graph of f on one side of the graph of f. It is a minimal point if the osculating hyperbola is locally below the graph; otherwise it is a maximal point.

It is now easy to prove the claim in part (1) of Theorem 3.1. Let x and y be minimum points and assume that there are no extremal points between them. Then the osculating hyperbolas between x and y do not intersect by the result of A. Kneser in [**Kn**] that we have already mentioned. It follows that the graph of f cannot be locally above the osculating hyperbolas both in x and y. Hence there is a point z between x and y in which the graph must lie locally below the osculating hyperbola.

We now begin with the definition of the intrinsic circle systems. We choose a point (a, b) on the graph of f. The branches of the admissible hyperbolas tangent to the graph of f in (a,b) fill up the first and the third quadrant of the complement of $x=a$ and $y=b$. Furthermore, they only meet in the point (a,b). Notice that we have a one parameter family of these hyperbolas. We assume the parameter to go from minus to plus infinity. Being a diffeomorphism, f only meets the lines $x=a$ and $y=b$ in the point (a,b). We can therefore find admissible hyperbolas that are tangent to the graph of f in (a,b), lying above it and not meeting it any other point. Let $\mathcal{H}$ be the set of all such admissible hyperbolas. Then $\mathcal{H}$ either corresponds to an open or closed half line of the parameter. Denote the hyperbola that corresponds to the endpoint of $\mathcal{H}$ by H_a.

If H_a is not a line, then we define the set
$$F_a^\bullet = \{x \in \mathbf{R} \mid (x, f(x)) \in H_a\}.$$

If H_a is a line, then we set
$$F_a^\bullet = \{x \in \mathbf{R} \mid (x, f(x)) \in H_a\} \cup \{\infty\}.$$

The set F_a° is defined analogously using hyperbolas that lie below the graph of f. The sets $F_a^\bullet$ and F_a° are clearly independent of the choice of the point at infinity. It follows that we can associate the sets $F_a^\bullet$ and F_a° to all points $a \in \mathbf{P}^1\mathbf{R}$, including the one that was chosen to be at infinity.

A more intrinsic way to define $F_a^\bullet$ is as follows. Let $\mathcal{P}$ denote the one-parameter family of projective transformations whose 1-jets at $a \in \mathbf{P}^1\mathbf{R}$ agree with the 1-jet of f at a. We assume that this family is parametrized by the real numbers. We consider $f \circ P_t^{-1}$ for $P_t \in \mathcal{P}$. It follows from the considerations above that there are numbers $t_0 \leq t_1$ such that if $t \notin [t_0, t_1]$, then $f \circ P_t^{-1}$ only has a fixed point in a. We assume the parameter chosen so that $f \circ P_t^{-1}$ moves points locally on the left of a away and brings those on the right closer, if $t < t_0$. Assume that the interval $[t_0, t_1]$ is the smallest possible with the above property. Then $F_a^\bullet$ is the fixed point set of $f \circ P_{t_0}^{-1}$ and F_a° is the fixed point set of $f \circ P_{t_1}^{-1}$.

THEOREM 3.2. *Let $f : \mathbf{P}^1\mathbf{R} \to \mathbf{P}^1\mathbf{R}$ be a diffeomorphism that is not a projective transformation. Then $(F_a^\bullet)_{a \in \mathbf{P}^1\mathbf{R}}$ and $(F_a^\circ)_{a \in \mathbf{P}^1\mathbf{R}}$ are compatible intrinsic circle systems.*

PROOF. We first prove that $(F_a^\bullet)_{a \in \mathbf{P}^1\mathbf{R}}$ is an intrinsic circle system. Let a, b be such that $b \in F_a^\bullet$. We can assume that $a, b \in \mathbf{R}$. Then $b \in H_a$. It clearly follows that $H_b = H_a$ and hence that $F_b^\bullet = F_a^\bullet$. Hence $(F_a^\bullet)_{a \in \mathbf{P}^1\mathbf{R}}$ satisfies (I1). Property (I2) follows from the fact that two distinct admissible hyperbolas intersect in at most two points, or equivalently that two distinct projective transformations agree in at most two points. Property (I3) is clear. We have thus proved that $(F_a^\bullet)_{a \in \mathbf{P}^1\mathbf{R}}$ is an intrinsic circle system. The proof that $(F_a^\circ)_{a \in \mathbf{P}^1\mathbf{R}}$ is an intrinsic circle system is completely analogous.

We now prove that the intrinsic circle systems are compatible. Property (C1) is immediate since f is not a projective transformation. Assume that $\text{rank}^\bullet(a) = 1$ and $a \in \mathbf{R}$. Then $\text{rank}^\circ(a) > 1$ since f is not a projective transformation. Also a is isolated in $F^\circ(a)$ for the same reason. We have thus proved (C2). This completes the proof. □

Each intrinsic circle system $F^\bullet$ (resp. F°) induces an equivalent relation (see I-§2). We denote the quotient spaces of $\mathbf{P}^1\mathbf{R}$ with respect to these relations by

$\mathbf{P}^1\mathbf{R}/F^\bullet$ and $\mathbf{P}^1\mathbf{R}/F^\circ$, respectively. We set

$$S(F^\bullet) := \{[p]^\bullet \in \mathbf{P}^1\mathbf{R}/F^\bullet; \text{rank}^\bullet([p]^\bullet) = 1\},$$
$$S(F^\circ) := \{[p]^\circ \in \mathbf{P}^1\mathbf{R}/F^\circ; \text{rank}^\circ([p]^\circ) = 1\},$$
$$T(F^\bullet) := \{[p]^\bullet \in \mathbf{P}^1\mathbf{R}/F^\bullet; \text{rank}^\bullet([p]^\bullet) \geq 3\},$$
$$T(F^\circ) := \{[p]^\circ \in \mathbf{P}^1\mathbf{R}/F^\circ; \text{rank}^\circ([p]^\circ) \geq 3\}.$$

Moreover, we set

$$s(F^\bullet) := \text{the cardinality of the set } S(F^\bullet),$$
$$s(F^\circ) := \text{the cardinality of the set } S(F^\circ),$$
$$t(F^\bullet) := \sum_{[p]^\bullet \in T(F^\bullet)} (\text{rank}^\bullet(p) - 2),$$
$$t(F^\circ) := \sum_{[p]^\circ \in T(F^\circ)} (\text{rank}^\circ(p) - 2).$$

Notice that $s(F^\bullet)$ (resp. $s(F^\circ)$) is the number of connected components of the set of clean maximal (resp. minimal) points of f. Hence the number $s(f)$ of connected components of clean extremal points of f equals $s(F^\bullet) + s(F^\circ)$.

We would like to interpret the meaning of the sets $T(F^\bullet)$ and $T(F^\circ)$ in terms of properties of the diffeomorphism f. The equivalence class of a is in $T(F^\bullet)$ if and only if there is a projective transformation P such that the fixed point set of the mapping $f \circ P^{-1}$ has at least three connected components, one of them containing a, and $f \circ P^{-1}$ moves all points in the complement of the fixed point set of $f \circ P^{-1}$ against the orientation of $\mathbf{P}^1\mathbf{R}$. Similarly, the equivalence class of a lies in $T(F^\circ)$ if and only if there is a projective transformation P such that the fixed point set of the mapping $f \circ P^{-1}$ has at least three connected components, one of them containing a, and $f \circ P^{-1}$ moves all points in the complement of the fixed point set of $f \circ P^{-1}$ with the orientation of $\mathbf{P}^1\mathbf{R}$. We set $t(f) = t(F^\bullet) + t(F^\circ)$. It follows from section 1 that $t(f)$ is finite if $s(f)$ is finite.

The following Bose type formula follows from I-Theorem 2.7. Notice that it has part (2) of Theorem 3.1 as in immediate corollary.

THEOREM 3.3. *Let $f : \mathbf{P}^1\mathbf{R} \to \mathbf{P}^1\mathbf{R}$ be a diffeomorphism that is not a projective transformation and assume that $s(f)$ is finite. Then $s(f) - t(f) = 4$.*

REMARK. It follows from Lemma 1.2 in [OT] that the sets $S(F^\bullet)$ and $S(F^\circ)$ are supported by a certain continuous function. So, by I-Theorem 2.2, $s(f) < \infty$ if and only if $t(f) < \infty$.

A further consequence of Section 1 is the following theorem.

THEOREM 3.4. *Let $f : \mathbf{P}^1\mathbf{R} \to \mathbf{P}^1\mathbf{R}$ be a diffeomorphism that is not a projective transformation. Then there are four points $a_1 \succ a_2 \succ a_3 \succ a_4$ on $\mathbf{P}^1\mathbf{R}$ such that a_1 and a_3 are clean maximal points of f and a_2 and a_4 are clean minimal points of f.*

We would now like to prove, as a corollary of Theorem 3.4, a result that is similar to part (2) of the theorem of A. Kneser and Segre that we quoted in Section 2: If p is a point on a simple closed curve γ on the unit sphere S^2 that is not a vertex, there are at least three distinct points p_1, p_2, p_3 on γ, all of which are different from p, with the property that the osculating planes at p_1, p_2, and p_3 pass through p. If p is a vertex, one can find two such points.

THEOREM 3.5. *Let $f : \mathbf{P}^1\mathbf{R} \to \mathbf{P}^1\mathbf{R}$ be a diffeomorphism and $a \in \mathbf{P}^1\mathbf{R}$. If a is not a clean extremal point of f, then there are three distinct points b_1, b_2 and b_3, all different from a and none of which is an extremal point, such that the osculating maps of f at b_1, b_2 and b_3 all agree with f in a. If a is a clean extremal point, then we can find at least two such points b_1 and b_2.*

PROOF. The idea of the proof is exactly the same as the one we used to prove Theorem 2.1. Assume that a is not a clean extremal point. By Theorem 3.4 there are four points $a_1 \succ a_2 \succ a_3 \succ a_4$ on $\mathbf{P}^1\mathbf{R}$ such that a_1 and a_3 are clean maximal points of f and a_2 and a_4 are clean minimal points of f. We choose the point at infinity between a_4 and a_1 and assume that it is different from a. We also assume that a lies between a_4 and a_1. The the graph of f lies between its osculating hyperbola at a_1 and its osculating hyperbola at a_2. As we move from a_1 to a_2 we must by continuity go through a point b_1 such that the osculating hyperbola of the graph of f at b_1 passes through the point $(a, f(a))$. We can use exactly the same argument as in the proof of Theorem 2.2 to show that b_1 can be chosen such that it is not an extremal point of f. We apply exactly the same argument to the intervals (a_2, a_3) and (a_3, a_4) to find the points b_2 and b_3. The case that a is a clean maximal point is similar. This proves the theorem. $\square$

Acknowledgments. The authors wish to thank S. Tabachnikov for his encouragement and very fruitful discussions during his stay at the Max-Planck-Institut, and V. D. Sedykh for his many excellent remarks on an earlier version of this paper. We are also very greatful to E. Heil for his historical remarks.

References

[A1] V. I. Arnold, *A ramified covering of $CP^2 \to S^4$, hyperbolicity and projective topology* (Russian), Sibirsk. Mat. Zh. **29** (1988), 36–47; English transl., Siberian Math. J. **29** (1988), 717–726.
[A2] _____, *Topological invariants of plane curves and caustics*, University Lecture Series, vol. 5, Amer. Math. Soc., Providence, RI, 1994.
[Bl] W. Blaschke, *Vorlesungen über Differentialgeometrie.* I, Verlag von Julius Springer, Berlin, 1921.
[Bo] G. Bol, *Projektive Differentialgeometrie,* 1. Teil, Vandenhoeck & Ruprecht, Göttingen, 1950.
[C] S. I. R. Costa, *On closed twisted curves,* Proc. Amer. Math. Soc. **109** (1990), 205–214.
[KR] R. V. Kadison and J. R. Ringrose, *Fundamentals of the theory of operator algebras,* vol. 1, Academic Press, New York, London, 1983.
[Kn] A. Kneser, *Bemerkungen über die Anzahl der Extreme der Krümmung auf geschlossenen Kurven und über verwandte Fragen in einer nicht-euklidischen Geometrie,* Festschrift Heinrich Weber zu seinem siebzigsten Geburtstag am 5. März 1912, B. G. Teubner, Leipzig and Berlin, 1912, pp. 170–180.
[J] S. B. Jackson, *A refinement of the Four-Vertex Theorem,* Technical Report, University of Maryland, 1969, 15pp.
[KU] O. Kobayashi and M. Umehara, *Geometry of scrolls,* Osaka J. Math. **33** (1996), 441–473.
[M] A. F. Möbius, *Über die Grundformen der Linien der dritten Ordnung,* Abhandlungen der Königl. Sächs. Gesellschaft der Wissenschaften, Math.-Phys. Klasse **I** (1852), 1–82; Also inA. F. Möbius, *Gesammelte Werke,* vol. II, Verlag von S. Hirzel, Leipzig, 1886, pp. 89–176.
[OT] V. Ovsienko and S. Tabachnikov, *Sturm theory, Ghys theorem on zeros of the Schwarzian derivative and flattening of Legendrian curves,* Selecta Math., New Series, **2** (1996), 297–307.

[Se] B. Segre, *Alcune proprietà differenziali in grande delle curve chiuse sghembe*, Rend. Math. (6) **1** (1968), 237–297.

[Sd] V. D. Sedykh, *The four-vertex theorem of a convex space curve* (Russian), Funktsional. Anal. i Prilozhen **26** (1992), no. 1, 35–41; English transl., Functional Anal. Appl. **26** (1992), 28–32.

[T] S. Tabachnikov, *On zeros of Schwarzian derivative*, to appear in V. Arnold's 60-th anniversary volume.

[U] M. Umehara, "A unified approach to the four vertex theorems. I", *Differential and symplectic topology of knots and curves* (S. Tabachnikov, ed.), Amer. Math. Soc. Transl. Ser. 2, 190, Amer. Math, Soc., Providence, RI, 1999.

[We] J. L. Weiner, *Global properties of spherical curves*, J. Diff. Geom. **12** (1977), 425–434.

Mathematisches Institut der Universität zu Köln, Weyertral 86-90, D-50931 Köln, Germany

Department of Mathematics, Graduate School of Science, Osaka University, Machikaneyama 1-16, Toyonaka, Osaka 560, Japan

E-mail address: umehara@math.wani.osaka-u.ac.jp

Topology of Two-Connected Graphs and Homology of Spaces of Knots

Victor A. Vassiliev

ABSTRACT. We propose a new method of computing cohomology groups of spaces of knots in $\mathbb{R}^n$, $n \geq 3$, based on the topology of configuration spaces and two-connected graphs, and calculate all such classes of order ≤ 3. As a byproduct we define the higher *indices*, which invariants of knots in $\mathbb{R}^3$ define at arbitrary singular knots. More generally, for any finite-order cohomology class of the space of knots in $\mathbb{R}^n$ we define its *principal symbol*, lying in the cohomology group of a certain finite-dimensional configuration space and characterizing our class modulo classes of smaller filtrations.

1. Introduction

The knots, i.e., smooth embeddings $S^1 \to \mathbb{R}^n$, $n \geq 3$, form an open dense subset in the space $\mathcal{K} \equiv C^\infty(S^1, \mathbb{R}^n)$. Its complement Σ is the *discriminant set*, consisting of maps, having self-intersections or singularities. Any cohomology class $\gamma \in H^i(\mathcal{K} \setminus \Sigma)$ of the space of knots can be described as the linking number with an appropriate chain of codimension $i+1$ in $\mathcal{K}$ lying in Σ.

In [**V2**] a method of constructing some of these (co)homology classes (in particular, knot invariants) was proposed. For $n = 3$ the 0-dimensional classes, arising from this construction, are exactly the "finite-type knot invariants", and for $n > 3$ this method provides a complete calculation of all cohomology groups of knot spaces in $\mathbb{R}^n$.

However, precise calculations by this method are very complicated. The strongest results obtained by now are as follows.

D. Bar-Natan calculated the $\mathbb{C}$-valued knot invariants of orders ≤ 9 (and used for this several weeks of computer work and the Kontsevich's realization theorem), see [**BN1**].

T. Stanford wrote a program, realizing the algorithm of [**V2**] for computing $\mathbb{Z}$-invariants, and found all such invariants of orders ≤ 7.

D. Teiblum and V. Turchin (also using a computer) found the first nontrivial 1-dimensional cohomology class of the space of noncompact knots in $\mathbb{R}^3$ (which is

---

1991 *Mathematics Subject Classification*. Primery 57M15, 57M25, 57M45.

Supported in part by RFBR (project 95-01-00846a), INTAS (project 4373) and Netherlands Organization for Scientific Research (NWO), project 47.03.005. Research at MSRI supported in part by NSF grant DMS-9022140.

of order 3) and proved that there are no other positive-dimensional cohomology classes of order 3. For a description of this class, see Section 4.4.2.

By the natural periodicity of cohomology groups of spaces of knots in $\mathbb{R}^n$ with different n (similar to the fact that all algebras $H^*(\Omega S^m)$ with m of the same parity are isomorphic up to a scaling of dimensions) these results can be extended to the cohomology of spaces of knots in $\mathbb{R}^n$ with arbitrary odd n: to any knot invariant of order i in $\mathbb{R}^3$ there corresponds an $(n-3)i$-dimensional cohomology class for any odd n, and the Teiblum–Turchin class is the origin of a series of $((n-3)3+1)$-dimensional cohomology classes in spaces of knots in $\mathbb{R}^n$, n odd.

Below we describe another in a sense opposite method of calculating these cohomology groups based on a different filtration of the discriminant variety. On the level of knot invariants in $\mathbb{R}^3$, this method is more or less equivalent to Bar-Natan's calculus of Chinese Character Diagrams and gives a partial explanation of its geometrical meaning. Using this method, it is possible to repeat "by hands" the calculation of Teiblum–Turchin (which was nontrivial even for a computer) and to obtain a similar result for the case of even n.

The idea of reducing the cohomology of spaces of nonsingular geometrical objects (such as, e.g., polynomials of a fixed degree in $\mathbb{C}^1$ or $\mathbb{R}^1$ without multiple roots) to the (Alexander dual) homology of the complementary discriminant set was proposed by Arnold in [**A1**]; see also [**A2**]. This idea proved to be very fruitful because the discriminant is a naturally stratified set (whose open strata consist of "equisingular" objects). Many homology classes of the discriminant can be calculated with the help of the filtration defined by this stratification.

The "reversed" filtrations of (some natural resolutions of) discriminant sets, introduced in [**V1**], greatly simplify these calculations and allow one to solve similar problems in the case of functions on multidimensional manifolds; see e.g., [**V5**], [**V6**]. Moreover, the spectral sequences defined by such filtrations are functorial with respect to the inclusion of (finite-dimensional) functional spaces, and thus give rise to stable sequences, calculating cohomology of complements of discriminants in infinite-dimensional spaces like the space of knots or of C^∞-functions without complicated singularities.

This reversion is a continuous generalization of the combinatorial formula of inclusion–exclusions. If we want to calculate the cardinality of a finite union of finite sets, then, instead of counting the points in "open strata" (i.e., for any k, in all k-fold intersections of these sets, from which all the $(k+1)$-fold intersections are removed), we first count separately all points in all sets. We then distract from the obtained sum the correction term, corresponding to all 2-fold intersections (not taking into account the fact that some of these double intersection points are also triple points), then add the correction terms corresponding to all triple points, etc. All results of [**V1**]–[**V7**] concerning the cohomology of complements of discriminants, were obtained by this method.

However, in the calculation of the first term E_1 of the spectral sequence from [**V2**] converging to the cohomology of knot spaces (in particular to knot invariants), an auxiliary spectral sequence was used, based on the "natural" substratification of the resolved discriminant. The main portion of hard calculations in the theory of finite-order knot invariants is the calculation in this auxiliary sequence (especially in those terms responsible for the 0-dimensional cohomology). In this paper we reverse this filtration as well. From the point of view of this new method the previous calculations are just the computation of cellular homology of certain configuration spaces, which can be studied also by more "theoretical" methods.

As a byproduct, we obtain the notion of the *index*, which any knot invariant assigns to any (finitely degenerate) singular knot in $\mathbb{R}^3$. In the standard theory of finite-order invariants this index was considered for the simplest points of Σ, i.e., for immersions having only transverse double points or (in a more implicit way) at most one triple point.

In the general situation this function is not numerical: it takes values in certain homology groups related to singular knots. Say, for an immersion having one generic k-fold self-intersection point this group is the $(2k-4)$-th homology group of the complex of *two-connected graphs*[1] with k nodes, in particular, according to [**BBLSW**] and [**T**], is $(k-2)!$-dimensional. In fact, the information provided by the index of such a singular knot is equivalent to the "totality of all extensions of our invariant to all Chinese Character Diagrams of order $k-1$ with exactly k legs"; see [**BN2**].

More generally, for any finite-order cohomology class of the space of knots we define its *principal symbol*, which lies in a cohomology group of a certain finite-dimensional configuration space and characterizes our class modulo classes of smaller filtrations.

Our first calculations lead to some essential problems in the homological combinatorics and representation theory (see [**BBLSW**], [**T**]), and the first answers indicate the existence of a rich algebraic structure behind it, which probably will allow one to guess many invariants and cohomology classes of arbitrary orders.

Notation. For any topological space X, $\bar{H}_*(X)$ denotes the *Borel–Moore homology group* of X, i.e., the homology group of its one-point compactification reduced modulo the added point.

Acknowledgments. I thank very much A. Vaintrob, who about January 1994 asked me whether it is possible to define generalized indices of complicated singular knots, and to A. Björner, V. Turchin and V. Welker for their interest and work on combinatorial aspects of the problem. Also I thank MSRI, where a considerable part of the work was done.

2. Complexes of connected and two-connected graphs

We consider only graphs without loops and multiple edges, but maybe with isolated nodes.

A graph with k nodes is *connected* if any two of its nodes are joined by a chain of its edges. A graph is l-*connected* if it is connected and after removing from it any j nodes, $j < l$, together with all incident edges, we obtain again a connected graph (with $k - j$ nodes).

The set of all graphs with given k nodes generates an (acyclic) simplicial complex. Namely, consider the simplex $\Delta(k)$, whose vertices are in one-to-one correspondence with all $\binom{k}{2}$ edges of the complete graph with these k nodes. The faces of this simplex are subgraphs of the complete graph; indeed, any face is characterized by the set of its vertices, i.e., by a collection of several edges of the complete graph. All faces of the simplex $\Delta(k)$ form an acyclic simplicial complex, which will also

---

[1]Unlike topological terminology, in combinatorics a graph with k nodes is called two-connected if it is connected (i.e., joins any two of these k nodes), and, moreover, removing from it any node together with all incident edges, we obtain also a connected graph.

be denoted by $\Delta(k)$. A generator of this complex is a graph with ordered edges, while permuting the edges we multiply such a generator by $\pm$ depending on the parity of the permutation. We will always choose the generator corresponding to the lexicographic order of edges induced by some fixed order of initial k nodes. The boundary of a graph is the formal sum of all graphs obtained from it by removing one of its edges, taken with coefficients 1 or -1.

Faces, corresponding to nonconnected graphs, form a subcomplex of the complex $\Delta(k)$. The corresponding quotient complex is denoted by $\Delta^1(k)$ and is called the *complex of connected graphs*. In a similar way, the complex $\Delta^l(k)$ of *l-connected graphs* is the quotient complex of $\Delta(k)$ generated by all faces corresponding to l-connected graphs. It is obviously isomorphic to the Borel–Moore homology group of the simplex $\Delta(k)$ (considered as a topological space) with all not l-connected faces removed.

THEOREM 1. (a) *For any k, the group $H_i(\Delta^1(k))$ is trivial for all*

$$i \neq k - 2, \quad \text{and} \quad H_{k-2}(\Delta^1(k)) \simeq \mathbb{Z}^{(k-1)!}.$$

(b) *A basis in the group $H_{k-2}(\Delta^1(k))$ consists of all snake-like (homeomorphic to a segment) graphs, one of whose endpoints is fixed.*

Statement (a) of this theorem is a corollary of a theorem of Folkman [**Fo**]; see, e.g., [**B**]. A proof of (b) and another proof of (a), based on the Goresky–MacPherson formula for the homology of plane arrangements, is given in [**V3**]. □

THEOREM 2. *For any k, the group $H_i(\Delta^2(k))$ is trivial if $i \neq 2k - 4$ and is isomorphic to*

$$\mathbb{Z}^{(k-2)!} \quad \text{if} \quad i = 2k - 4.$$

This theorem was proved independently and in different ways by Eric Babson, Anders Björner, Svante Linusson, John Shareshian, and Volkmar Welker, on one hand, and almost simultaneously (only a day later) by Victor Turchin on the other; see [**BBLSW**], [**T**]. □

EXAMPLE 1. Suppose that $k = 3$ and the original nodes are numbered by 1, 2 and 3. The simplex $\Delta(3)$ is a triangle, whose vertices are called $(1,2)$, $(1,3)$ and $(2,3)$. Among its 7 faces only four correspond to connected graphs, namely, all faces of dimension 1 or 2. In particular, the homology group $H_i(\Delta^1(3))$ is trivial if $i \neq 1$ and is isomorphic to $\mathbb{Z}^2$ if $i = 1$.

The unique face of $\Delta(3)$ corresponding to a two-connected graph is the triangle itself. Thus $H_i(\Delta^2(3))$ is trivial if $i \neq 2$ and is isomorphic to $\mathbb{Z}$ if $i = 2$.

EXAMPLE 2. The simplex $\Delta(4)$ has $\binom{4}{2} = 6$ vertices. The corresponding complex of two-connected graphs consists of the simplex itself, all 6 faces of dimension 4, and 3 faces of dimension 3 corresponding to all cycles of length 4. It is easy to calculate that $H_i(\Delta^2(4)) = 0$ for $i \neq 4$ and $H_4(\Delta^2(4)) \simeq \mathbb{Z}^2$. Namely, this homology group is generated by three 4-chains any of which is the difference of two graphs with 5 edges obtained from the complete graph by removing edges, connecting complementary pairs of points; see Figure 1.

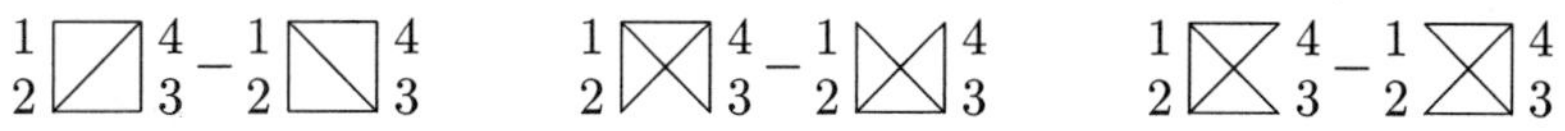

FIGURE 1. Basic chains for two-connected graphs with 4 vertices.

Such basic chains are numbered by unordered partitions of four points into two pairs and satisfy one relation: the sum of all three chains is equal to the boundary of the complete graph.

3. Simplest examples of indices of singular knots

For any immersion $\phi : S^1 \to \mathbb{R}^3$ with exactly i transverse crossings we can consider all 2^i possible small resolutions of this singular immersion, replacing any of its self-intersection points (see Figure 2a) by either undercrossing or overcrossing. Any such local resolution can be invariantly called positive (see Figure 2b) or negative (Figure 2c), see [**V2**]. The sign of the entire resolution is defined as $(-1)^{\#-}$, where $\#-$ is the number of *negative* local resolutions in it.

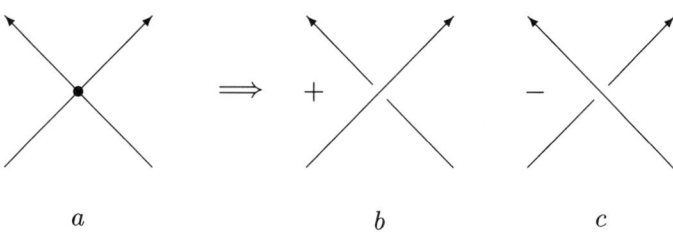

FIGURE 2. Local perturbations of a self-intersection point

Given a knot invariant, the index (or "i-th jump") of our singular immersion is defined as the sum of values of this invariant at all positive resolutions minus the similar sum over negative resolutions. Invariants of order $\leq j$ can be defined as those taking zero index at all immersions with $> j$ transverse crossings.

Further, let $\phi : S^1 \to \mathbb{R}^3$ be an immersion with l transverse double self-intersections and one generic triple point. There are 6 different perturbations of this triple point, splitting it into a pair of double points; see Figure 3 (= Figure 15 in [**V2**]). Thus we get 6 immersions with $l+2$ double crossings each, and, given a knot invariant, 6 indices $I(1), \ldots, I(6)$ of order $l + 2$ of these double crossings. These indices are dependent: they satisfy the *four-term relations*

(1) $$I(1) - I(4) = I(2) - I(5) = I(6) - I(3).$$

The common value of these three differences is a characteristic assigned by our invariant to the initial singular immersion with a triple point and is also called its index; see [**V2**].

An equivalent definition of the order of an invariant is as follows. An invariant is of order j if all such indices assigned by it to all immersions with $\geq j - 1$ double and one triple generic self-intersection points vanish.

In Section 5.3 we define similar indices for maps $S^1 \to \mathbb{R}^3$ having an arbitrary finite number of multiple self-intersections or singular points.

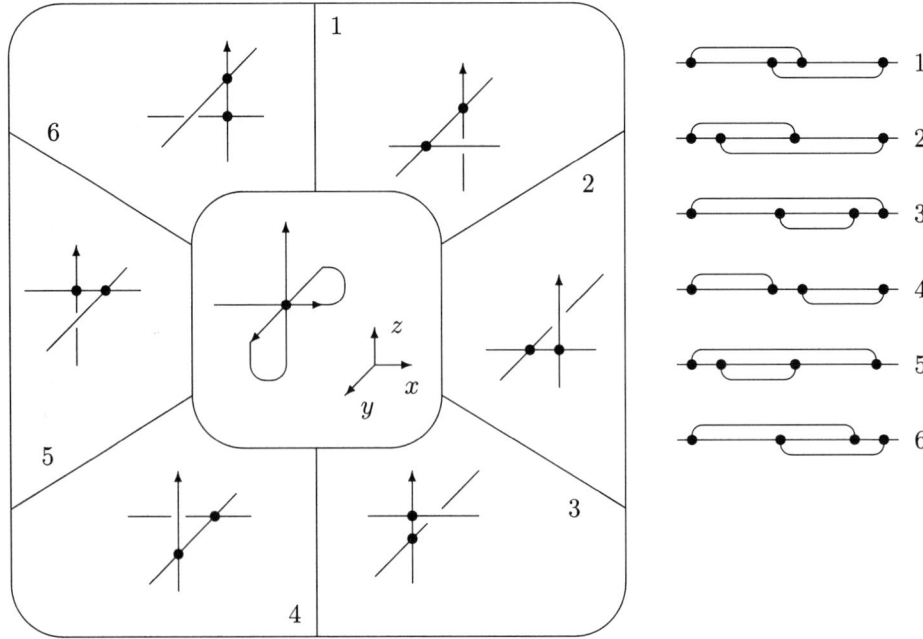

FIGURE 3. Splittings of a triple point

4. The discriminant set and its resolution

In this section we recall basic facts from [**V2**] concerning the topological structure of the discriminant set in the space $\mathcal{K}$ of all smooth maps $S^1 \to \mathbb{R}^n$, $n \geq 3$.

For simplicity we will regard this space as a space of very large but finite dimension ω. A partial justification of this assumption uses finite-dimensional approximations of $\mathcal{K}$; see [**V2**]. Below by the single quotes ' ... ' we indicate nonrigorous assertions that use this assumption and need a reference to [**V2**] for such a justification.

4.1. Simplicial resolution of the discriminant. The *resolution* σ of the discriminant set Σ is constructed as follows. Denote by Ψ the space of all unordered pairs (x, y) of points of S^1 (allowing $x = y$); it is easy to see that Ψ is diffeomorphic to the closed Möbius band. Consider a generic embedding I of Ψ in a space $\mathbb{R}^N$ of $(very)^2$ large dimension $N \gg \omega^2$. For any discriminant map $\phi : S^1 \to \mathbb{R}^n$ consider all points $(x, y) \subset \Psi$ such that either $x \neq y$ and $\phi(x) = \phi(y)$, or $x = y$ and $\phi'(x) = 0$. Denote by $\Delta(\phi)$ the convex hull in $\mathbb{R}^N$ of images of all such points under the embedding I. If I is generic, then this convex hull is a simplex, whose vertices coincide with all these images. The space σ is defined as the union of all simplices of the form $\phi \times \Delta(\phi) \subset \mathcal{K} \times \mathbb{R}^N$. The restriction on σ of the obvious projection $\mathcal{K} \times \mathbb{R}^N \to \mathcal{K}$ is proper and induces a 'homotopy equivalence' $\pi : \bar\sigma \to \bar\Sigma$ of one-point compactifications of spaces σ and Σ. By the 'Alexander duality', the homology groups $\bar{H}_*(\sigma) \equiv \bar{H}_*(\Sigma)$ of these compactifications 'coincide' (up to a change of dimensions) with the cohomology groups of the space of knots:

(2) $$H^i(\mathcal{K} \setminus \Sigma) \simeq \bar{H}_{\omega-i-1}(\Sigma) \equiv \bar{H}_{\omega-i-1}(\sigma).$$

The calculation of these groups is based on the natural stratification of Σ and σ. Let A be a nonordered finite collection of natural numbers, $A = (a_1, a_2, \ldots, a_{\#A})$, any of which is not less than 2, and b a nonnegative integer. Set $|A| = a_1 + \cdots + a_{\#A}$. An (A, b)-*configuration* is a collection of $|A|$ distinct points in S^1 separated into groups of cardinalities $a_1, \ldots, a_{\#A}$, plus a collection of b distinct points in S^1 (some of which can coincide with the above $|A|$ points). For brevity, $(A, 0)$-configurations are simply called A-configurations. A map $\phi : S^1 \to \mathbb{R}^n$ *respects* an (A, b)-configuration if it glues together all points inside any of its groups of cardinalities $a_1, \ldots, a_{\#A}$, and its derivative ϕ' is equal to 0 at all the b points of this configuration. For any (A, b)-configuration the set of all maps preserving this configuration is an affine subspace in $\mathcal{K}$ of codimension $n(|A| - \#A + b)$; the number $|A| - \#A + b$ is called the *complexity* of the configuration. Two (A, b)-configurations are *equivalent* if they can be transformed into one another by an orientation-preserving homeomorphism $S^1 \to S^1$.

For any (A, b)-configuration J, $A = (a_1, \ldots, a_{\#A})$, the corresponding simplex $\Delta(J) \subset \mathbb{R}^N$ is defined as the simplex $\Delta(\phi)$ for an arbitrary *generic* map ϕ respecting J; it has exactly $\binom{a_1}{2} + \cdots + \binom{a_{\#A}}{2} + b$ vertices.

For any class **J** of equivalent (A, b)-configurations, the corresponding **J**-*block* in σ is defined as the union of all pairs $\{\phi, x\}$, where ϕ is a map $S^1 \to \mathbb{R}^n$, respecting some (A, b)-configuration J of this equivalence class, and x is a point of the simplex $\Delta(J)$.

4.2. Main spectral sequence.

4.2.1. *Main filtration in the resolved discriminant.* The *main filtration* in σ is defined as follows: its term σ_i is the union of all **J**-blocks over all equivalence classes **J** of (A, b)-configurations of complexities $\leq i$.

This filtration is the unique useful filtration in all of σ and will not be revised. Indeed, its term $\sigma_i \setminus \sigma_{i-1}$ is the space of an affine bundle of dimension $\omega - ni$ over a finite-dimensional base, which can easily be described: the one-point compactification of this base is a finite cell complex; see [**V2**]. The spectral sequence, 'calculating' the groups (2) and induced by this filtration, satisfies the condition

(3) $$E^1_{p,q} = 0 \quad \text{for} \quad p(n-2) + q > \omega - 1.$$

(By the definition of the spectral sequence, $E^1_{p,q} = \bar{H}_{p+q}(\sigma_i \setminus \sigma_{i-1})$.)

The Alexander dual *cohomological* spectral sequence $E^{p,q}_r$ is obtained from this one by the formal change of indices,

(4) $$E^{p,q}_r \equiv E^r_{-p, \omega - 1 - q};$$

it 'converges' to some subgroups of groups $H^{p+q}(\mathcal{K} \setminus \Sigma)$ (if $n > 3$, then to all of these groups), and the support of its term E_1 belongs to the wedge

(5) $$p \leq 0, p(n-2) + q \geq 0.$$

This filtration induces an infinite filtration in (some subgroup of) the cohomology group $H^*(\mathcal{K} \setminus \Sigma) \simeq \bar{H}_{\omega - * - 1}(\sigma)$: an element of this group has order i if it can be defined as the linking number with a cycle lying in the term σ_i. In particular, for $* = 0$ we get a filtration in the space of knot invariants; it is easy to see that this filtration coincides with the elementary characterization of finite-order invariants given in Section 3.

4.2.2. *Kontsevich's realization theorem.* M. Kontsevich [**K2**] proved that the spectral sequence (4) (over $\mathbb{C}$) degenerates at the first term $E_1^{p,q} \equiv E_\infty^{p,q}$.

If $n = 3$, then for the groups $E^{p,q}$ with $p + q = 0$ which provide knot invariants, this follows from his integral realization of finite-order invariants described in [**K1**], [**BN2**]. Moreover, for arbitrary n and $p < 0$ exactly the same construction proves the degeneration of the "leading term" of the column $E^{p,*}$:

(6) $$E_1^{p,p(2-n)}(\mathbb{C}) \simeq E_\infty^{p,p(2-n)}(\mathbb{C}).$$

In several other cases (in particular if $-p$ is sufficiently small with respect to n) this follows also from dimensional reasons. However, generally for greater values of q the proof is more complicated.

4.2.3. *J-blocks and complexes of connected graphs.* Let J be an (A, b)-configuration of complexity i. By construction, the corresponding **J**-block is a fiber bundle, whose base is the space of (A, b)-configurations J' equivalent to J, and the fiber over such a configuration J' is the direct product of an affine space of dimension $\omega - ni$ (consisting of all maps respecting this configuration) and the simplex $\Delta(J')$. Consider this **J**-block as the space of the fiber bundle, whose base is the $(\omega - ni)$-dimensional affine bundle over the previous base, and the fibers are simplices $\Delta(J')$. Some points of these simplices belong not only to the i-th term σ_i of our filtration, but even to the $(i-1)$-th term. These points form a simplicial subcomplex of the simplex $\Delta(J')$. Let us describe it. Any face of $\Delta(J')$ can be depicted by $\#A$ graphs with $a_1, \ldots, a_{\#A}$ vertices and b signs $\pm$. Indeed, to any vertex of $\Delta(J')$ there corresponds either an edge connecting two points inside some of our $\#A$ groups of points or one of the last b points of the (A, b)-configuration J'.

PROPOSITION 1 (see [**V2**]). *A face of $\Delta(J')$ belongs to σ_{i-1} if and only if either at least one of the corresponding $\#A$ graphs is nonconnected or at least one of b points does not participate in its picture (= participates with sign $-$).* □

Denote by $\Delta^1(J')$ the quotient complex of $\Delta(J')$ by the union of all faces belonging to σ_{i-1}. Its homology group $H^*(\Delta(J'))$ obviously coincides with the Borel–Moore homology group of the topological space $\Delta(J') \setminus \sigma_{i-1}$. Theorem 1 immediately implies the following statement.

PROPOSITION 2. *The quotient complex $\Delta^1(J')$ is acyclic in all dimensions other than $|A| - \#A + b - 1 \equiv i - 1$, whereas*

(7) $$H_{i-1}(\Delta^1(J')) \simeq \otimes_{j=1}^{\#A} \mathbb{Z}^{(a_j - 1)!}.$$

4.3. Noncompact knots. Simultaneously with the usual knots, i.e., embeddings $S^1 \to \mathbb{R}^n$, we will consider *noncompact knots*, i.e., embeddings $\mathbb{R}^1 \to \mathbb{R}^n$ coinciding with the standard linear embedding outside some compact subset in $\mathbb{R}^1$; see [**V2**]. The space of all smooth maps with this behavior at infinity will also be denoted by $\mathcal{K}$, and the discriminant $\Sigma \subset \mathcal{K}$ again is defined as the set of all such maps, having singularities and self-intersections. In this case the configuration space Ψ, participating in the construction of the resolution σ, is not the Möbius band but the closed half-plane $\mathbb{R}^2 / \{(t, t') \equiv (t', t)\}$, which we usually will realize as the half-plane $\{(t, t') | t \leq t'\}$.

There is an obvious one-to-one correspondence between isotopy classes of standard and noncompact knots in $\mathbb{R}^3$, and spaces of invariants provided by the above spectral sequences in both theories naturally coincide. On the other hand, the CW-structure on resolved discriminants in spaces of noncompact knots is easier; this is the reason why in [**V2**] only the noncompact knots were considered. In Section 5 we develop some new techniques for calculating cohomology groups of spaces of compact or noncompact knots in $\mathbb{R}^n$, and in 5.1 and Section 6 (respectively, Section 7) apply it to the calculation of cohomology classes of order ≤ 3 of the space of noncompact knots in $\mathbb{R}^n$ (respectively, to the classes of order ≤ 2 of the space of compact knots). Everywhere below M^1 denotes either S^1 or $\mathbb{R}^1$.

4.4. Ancient auxiliary filtration in the term $\sigma_i \setminus \sigma_{i-1}$ of the main filtration. By definition, the term $E^1_{i,q}$ of the main spectral sequence is isomorphic to $\bar{H}_{i+q}(\sigma_i \setminus \sigma_{i-1})$. To calculate these groups, the *auxiliary filtration* in the space $\sigma_i \setminus \sigma_{i-1}$ was defined in [**V2**]. Namely, its term G_α is the union of all **J**-blocks, such that **J** is an equivalence class of (A,b)-configurations of complexity i, consisting of $\leq \alpha$ geometrically distinct points. In particular, $G_{2i} = \sigma_i \setminus \sigma_{i-1}$ and $G_{i-1} = \emptyset$. By Proposition 2, $\bar{H}_j(G_\alpha) = 0$ if $j \geq \omega - i(n-1) + \alpha$.

4.4.1. *Example: Knot invariants in* $\mathbb{R}^3$. If $n = 3$, then $\bar{H}_j(G_{2i-2}) = 0$ for $j \geq \omega - 2$. In particular, calculating the group $\bar{H}_{\omega-1}(\sigma_i \setminus \sigma_{i-1})$, we can ignore all **J**-blocks with auxiliary filtration $\leq 2i - 2$:

$$(8) \qquad \bar{H}_{\omega-1}(\sigma_i \setminus \sigma_{i-1}) \simeq \bar{H}_{\omega-1}((\sigma_i \setminus \sigma_{i-1}) \setminus G_{2i-2}).$$

The remaining part $(\sigma_i \setminus \sigma_{i-1}) \setminus G_{2i-2}$ consists of **J**-blocks satisfying the following definition.

DEFINITION. An (A,b)-configuration J (and the corresponding **J**-block) of complexity i is *simple* if one of the three types is satisfied:[2]

 I. $A = (2^{\times i})$, $b = 0$;
 II. $A = (2^{\times(i-1)})$, $b = 1$, and the last point does not coincide with any of $2i - 2$ points defining the A-part of the configuration;
 III. $A = (3, 2^{\times(i-2)})$, $b = 0$.

The complex $\Delta^1(J)$ corresponding to a configuration J of type I or II, consists of unique cell of dimension $i - 1$ (i.e., of the simplex $\Delta(J)$ itself), while for $A = (3, 2, \ldots, 2)$ it consists of one i-dimensional simplex and certain three faces of dimension $i - 1$.

A chain complex calculating groups (8) (and, moreover, all the groups $\bar{H}_*(\sigma_i \setminus \sigma_{i-1})$) was written out in [**V2**]; its part calculating the top-dimensional group (8) can be described in the following terms.

An (A,b)-configuration of type I (or the equivalence class of such configurations) can be depicted by a *chord diagram*, i.e., a collection of i determining i pairs of distinct points of S^1 or $\mathbb{R}^1$. (A,b)-configurations of type III are depicted in [**V2**] by similar diagrams with $i+1$ chords, three of which form a triangle and the remaining $i - 2$ have no common endpoints. In [**BN2**] this configuration is depicted by the same collection of $i - 2$ chords and a **Y**-like star connecting the points of the triple with a point not on the line (or circle). Configurations of types I and III are called $[i]$- and $\langle i \rangle$-configurations, respectively; see [**V2**].

An element of the group (8) is a linear combination of several **J**-blocks of type I and interior parts (swept out by open i-dimensional simplices) of blocks of type III. This linear combination should satisfy the homology condition near all strata of

---

[2] We denote by $2^{\times m}$ the expression $2, \ldots, 2$ with 2 repeated m times.

codimension 1 in σ_i: any such stratum should participate in the algebraic boundary of this combination with coefficient 0. Near such a stratum corresponding to an (A,b)-configuration J of type II, there is unique stratum of maximal dimension. This is the stratum of type I, whose (A,b)-configuration can be obtained from that of J by replacing its unique singular point by a small chord connecting two points close to it. The corresponding homological condition is as follows. Any stratum of maximal dimension whose chord diagram can be obtained in this way (i.e., has a chord whose endpoints are not separated by endpoints of other chords) should participate in the linear combination with zero coefficient.

Further, any configuration of type III defines three strata of codimension 1 in σ_i (some of which can coincide) swept out by three $(i-1)$-dimensional faces of corresponding simplices $\Delta(J')$. Such a stratum can be described by an $\langle i \rangle$-configuration, in which one chord, forming the triangle, is erased. This stratum is incident to three strata of maximal dimension: one stratum swept out by open i-dimensional simplices from the same **J**-block of type III defined by the same $\langle i \rangle$-configuration, and two strata of type I corresponding to chord diagrams obtained from our configuration by splitting the endpoints of our noncomplete triangle:

The homological condition is as follows: the difference of coefficients with which two latter strata participate in the linear combination should be equal to the coefficient of the stratum swept out by i-simplices. In particular, all three such differences corresponding to erasing any of three edges of the chord triangle should be equal to one another. These *three* equalities (among which only two are independent) are called the *4-term relations*; see Section 3.

The common value of these three differences is a characteristic of our $(3,2,\ldots,2)$-configuration (and of the element α of the homology group (8)): it is equal to the coefficient with which the main stratum of the **J**-block of type III participates in the cycle α. It is natural to call it the *index* of this cycle at this configuration.

In a similar way, any element α of the group (8) assigns an index to any i-chord diagram: this is the coefficient, with which the corresponding **J**-block of type I participates in the cycle realizing α.

The group (8) is thus canonically isomorphic to the group of all $\mathbb{Z}$-valued functions on the set of i-chord diagrams, which (a) take zero value on all diagrams having chords not crossed by other chords of the diagram, and (b) satisfy the 4-term relations defined by all possible $\langle i \rangle$-configurations.

When i grows, this system of equations grows exponentially, and, which is even worse, the answers do not satisfy any transparent rule; see [**BN1**].

4.4.2. *The Teiblum–Turchin cocycle.* In this subsection we consider the resolution of the discriminant in the space of noncompact knots $\mathbb{R}^1 \to \mathbb{R}^3$.

Let **J** be an arbitrary equivalence class of (A,b)-configurations of complexity i in $\mathbb{R}^1$. Denote by $\tilde{B}(\mathbf{J})$ the intersection of the corresponding **J**-block with $\sigma_i \setminus \sigma_{i-1}$. This set $\tilde{B}(\mathbf{J})$ consists of several cells which are in one-to-one correspondence with all possible collections of $\#A$ connected graphs with $a_1, \ldots, a_{\#A}$ vertices, respectively; for some examples see Figure 4. Such cells corresponding to all equivalence classes **J** of (A,b)-configurations of complexity i define a cell decomposition of the quotient space σ_i / σ_{i-1}.

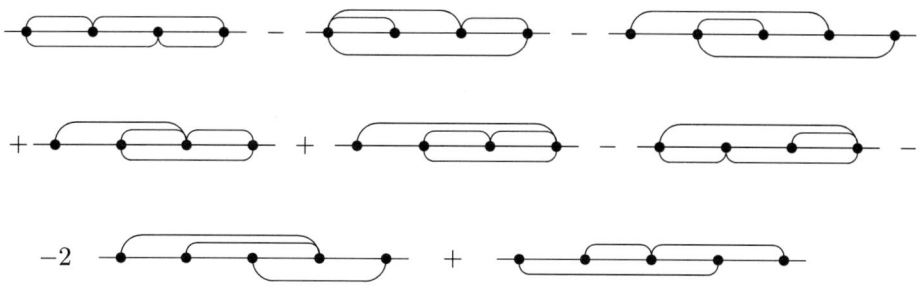

FIGURE 4. The Teiblum–Turchin cocycle

THEOREM (D. Teiblum, V. Turchin). *The linear combination of suitably oriented (see* [**V2**]) *cells of $\sigma_3 \setminus \sigma_2$ shown in Figure 4 defines a nontrivial $(\omega - 2)$-dimensional Borel–Moore homology class of $\sigma_3 \setminus \sigma_2$ (and hence also an $(\omega - 2 - 3(n-3))$-dimensional homology class of the term $\sigma_3 \setminus \sigma_2$ of the similar resolution of the space of singular noncompact knots in $\mathbb{R}^n$ for any odd $n \geq 3$).*

On the other hand, the group $\bar{H}_{\omega-3-3(n-3)}(\sigma_2)$ is trivial (see [**V2**]); hence this class can be extended to a well-defined class in the group $\bar{H}_{\omega-2-3(n-3)}(\sigma)$ and in $H^{1+3(n-3)}(\mathcal{K} \setminus \Sigma)$. If $n > 3$, then by dimensional reasons this class is nontrivial; however its nontriviality for $n = 3$ and any explicit realization remain unknown.

5. New auxiliary filtration

For any (A, b)-configuration J, denote by $\rho(J)$ the number of *geometrically distinct* points in it. In particular, for any A-configuration J, $\rho(J) = |A|$. The greatest possible value of $\rho(J)$ over all A- or (A, b)-configurations of complexity i is equal to $2i$.

DEFINITION. The *stickness* of an A-configuration J of complexity i is the number $2i - \rho(J)$. The *reversed auxiliary filtration* $\Phi_0 \subset \cdots \subset \Phi_{i-1}$ in the term $\sigma_i \setminus \sigma_{i-1}$ of the main filtration is defined as follows. Its term Φ_α is the closure of the union of **J**-blocks over all equivalence classes **J** of A-configurations of complexity i and stickness $\leq \alpha$.

If J is an (A, b)-configuration with $b > 0$, then the corresponding **J**-block belongs to the closure of an $\tilde{\mathbf{J}}$-block, where $\tilde{J}$ is an $\tilde{A}$-configuration, $\tilde{A} = A \cup 2^{\times b}$. Thus $\Phi_{i-1} = \sigma_i \setminus \sigma_{i-1}$.

So, to any finite-order cohomology class v of the space of knots in $\mathbb{R}^n$ (in particular, to any invariant of knots in $\mathbb{R}^3$) there correspond two numbers: the first, $i(v)$, is its order, and the second, $r(v)$, is its reversed filtration, i.e., the minimal reversed filtration in $\sigma_{i(v)} \setminus \sigma_{i(v)-1}$ of realizing its cycles in $\sigma_{i(v)}$. This pair of numbers is called the *bi-order* of the cohomology class (in particular, of the invariant).

5.1. First examples. *Notation.* For any topological space X, the k-th *configuration space* $B(X, k)$ is the space of all subsets of cardinality k in X. We denote by $\pm\mathbb{Z}$ the local system of groups on the configuration space $B(X, k)$, locally isomorphic to $\mathbb{Z}$ and such that elements of $\pi_1(B(X, k))$, defining odd permutations of k points, act on its fibers as multiplication by -1. The Borel–Moore homology group

$\bar{H}_*(B(X,k), \pm \mathbb{Z})$ is the homology group of locally finite chains with coefficients in this local system; see [**V6**].

In the following two examples we consider the discriminant in the space $\mathcal{K}$ of noncompact (maybe singular) knots.

EXAMPLE 3. If $i = 1$, then the entire space $\sigma_1 \equiv \Phi_0$ is the space of an $(\omega - n)$-dimensional affine fiber bundle over the manifold Ψ diffeomorphic to the closed half-plane. In particular, all its Borel–Moore homology groups are trivial and there are no nonzero cohomology classes of order 1 for any n.

EXAMPLE 4. Our filtration of the set $\sigma_2 \setminus \sigma_1$ consists of two terms $\Phi_0 \subset \Phi_1$. Its term Φ_0 is the space of a fiber bundle whose base is the configuration space $B(\Psi, 2)$ and the fiber is the product of an open interval and an affine space of codimension $2n$ in the space $\mathcal{K}$. The generator of the group $\pi_1(B(\Psi, 2)) \simeq \mathbb{Z}$ changes the orientation of the first factor of the fiber and multiplies the orientation of the second by $(-1)^n$. In particular, the Borel–Moore homology group of the term Φ_0 is isomorphic (up to the shift of dimensions) to the group $\bar{H}_*(B(\Psi, 2))$ if n is odd and to $\bar{H}_*(B(\Psi, 2), \pm \mathbb{Z})$ if n is even. It is easy to calculate (see also Lemma 1 below) that both of these groups are trivial in all dimensions.

The term $\Phi_1 \setminus \Phi_0$ is the space of the fiber bundle whose base is the configuration space $B(\mathbb{R}^1, 3)$ of triples of points in the line and the fiber is the product of an affine subspace of codimension $2n$ in $\mathcal{K}$ and the interior part of a triangle. The Borel–Moore homology of this term is obviously isomorphic to $\mathbb{Z}$ in the dimension $\omega - 2n + 5$ and is trivial in all other dimensions. Thus the group of cohomology classes of order 2 is isomorphic to $\mathbb{Z}$ in dimension $2n - 6$ and is trivial in all other dimensions.

A COMPARISON. Calculating the homology of $\sigma_2 \setminus \sigma_1$ with the help of the ancient auxiliary filtration, we assign the edges of these triangles to the same term as their interior parts; as a consequence, they separate the space $B(\Psi, 2)$ into several cells according to combinatorial types of chord diagrams. In the new approach all these cells are joined together and form a manifold $B(\Psi, 2)$ with simple topology.

In a similar way, for any i the space $\sigma_i \setminus \sigma_{i-1}$ consists of i terms $\Phi_0 \subset \cdots \subset \Phi_{i-1}$ of the reversed filtration, and the ultimate term $\Phi_{i-1} \setminus \Phi_{i-2}$ is the space of a fiber bundle whose base is the configuration space $B(M^1, i+1)$ and the fiber is the direct product of an affine subspace of codimension ni in the space $\mathcal{K}$ and an $\binom{i+1}{2}$-vertex simplex, from which all faces belonging to Φ_{i-2} are removed. These faces are exactly those corresponding to not two-connected graphs; see Theorem 3 below.

Similarly, to any immersion $M^1 \to \mathbb{R}^n$ with $j < \infty$ self-intersection points of finite multiplicities $a_1, \ldots, a_j$, we associate the homology group of the corresponding component of the term $\Phi_\alpha \setminus \Phi_{\alpha-1}$ of our filtration in $\sigma_i \setminus \sigma_{i-1}$, where $i = (\sum_{m=1}^{j} a_m - j)$, $\alpha = \sum_{m=1}^{j} a_m - 2j$. This group is isomorphic (up to a shift of dimensions) to the tensor product of homology groups of j complexes of two-connected graphs with $a_1, \ldots, a_j$ nodes.

5.2. Structure of the reversed filtration.

5.2.1. *A-sets.* The set $\Phi_\alpha \setminus \Phi_{\alpha-1}$ of the new filtration in $\sigma_i \setminus \sigma_{i-1}$ consists of several components numbered by unordered decompositions of the number $2i - \alpha$ into $i - \alpha$ integer summands, each being ≥ 2. Namely, to any such decomposition

(9) $\quad A = (a_1, a_2, \ldots, a_{\#A}), \quad \#A = i - \alpha, \quad \sum a_l = 2i - \alpha, \quad a_l \geq 2,$

there corresponds the closure of the union of all **J**-blocks defined by all equivalence classes **J** of A-configurations. Let us describe this closure explicitly.

DEFINITION. An *A-collection* is a collection of $|A|$ points in M^1 separated into $\#A$ groups of cardinalities $a_1, \ldots, a_{\#A}$, such that the points inside each group of cardinality > 2 are pairwise distinct.

A map $\phi : M^1 \to \mathbb{R}^n$ *respects* an A-collection if $\phi(x) = \phi(y)$ for any two different points x, y of the same group and $\phi'(x) = 0$ for any point x such that (x, x) is a group of cardinality $a_i = 2$ of this collection.

The set of all maps $\phi : M^1 \to \mathbb{R}^n$, respecting an A-collection, is an affine subspace in $\mathcal{K}$ of codimension $\leq n(|A| - \#A)$. Indeed, any group of cardinality a_j gives us $n(a_j - 1)$ independent restrictions on ϕ (although the union of all $n\sum(a_j - 1) \equiv n(|A| - \#A)$ such conditions can be dependent).

An A-collection is an *A-set* if these conditions are independent, i.e., the codimension of the space of maps $\phi : M^1 \to \mathbb{R}^n$ respecting this collection is equal to $n(|A| - \#A)$.

For example, $(2, 2, 2)$-sets are any subsets of cardinality 3 in Ψ not of the form $((x < y), (x < z), (y < z))$.

More generally, given an A-collection, consider the graph with $|A|$ vertices corresponding to its points and with edges of two types: all points inside any group connected by black edges, and geometrically coincident points of different groups connected by white edges. If there is a degenerate group of the form (x, x) and the point x belongs also to other group(s), then it is depicted by two points connected by a black edge, and only one of these points is connected by white edges with corresponding points of other groups.

PROPOSITION 3. *The following three conditions are equivalent:*

1) *our A-collection is an A-set;*

2) *any cycle of the corresponding graph without repeating nodes contains no white edges;*

3) *the graph obtained from the original one by contracting all black edges is a forest (i.e., a disjoint union on several trees).*

In particular, two different groups of an A-set cannot have two common points.

Proof. The equivalence of conditions 2) and 3) is obvious. Let us prove the implication 1) $\Rightarrow$ 2). Suppose that we have a cycle such as indicated in condition 2) but containing white edges. Removing an arbitrary vertex incident to a white edge of the cycle, from the corresponding group of the A-collection (say, from the group of cardinality a_l) we obtain an $\tilde{A}$-collection, $\tilde{A} = (a_1, \ldots, a_l - 1, \ldots, a_{\#A})$, defining the same subspace in $\mathcal{K}$. Thus the codimension of this subspace is no greater than $n(|A| - \#A - 1)$ and condition 1) is not satisfied. Conversely, let us prove that 3) $\Rightarrow$ 1). Any nonempty forest has a 1-valent node. In our case this node is obtained by contracting a group of ≥ 2 nodes of the original graph. Hence this group contains a node not incident to white edges. Removing the corresponding point from the A-collection (and cancelling the group containing this point if it is a group of cardinality 2) we obtain an $\tilde{A}$-collection with $|\tilde{A}| - \#\tilde{A} = |A| - \#A - 1$, which again satisfies conditions 2) and 3).

On the other hand, erasing this point we lose exactly n linear conditions on the corresponding subspace in $\mathcal{K}$, which are independent on the others. Thus condition 1) follows by induction over the complexity $|A| - \#A$ of our A-collection. $\square$

An A-set and an A'-set are *related* if the sets of maps $M^1 \to \mathbb{R}^n$ respecting these sets coincide. For example, the (3)-set $((x < y < z))$ and the $(2,2)$-set $((x,y),(x,z))$ are related. There is a partial order on any class of related sets. The set $\tilde{\Upsilon}$ is a *completion* of the related set Υ if $\tilde{\Upsilon}$ is obtained from Υ by replacing two intersecting groups with their union or by a sequence of such operations.

The space of all A-sets with given A is denoted by $C(A)$.

An open dense subset in $C(A)$ consists of A-configurations; see Section 4.1. For example, such a subset in $C(2,2,2)$ consists of all points $((x,x'),(y,y'),(z,z')) \in B(\Psi, 3)$ such that two of six numbers $x, x', \ldots, z'$ are distinct.

5.2.2. *The complex $\Lambda(\Upsilon)$.* Given an A-set Υ in M^1, $\Upsilon = (\Upsilon_1, \ldots, \Upsilon_{\#A})$, consider the simplex $\Delta(\Upsilon)$ with $\sum_{j=1}^{\#A} \binom{a_j}{2}$ vertices corresponding to all pairs $(x,y) \in \Psi$ such that x and y belong to one group of Υ. A face of this simplex can be encoded by a collection of $\#A$ graphs whose edges span the pairs corresponding to vertices of this face.

Such a face is called 2-*connected* if all these graphs are two-connected.

We will identify the nodes of these graphs with the corresponding points in M^1.

Denote by $\Lambda(\Upsilon)$ the union of interior points of all two-connected faces of our simplex.

Theorem 2 immediately implies the following statement.

COROLLARY 1. (1) *The group $\bar{H}_l(\Lambda(\Upsilon))$ is trivial for all $l \neq 2|A| - 3\#A - 1$ and is isomorphic to $\otimes_{j=1}^{\#A} \mathbb{Z}^{(a_j-2)!}$ for $l = 2|A| - 3\#A - 1$; this isomorphism is defined canonically up to permutations of factors $\mathbb{Z}^{(a_j-2)!}$, corresponding to coinciding numbers a_j.*

(2) For any decomposition of the set Υ into two nonintersecting subsets Υ', Υ'', each being the union of several groups of Υ, we have

(10) $$\bar{H}_{*-1}(\Lambda(\Upsilon)) \simeq \bar{H}_*(\Lambda(\Upsilon')) \otimes \bar{H}_*(\Lambda(\Upsilon'')),$$

where the lower index $*-1$ denotes the shift of grading. □

In other words, if we introduce the notation

$$\mathcal{H}(\Upsilon) \equiv \bar{H}_{2|A|-3\#A-1}(\Lambda(\Upsilon))$$

for any A and any A-set Υ, then $\mathcal{H}(\Upsilon) \simeq \mathcal{H}(\Upsilon') \otimes \mathcal{H}(\Upsilon'')$.

Namely, if we have two cycles $\gamma' \subset \Lambda(\Upsilon') \subset \Delta(\Upsilon')$, $\gamma'' \subset \Lambda(\Upsilon'') \subset \Delta(\Upsilon'')$, then the corresponding *join* cycle $\gamma' * \gamma'' \subset \Delta(\Upsilon) \equiv \Delta(\Upsilon') * \Delta(U''')$ is realized as the union of all open intervals connecting the points of γ' and γ''. Similarly, if γ^j, $j = 1, \ldots, \#A$, are cycles defining some elements of groups $\bar{H}_*(\Lambda(\Upsilon_j)) \simeq \mathbb{Z}^{(a_j-2)!}$, then their join $\gamma^1 * \cdots * \gamma^{\#A}$ is swept out by open $(\#A - 1)$-dimensional simplices whose vertices are the points of these cycles γ^j. The corresponding homology map $\otimes \bar{H}_*(\Lambda(\Upsilon_j)) \to \bar{H}_{*+\#A-1}(\Lambda(\Upsilon))$ is an isomorphism and depends on the choice of orientation of these simplices (= the ordering of their vertices or, equivalently, of the groups $\Upsilon_j \subset \Upsilon$ up to even permutations).

Let A be a multi-index (9) of complexity i, and Υ an A-set in M^1. The simplex $\Delta(\Upsilon)$ can be realized as the simplex in $\mathbb{R}^N$ spanned by all points $I(x,y)$, $(x,y) \in \Psi$, such that x and y belong to the same group of Υ. For any map $\phi \in \mathcal{K}$ respecting Υ, the simplex $\phi \times \Delta(\Upsilon) \subset \mathcal{K} \times \mathbb{R}^N$ belongs to the term σ_i of the main filtration, and its intersection with the space $\sigma_i \setminus \sigma_{i-1}$ belongs to the term Φ_α of the reversed filtration of this space, where $\alpha = 2i - \rho(\Upsilon)$.

PROPOSITION 4. *If $\bar{\Upsilon}$ is a completion of Υ and $\bar{\Upsilon} \neq \Upsilon$, then the simplex $\Delta(\Upsilon)$ is a not two-connected face of $\Delta(\bar{\Upsilon})$.*

Proof. Consider the two-color graph of the A-set Υ mentioned in Proposition 3. The graph representing the face $\Delta(\Upsilon) \subset \Delta(\bar{\Upsilon})$ is obtained from it by contracting all white edges. Since $\Upsilon \neq \bar{\Upsilon}$, there is at least one white edge connecting two points of the same group of $\bar{\Upsilon}$. Removing from $\Delta(\Upsilon)$ the vertex obtained from this edge, we get a graph, which by Proposition 3 is not connected. □

THEOREM 3. *For any multi-index A of the form (9) and any A-set Υ, a point of the simplex $\phi \times \Delta(\Upsilon)$ belongs to $\Phi_\alpha \setminus \Phi_{\alpha-1}$ if it is an interior point of a two-connected face and belongs to $\Phi_{\alpha-1}$ otherwise.*

Proof. First we prove the "only if" part. Consider a face of the simplex $\Delta(\Upsilon)$ such that one of the corresponding graphs $g_1, \ldots, g_{\#A}$, say g_m, is connected but not two-connected. We need to prove that this face belongs to $\Phi_{\alpha-1}$. Since $\Phi_{\alpha-1}$ is closed in Φ_α, it is sufficient to prove this property for all A-sets Υ from an arbitrary dense subset in $C(A)$. In particular we can assume that Υ is an A-configuration (see Section 4.1), i.e., all its $|A|$ points are pairwise distinct.

Let $y \in M^1$ be a node of our graph g_m such that by removing it we split g_m into two nonempty graphs g'_m, g''_m with a'_m and a''_m nodes, respectively, $a'_m + a''_m = a_m - 1$. Let A' be the multi-index $(a_1, \ldots, a_{m-1}, a'_m + 1, a''_m + 1, a_{m+1}, \ldots, a_{\#A})$ and Υ' an A'-set, related to Υ, whose groups of cardinalities $a_1, \ldots, a_{m-1}, a_{m+1}, \ldots, a_{\#A}$ coincide with these for Υ, and the last two coincide with the sets of vertices of g'_m, g''_m, augmented by y. Then our face of the simplex $\phi \times \Delta(\Upsilon) \simeq \Delta(\Upsilon)$ lies also in the simplex $\phi \times \Delta(\Upsilon')$. This simplex belongs to the closure of the set swept out by similar simplices $\phi' \times \Delta(\Upsilon'(\tau))$, $\tau \in (0, \varepsilon]$, where $\Upsilon'(\tau)$ are A'-sets with $\rho(\Upsilon'(\tau)) > \rho(\Upsilon') = \rho(\Upsilon)$, namely, some of their $|A'| - 2$ points coincide with these for Υ', and only two points y of the m-th and $(m+1)$-th groups are replaced by $y - \varepsilon$ and $y + \varepsilon$, respectively. Thus our face belongs to a lower term of the reversed filtration.

Conversely, for any multi-index A of the form (9), the closure of the union of all **J**-blocks over all A-configurations J consists of simplices of the form $\phi \times \Delta(\Upsilon)$, where Υ is an A-set, and ϕ a map respecting it. By Proposition 4, if Υ is such an A-set and $\bar{\Upsilon}$ a completion of Υ, $\bar{\Upsilon} \neq \Upsilon$, then such a simplex is a not two-connected face of $\Delta(\bar{\Upsilon})$. □

COROLLARY 2. *The component of $\Phi_\alpha \setminus \Phi_{\alpha-1}$, corresponding to the multi-index A of the form (9), is the space of a fiber bundle over the space of all A-sets, whose fiber over the A-set Υ is the direct product of the complex $\Lambda(\Upsilon)$ and the affine subspace of codimension ni in $\mathcal{K}$, consisting of all maps ϕ, respecting Υ.* □

Denote this component by $S(A)$.

COROLLARY 3. *For any A, the group $\bar{H}_*(S(A))$ is trivial in all dimensions greater than $\omega - 1 - (n-3)i$.*

This follows immediately from Theorems 2 and 3 and implies formulas (3) and (5).

5.2.3. *An upper estimate of the number of invariants of bi-order $(i, i-1)$.* Recall that the greatest possible value of the reversed filtration of a cohomology class (in particular, of a knot invariant in $\mathbb{R}^3$) of order i is equal to $i - 1$.

PROPOSITION 5. *For any i, the number of linearly independent knot invariants of bi-order $(i, i-1)$ (modulo invariants of lower bi-orders) is estimated from above by the number $(i-1)! \equiv \dim H_{2i-2}(\Delta^2(i+1))$.*

This follows from Theorem 2 and the fact that the unique multi-index A of the form (9) with given i and stickness $i-1$ consists of one number $(i+1)$. □

This estimate is not sharp. Indeed, the group $\mathbb{Z}_{i+1}$ of cyclic permutations of vertices acts naturally on the complex $\Delta(i+1)$, hence also on the group $H_*(\Delta^2(i+1))$. It is easy to see that an element of this group, corresponding to a knot invariant of bi-order $(i, i-1)$, should be invariant under this action, and we obtain the following improvement of Proposition 5.

PROPOSITION 5′. *The number of linearly independent knot invariants of bi-order $(i, i-1)$ (modulo invariants of lower bi-orders) does not exceed the dimension of the subgroup in $H_{2i-2}(\Delta^2(i+1))$, consisting of $\mathbb{Z}_{i+1}$-invariant elements.* □

The exact formula for this dimension was found in [**BBLSW**], Corollary 4.7.

5.3. Index of a knot invariant at a complicated singular knot.

Let A be a multi-index $(a_1, \ldots, a_{\#A})$ of complexity i, J an A-configuration, consisting of groups $J_1, \ldots, J_{\#A}$, and ϕ an immersion $M^1 \to \mathbb{R}^3$ respecting J but not respecting more complicated configurations. Define the group $\Xi_*(J)$ as the tensor product

$$(11) \qquad \otimes_{m=1}^{\#A} \bar{H}_*(\Lambda(J_m)),$$

where $\Lambda(J_m)$ is the complex of two-connected graphs whose nodes are identified with points of the group J_m.

By Theorem 2 this graded group has the unique nontrivial term, which is of dimension $2|A| - 4\#A$ and is isomorphic to $\otimes_{m=1}^{\#A} \mathbb{Z}^{(a_j-2)!}$. In particular, there is an isomorphism

$$(12) \qquad \Xi_*(J) \equiv \bar{H}_{*+\#A-1}(\Lambda(J)).$$

By the *join* construction (see subsection 5.2.2) this isomorphism is defined canonically up to a sign, which can be specified by any ordering of our groups J_m and is different for orders of different parity.

Any knot invariant V of order i defines an element of this group $\Xi_*(J)$, the *index* $\langle V|\phi \rangle$ of the singular knot ϕ; this index generalizes similar indices of not very complicated singular knots considered in Section 3 and subsection 4.4.1. First we define it in the case of noncompact knots $\mathbb{R}^1 \to \mathbb{R}^3$. Consider any sufficiently small open contractible neighborhood U of our point ϕ in the space of discriminant maps equisingular to ϕ (i.e., respecting equivalent configurations). The complete preimage of U in the space σ_i is homeomorphic to the direct product of U and the simplex $\Delta(\phi)$. Remove from this simplex all faces lying in lower terms of the main and reversed auxiliary filtrations. The remaining domain will be homeomorphic to $U \times \Lambda(J)$. The intersection of our invariant (i.e., of the cycle $\gamma \in \bar{H}_{\omega-1}(\sigma_i)$ realizing it) with this domain defines an element of the group $\bar{H}_{\omega-1}(U \times \Lambda(J))$, hence (via the Künneth isomorphism) a class in $\bar{H}_{2|A|-3\#A-1}(\Lambda(J))$ and, by the isomorphism (12), also in the group $\Xi_*(J)$; this class is the desired index $\langle V|\phi \rangle$.

To make this definition precise, we need to specify the orientation of U participating in the construction of the Künneth isomorphism. This orientation consists of (a) an orientation of the space **J** of configurations equivalent to J, and (b) an orientation of the space $\chi(J)$ of maps $\mathbb{R}^1 \to \mathbb{R}^3$ respecting J. To define the first orientation, we order all groups J_m of J in an arbitrary way and order points of any group by their order in $\mathbb{R}^1$. Thus we get an order of all points of J (first points of the first group in increasing order, then of the second, etc.), and thus also the orientation of the configuration space **J**.

Further, we assume that an orientation of the functional space $\mathcal{K}$ is fixed. Then orientations of the subspace $\chi(J)$ can be identified with its coorientations, i.e., orientations of the normal bundle. Let X, Y, Z be coordinates in $\mathbb{R}^3$, so that the map ϕ consists of three real functions $X(t), Y(t), Z(t)$. Then a coorientation of $\chi(J)$ can be specified by the differential form

$[d(X(\text{the second point } t_{1,2} \text{ of the first group } J_1 \subset J)$
$- X(\text{the first point } t_{1,1} \text{of this group}))$
$\wedge d(Y(t_{1,2}) - Y(t_{1,1})) \wedge d(Z(t_{1,2}) - Z(t_{1,1}))] \wedge \cdots \wedge [d(X(t_{1,a_1}) - X(t_{1,a_1-1}))$
$\wedge d(Y(t_{1,a_1}) - Y(t_{1,a_1-1})) \wedge d(Z(t_{1,a_1}) - Z(t_{1,a_1-1}))]$
$\wedge (\text{the same for the second group } J_2) \wedge \cdots \wedge (\text{the same for } J_{\#A}).$

LEMMA 1. (1) *The orientation of U, defined by this pair of orientations of **J** and $\chi(J)$, depends on the choice of the order of groups $J_m \subset J$; namely, an odd permutation of these groups multiplies it by -1.*

(2) *This orientation will be preserved if we change the increasing order of points in an arbitrary group J_m by means of any cyclic permutation.*

The proof is elementary. □

The first statement of Lemma 1 implies that our index $\langle V | \phi \rangle$ does not depend on the choice of the order of groups $J_m \subset J$ (formally participating in its construction). Indeed, changing this order by an odd permutation, we multiply by -1 both the Künneth isomorphism and the isomorphism (12).

The second statement allows us to define a similar index also in the case of compact knots $S^1 \to \mathbb{R}^3$.

Moreover, assume that our discriminant point ϕ is generic in its stratum of complexity i, i.e., it does not belong to the closure of the set of maps respecting configurations of complexities $> i$. (Typical examples of nongeneric points are immersions with self-tangency points or triple points, at which three local branches are coplanar.) Then *any* knot invariant (of an arbitrary order) defines in the same way the generalized index taking values in the same homology group. Indeed, in this case the complete preimage of our neighborhood U in the entire $\sigma \setminus \sigma_{i-1}$ (and not only in $\sigma_i \setminus \sigma_{i-1}$) has the same structure of the direct product $U \times \Lambda(J)$.

For any class **J** of equivalent A-configurations, all groups $\Xi_*(J)$, $J \in \mathbf{J}$, are canonically isomorphic to one another, thus we can define the group $\Xi_*(\mathbf{J})$ to be any of them.

LEMMA 2. *Let V be an invariant of order i of knots in $\mathbb{R}^3$, and **J** a class of equivalent A-configurations in M^1 of complexity $|A| - \#A = i$. Then all indices $\langle V | \phi \rangle \in \Xi_*(\mathbf{J})$ for all generic immersions ϕ respecting A-configurations $J \in \mathbf{J}$, coincide.* □

This (obvious) lemma allows us to define the index $\langle V | \mathbf{J} \rangle \in \Xi_*(\mathbf{J})$ as the common value of these indices $\langle V | \phi \rangle$.

These indices satisfy the natural homological condition (the analog of the *STU*-relation of [**BN2**]). Let J, J' be two A-configurations from neighboring equivalence classes, i.e., J' can be obtained from J by the permutation of exactly two neighboring points $t_1 < t_2$ of different groups $J_m, J_{m+1} \subset J$. Then groups $\Xi_*(\mathbf{J}), \Xi_*(\mathbf{J}')$ are obviously identified. Let $A!$ be the multi-index obtained from A by replacing two numbers a_m, a_{m+1} with one number $a_m + a_{m+1} - 1$, and $J!$ the $A!$-configuration obtained from J by contracting the segment $[t_1, t_2]$ into a point τ, which will thus belong to the group of cardinality $a_m + a_{m+1} - 1$. The index $\langle V | J! \rangle$ is an element of the group $\Xi_*(J!)$, i.e., a linear combination of collections of $\#A!$ two-connected graphs (with ordered edges). Its boundary $\partial \langle V | J! \rangle$ is a linear combination of similar collections of graphs, exactly one of which is not two-connected. Denote by $\partial_\tau \langle V | J! \rangle$ the part of this linear combination spanned by all collections such that by removing from any of them the point τ, we split the corresponding graph Γ_m with $a_m + a_{m+1} - 1$ vertices into two disjoint graphs whose nodes are the points of $J_m \setminus t_1$ and $J_{m+1} \setminus t_2$. Any such graph Γ_m is the union of two graphs with a_m and a_{m+1} nodes and common node τ, thus the space of all such linear combinations can also be identified with any of groups $\Xi_*(\mathbf{J}), \Xi_*(\mathbf{J}')$. The promised homological condition is that the chain in σ_i that is Alexander dual to V, actually is a cycle, and in particular its algebraic boundary near the common boundary of $\mathbf{J}$-, $\mathbf{J}'$- and $\mathbf{J}!$-blocks is equal to zero. This condition is as follows:

$$(13) \qquad \langle V | \mathbf{J} \rangle - \langle V | \mathbf{J}' \rangle = \pm \partial_\tau \langle V | \mathbf{J}! \rangle,$$

where the coefficient $\pm$ depends on the choice of local orientations of all participating strata.

The system of equations (13) corresponding to all points τ of the configuration $\mathbf{J}!$, is strong enough to determine the index $\langle V | \mathbf{J}! \rangle$ if we know all indices $\langle V | \mathbf{J} \rangle$ for all configuration classes $\mathbf{J}$ of stickness smaller than that of $\mathbf{J}!$. Indeed, any knot invariant V of order i is determined (up to terms of lower orders) by its indices at chord diagrams, i.e., the configurations of stickness 0.

This equality also allows us to give the following characterization of the reversed filtration. Let $\mathbf{J}, \mathbf{J}'$ be two classes of equivalent A-configurations with the same A. An arbitrary correspondence between their groups $J_m, J'_{m'}$ of equal cardinalities allows us to identify the groups $\Xi_*(\mathbf{J})$ and $\Xi_*(\mathbf{J}')$. This identification is not unique if some of numbers a_m of the multi-index A coincide. Moreover, in the last case we can define a group of automorphisms acting on $\Xi_*(\mathbf{J})$ and generated by all possible permutations of groups of the same cardinalities a_m that preserve the order (cyclic order, if $M^1 = S^1$) of points inside any group; see Corollary 1 above.

PROPOSITION 6. *For any knot invariant V of order i, the following conditions are equivalent:*

(1) *V has reversed filtration $\leq \alpha$;*

(2) *for any multi-index A with $|A| - \#A = i$, $i - \#A = \alpha$, all the indices $\langle V | \mathbf{J} \rangle \in \Xi_*(\mathbf{J})$ defined by V at all equivalence classes $\mathbf{J}$ of all A-configurations are invariant under all the identifications and automorphisms described in the previous paragraph.*

Proof. $(1) \Rightarrow (2)$. If $\bar{V}$ is the class in $\bar{H}_{\omega-1}(\sigma_i)$ representing V, then for any A as above its restriction on $\Phi_\alpha \setminus \Phi_{\alpha-1}$ defines a class in $\bar{H}_{\omega-1}(S(A))$. The indices $\langle V | \mathbf{J} \rangle$ for all classes $\mathbf{J}$ of equivalent A-configurations are defined by restrictions of this class to all corresponding $\mathbf{J}$-blocks. In particular, these indices for neighboring classes of A-configurations (i.e., for those obtained one from another by a change of orders of only two points from different groups) should coincide; otherwise our cycle in $S(A)$ would have a nontrivial boundary at the border between these two blocks.

The implication (2) $\Rightarrow$ (1) means that there are no nontrivial elements of the group $\bar{H}_{\omega-1}(\sigma_i \setminus \sigma_{i-1})$ represented by cycles with support in $(\sigma_i \setminus \sigma_{i-1}) \setminus \Phi_\alpha$. This follows from the fact that any invariant V of order i is determined up to invariants of lower orders by its indices at all i-chord diagrams (i.e., at all **J**-blocks with $A = (2, \ldots, 2)$). $\square$

5.4. The symbol of a cohomology class of a finite order.
Again, let n be an arbitrary natural number greater than 2 and $\mathcal{K}$ the space of noncompact knots $\mathbb{R}^1 \to \mathbb{R}^n$. Let $V \in H^a(\mathcal{K} \setminus \Sigma)$ be any cohomology class of order i, i.e., a class Alexander dual to an $(\omega - a - 1)$-dimensional cycle $\bar{V} \in \sigma_i$. Let α be the reversed filtration of this class, i.e., we can choose this cycle $\bar{V}$ in such a way that it lies in $\Phi_\alpha \cup \sigma_{i-1}$. Then for an arbitrary multi-index A with $|A| - \#A = i$ and $\#A = i - \alpha$, the *symbol* $s(A, V)$ is defined as follows.

Let us denote by $\Xi(A)$ the local system of groups on the configuration space $C(A)$ whose fiber over an A-set Υ is identified with $\Xi_*(\Upsilon)$. This local system is not constant if some of numbers a_m coincide: a loop in $C(A)$ permuting some groups of the same cardinality acts nontrivially on the fiber.

Further, let $\tilde{\Xi}(A)$ be a similar local system with a slightly more complicated monodromy action. Namely, any loop in $C(A)$ permuting exactly two groups of points, acts exactly as in the system $\Xi(A)$ (i.e., permutes the corresponding factors $\bar{H}_*(\Lambda(J_m))$ in the tensor product (11)) if the cardinality of any of these groups is odd, and additionally multiplies such an operator by -1 if all these cardinalities are even.

The symbol $s(A, V)$, which we are going to define, is an element of the group $H^a(C(A), \Xi(A))$ if n is odd and of $H^a(C(A), \tilde{\Xi}(A))$ if n is even.

The construction repeats for the index of a knot invariant given in subsection 5.3. Namely, we consider the class of the realizing V dual cycle $\bar{V}$ in the group $\bar{H}_{\omega-a-1}(\Phi_\alpha \setminus \Phi_{\alpha-1})$, then, using Corollary 2 of Theorem 3, Thom isomorphism for the fiber bundle mentioned in this Corollary, and Poincaré duality, we reduce this group to the cohomology group of $C(A)$ with coefficients in a local system associated to the homology bundle of the fiber bundle $\Phi_\alpha \setminus \Phi_{\alpha-1} \to C(A)$, which is nothing other than the local system $\Xi(A)$ or $\tilde{\Xi}(A)$.

For instance, if $n = 3$ and $a = 0$, i.e., V is just a knot invariant, then this symbol coincides with the totality of indices of V corresponding to all classes of A-configurations. The invariance of these indices mentioned in Proposition 6, follows from the fact that the 0-dimensional cocycle of a local system is just a global section of this system and thus is invariant under its monodromy action.

5.5. Multiplication.
5.5.1. *Multiplication of knot invariants in $\mathbb{R}^3$.* We have two filtrations in the space of finite-order invariants of knots in $\mathbb{R}^3$: the order i and, for any fixed i, the degree α with respect to the reversed auxiliary filtration in $\sigma_i \setminus \sigma_{i-1}$. It is well known that the order is multiplicative. In fact, the same is true also for the second number.

THEOREM 4. *Let V', V'' be two knot invariants of orders i', i''. Then for any multi-index $A = (a_1, \ldots, a_{\#A})$ of complexity $i = i' + i''$ and any equivalence class $\mathbf{J}$ of A-configurations, the index $\langle V' \cdot V'' | \mathbf{J} \rangle \in \Xi_*(\mathbf{J})$ is equal to*

$$\sum_{J' \cup J'' = J} \langle V' | J' \rangle \otimes \langle V'' | J'' \rangle, \tag{14}$$

where summation is taken over all decompositions of any A-configuration J representing $\mathbf{J}$ into disjoint unions of an A'-configuration J' and A''-configuration J'' of complexities i' and i'', such that any group of J belongs to either J' or J''.

Proof. For $A = (2^{\times i})$ this is just the Kontsevich decomposition formula for the index of $V' \cdot V''$ on an i-chord diagram: for any $(2, \ldots, 2)$-configuration τ,

$$\langle (V' \cdot V''), (\tau) \rangle = \sum_{\tau' \cup \tau'' = \tau} V'(\tau') V''(\tau''), \tag{15}$$

summation over all $\binom{i}{i'}$ ordered decompositions of the i-chord diagram τ into an i'-chord diagram τ' and an i''-chord diagram τ''; see, e.g., [**BN2**].

For an arbitrary multi-index A of the same complexity i formula (14) follows by induction over the stickness, while the induction step is formula (13). □

COROLLARY 4. *The bi-order (i, α) is multiplicative: if V' and V'' are two invariants of bi-orders (i', α') and (i'', α''), respectively, then $V' \cdot V''$ is an invariant of bi-order $(i' + i'', \alpha' + \alpha'')$.*

Indeed, in this case formula (14) gives the same answer for all A-configurations J of complexity $i' + i''$ and stickness $\alpha' + \alpha''$; thus our Corollary follows from Proposition 6. □

5.5.2. *Multiplication conjecture for higher cohomology of spaces of knots.* For any multi-indices A', A'' and $A \equiv A' \cup A''$, there are natural operations

$$H^*(C(A'), \Xi(A')) \otimes H^*(C(A''), \Xi(A'')) \to H^*(C(A), \Xi(A)) \tag{16}$$

and

$$H^*(C(A'), \tilde{\Xi}(A')) \otimes H^*(C(A''), \tilde{\Xi}(A'')) \to H^*(C(A), \tilde{\Xi}(A)), \tag{17}$$

such that for any cohomology classes $V', V'' \in H^*(\mathcal{K} \setminus \Sigma)$ of bi-orders (i', α') and (i'', α''), respectively, the class $V' \cdot V''$ has bi-order $(i' + i'', \alpha' + \alpha'')$. Further, its principal symbol $s(A, V' \cdot V'')$ can be expressed through the symbols of V' and V'' by a formula similar to (14), in which the operation of the tensor multiplication is replaced by the operation (16) if n is odd and by operation (17) if n is even; see [**Fu**], Section 8.

5.6. Configurations containing groups of 2 points do not contribute to the cohomology of lowest possible dimension of the space of noncompact knots in $\mathbb{R}^n$. Consider the *reversed spectral sequence* $\mathcal{E}_{p,q}^r$, calculating the group $\bar{H}_*(\sigma_i \setminus \sigma_{i-1})$ of the discriminant in the space of noncompact knots $\mathbb{R}^1 \to \mathbb{R}^n$ and generated by the reversed filtration $\{\Phi_\alpha\}$. By definition, the term $\mathcal{E}_{\alpha,q}^1$ of this spectral sequence is isomorphic to $\bar{H}_{\alpha+q}(\Phi_\alpha \setminus \Phi_{\alpha-1})$ and thus splits into the direct sum of similar groups $\bar{H}_{\alpha+q}(S(A))$ over all indices A of the form (9) with $\#A = i - \alpha$.

In particular, by Corollary 3 it is trivial if $\alpha + q > \omega - 1 - (n-3)i$.

For instance, for any i the term Φ_0 of our filtration consists of unique stratum $S(A)$ with $A = (2, \ldots, 2)$.

PROPOSITION 7. *For any i, two top terms $\mathcal{E}^1_{0,\omega-1-(n-3)i}$ and $\mathcal{E}^1_{0,\omega-2-(n-3)i}$ of the column $\mathcal{E}^1_{0,*} \equiv \mathcal{E}^1_{0,*}(i)$ of the reversed auxiliary spectral sequence calculating $\bar{H}_*(\sigma_i \setminus \sigma_{i-1})$ are trivial.*

This proposition is a special case of the following one.

PROPOSITION 8. *For any multi-index A of the form (9) such that at least one (respectively, at least two) of numbers a_i equals 2, the group $\bar{H}_{\omega-1-(n-3)i}(S(A))$ (respectively, both groups $\bar{H}_{\omega-1-(n-3)i}(S(A))$, $\bar{H}_{\omega-2-(n-3)i}(S(A))$) is trivial.*

EXAMPLE 5. Let $n = 3$. Calculating the group $\bar{H}_{\omega-1}(\sigma_i)$ of knot invariants of order i, we can take into account only strata $S(A)$ with at most one element 2 in the multi-index A. Moreover, such strata with exactly one 2 in A can provide only relations in this group, and not generators.

LEMMA 3. *For any natural k, $\bar{H}_*(B(\Psi, k)) \simeq 0$ and $\bar{H}_*(B(\Psi, k), \pm\mathbb{Z}) \simeq 0$.*

(For the definition of the local system $\pm\mathbb{Z}$, see subsection 5.1.)

Proof of Lemma 3. We use the decomposition of $B(\Psi, k)$ similar to the cell decomposition of the space $B(\mathbb{R}^2, k)$ used in [**Fu**]. Namely, to any decomposition $k = k_1 + \cdots + k_m$ of the number k into natural numbers k_i we assign the set of all k-subsets of Ψ, consisting of points $((t_1 \leq t'_1), \ldots, (t_k \leq t'_k))$, such that the smallest value of numbers t_l appears in exactly k_1 pairs $(t_l \leq t'_l)$, the next smallest value in k_2 pairs, etc. Denote this set by $e(k_1, \ldots, k_m)$. Filter the space $B(\Psi, k)$, assigning to the l-th term of the filtration the union of all such sets of dimension $\leq l$. It is easy to see that any such set is diffeomorphic to the direct product of a closed octant in $\mathbb{R}^m$ and an open k-dimensional cell, thus its Borel–Moore homology group is trivial, as well as the similar group with coefficients in (the restriction to this set of) the local system $\pm\mathbb{Z}$. Therefore, the spectral sequences calculating both groups $\bar{H}_*(B(\Psi, k)), \bar{H}_*(B(\Psi, k), \pm\mathbb{Z})$ and generated by our filtration vanish in the first term. □

Proof of Proposition 7. If $A = (2^{\times i})$, then $C(A)$ is an open subset in the configuration space $B(\Psi, i)$, whose complement $B(\Psi, i) \setminus C(A)$ has codimension 3. In particular, by Lemma 3 $\bar{H}_j(C(A)) = 0$ for $j = 2i$ or $2i-1$. It is easy to calculate that for odd n the fiber bundle $S(A) \to C(A)$ is orientable and for even n it changes its orientation together with the local system $\pm\mathbb{Z}$. Thus we have the Thom isomorphism $\bar{H}_{*+\omega-(n-2)i-1}(S(A)) \simeq \bar{H}_*(C(A))$ (for odd n) and $\simeq \bar{H}_*(C(A), \pm\mathbb{Z})$ (for even n). □

LEMMA 4. *For any finite subset $\theta \subset \Psi$, both groups $\bar{H}_l(B(\Psi \setminus \theta, k))$, $\bar{H}_l(B(\Psi \setminus \theta, k), \pm\mathbb{Z})$ are trivial for $l \neq k$, and for $l = k$ they are free Abelian groups of rank equal to the number of functions $\chi : \theta \to \mathbb{Z}_+$ such that $\sum_{z \in \theta} \chi(z) = k$.*

Proof. Consider any direction in the plane $\mathbb{R}^2 \supset \Psi$ transversal to the diagonal $\partial\Psi = \{t = t'\}$ and such that no two points of θ are connected by a vector of this direction. Let $\pi : \Psi \to \partial\Psi$ be the projection along this direction. To any point $Z = (z_1, \ldots, z_k) \subset \Psi \setminus \theta$ of the space $B(\Psi \setminus \theta, k)$ the following data are assigned:

(1) the topological type of the configuration in $\partial\Psi$ formed by the points $\pi(z_j)$ (counted with multiplicities) and the set $\pi(\theta)$;
(2) for any point $w \in \theta$, the number of points $z_j \in Z$ such that $\pi(z_j) = \pi(w)$ and z_j is separated by w from $\partial\Psi$ in the line $\pi^{-1}(\pi(w))$.

For any such collection of data, the subset in $B(\Psi \setminus \theta, k)$ formed by configurations Z with these data, is homeomorphic to $\mathbb{R}_+^l \times \mathbb{R}^s$, where l is the number of lines $\pi^{-1}(\cdot)$ containing points of Z not separated from $\partial \Psi$ by the points of θ, and $l+s$ is equal to the dimension of this subset, i.e., to $k+$(the number of geometrically distinct points of $\pi(Z)$ for any Z from this subset). In particular, the Borel–Moore homology group of such a subset is trivial if $l > 0$, and the set of configurations with $l = 0$ consists of several k-dimensional cells, which are in obvious correspondence with the functions $\theta \to \mathbb{Z}_+$ described in Lemma 4. □

Proof of Proposition 8. If exactly one of numbers a_j is equal to 2, then $S(A)$ is a connected manifold with nonempty boundary, in particular its Borel–Moore homology group of top dimension is trivial.

Now suppose that there are at least two such numbers. Let $A_{>2}$ be the same multi-index A, from which all numbers a_j equal to 2 are removed. Consider the projection

$$(18) \qquad C(A) \to C(A_{>2})$$

erasing from any A-set all its groups of cardinality 2. For any point $\nu \in C(A_{>2})$ denote by $\theta(\nu)$ the set of all pairs (x, y) lying in some of groups of ν. By the Leray spectral sequence of the composite fiber bundle $S(A) \to C(A) \to C(A_{>2})$, we need only to prove the following lemma.

LEMMA 5. *If the number r of twos in A is greater than 1, then for any $\nu \in C(A_{>2})$ the group $\bar{H}_*$ of the fiber of the projection (18) over the point ν is trivial in dimensions $2r$ and $2r - 1$.*

Proof. Any such fiber consists of all configurations $Z \in B(\Psi, r)$ not containing the points of the finite subset $\theta(\nu) \subset \Psi$ and satisfying some additional restrictions, which forbid certain subvariety of codimension ≥ 3 in $B(\Psi, r)$. Thus Lemma 5 follows from Lemma 4. □

6. Order 3 cohomology of spaces of noncompact knots

Here we calculate the column $E_{3,*}^\infty$ of the main spectral sequence converging to the finite-order cohomology of the space of noncompact knots $\mathbb{R}^1 \to \mathbb{R}^n$. In fact, it is sufficient to calculate the column $E_{3,*}^1$ of its initial term E^1, i.e., the group $\bar{H}_*(\sigma_3 \setminus \sigma_2)$ of the corresponding discriminant variety. This group is described in the following statement.

THEOREM 5. *For any $n \geq 3$, all groups $\bar{H}_j(\sigma_3 \setminus \sigma_2)$ are trivial except the groups with $j = \omega - 1 - 3(n-3)$ and $\omega - 2 - 3(n-3)$, which are isomorphic to $\mathbb{Z}$.*

This statement for $n = 3$ is not new: the group $\bar{H}_{\omega-1}(\sigma_3 \setminus \sigma_2) \simeq \mathbb{Z}$ (of knot invariants of order 3) was calculated in [**V2**], and all other groups $\bar{H}_j(\sigma_3 \setminus \sigma_2)$ in an unpublished work of D. M. Teiblum and V. E. Turchin; see subsection 4.4.2. Their calculation was based on the cellular decomposition of $\sigma_3 \setminus \sigma_2$, constructed in [**V2**], and was nontrivial even for the computer.

COROLLARY 5. *The column $E_{3,*}^\infty$ of the main spectral sequence coincides with $E_{3,*}^1$; namely, it consists of exactly two nontrivial terms $E_{3,\omega-1-3(n-2)}^\infty \simeq \mathbb{Z}$ and $E_{3,\omega-2-3(n-2)}^\infty \simeq \mathbb{Z}$.*

Proof of Corollary 5. The fact that all differentials $d^r : E^r_{3,*} \to E^r_{3-r,*+r-1}$, $r \geq 1$, are trivial, follows immediately from the construction of columns $E^1_{2,*}$ and $E^1_{1,*}$; see Examples 3 and 4 in subsection 5.1.

On the other hand, the inequality (3) implies that there are also no nontrivial differentials d^r acting *into* $E^r_{3,\omega-1-3(n-2)}$; if $n > 3$, then the same is true also for $E^r_{3,\omega-2-3(n-2)}$. Finally, if $n = 3$, then the similar triviality of the homomorphisms $d^r : E^r_{3+r,\omega-4-r} \to E^r_{3,\omega-5}$ follows from the Kontsevich realization theorem. □

The proof of Theorem 5 occupies the rest of Section 6.

6.1. Term $\mathcal{E}^1$ of the reversed spectral sequence.

The term $\sigma_3 \setminus \sigma_2$ of the main filtration consists of three terms of the reversed auxiliary filtration: $\Phi_0 \subset \Phi_1 \subset \Phi_2 \equiv \sigma_3 \setminus \sigma_2$.

The sets Φ_0, $\Phi_1 \setminus \Phi_0$ and $\Phi_2 \setminus \Phi_1$ consist of one component $S(A)$ each, with A equal to $(2,2,2)$, $(3,2)$ and (4), respectively.

LEMMA 6. *For any n, the following holds:*
(1) $\bar{H}_j(\Phi_0) = 0$ *for all* $j \neq \omega - 3n + 6$, *and* $\bar{H}_{\omega-3n+6}(\Phi_0) = \mathbb{Z}$;
(2) $\bar{H}_j(\Phi_1 \setminus \Phi_0) = 0$ *for all* $j \neq \omega - 3n + 7$, *and* $H_{\omega-3n+7}(\Phi_1 \setminus \Phi_0) = \mathbb{Z}^3$;
(3) $\bar{H}_j(\Phi_2 \setminus \Phi_1) = 0$ *for all* $j \neq \omega - 3n + 8$, *and* $\bar{H}_{\omega-3n+8}(\Phi_2 \setminus \Phi_1) = \mathbb{Z}^2$.

Proof. (1) The space $C(2,2,2)$ of all $(2,2,2)$-sets is an open subset in the configuration space $B(\Psi, 3)$, and the difference $\mathbf{T} \equiv B(\Psi, 3) \setminus C(2,2,2)$ is the set of all triples of the form $((x,y), (x,z), (y,z)) \subset \Psi$ with $x < y < z$. This set is obviously a closed subset in $B(\Psi, 3)$ diffeomorphic to a 3-dimensional cell. Thus by Lemma 3 and the exact sequence of the pair $(B(\Psi, 3), \mathbf{T})$, both groups $\bar{H}_l(C(2,2,2))$ and $\bar{H}_l(C(2,2,2), \pm \mathbb{Z})$ are trivial for any l other than 4 and are isomorphic to $\mathbb{Z}$ for $l = 4$. Fibers of the bundle $S(2,2,2) \to C(2,2,2)$ are products of the open triangle $\Lambda(2,2,2)$ and an affine space of dimension $\omega - 3n$. It is easy to calculate that for odd n this bundle is orientable, and for even n its orientation bundle coincides with $\pm\mathbb{Z}$, thus statement 1 follows from the Thom isomorphism.

(2) The space $C(3,2)$ is the space of a fiber bundle whose base is the configuration space $B(\mathbb{R}^1, 3)$ and the fiber over the triple $(x < y < z) \subset \mathbb{R}^1$ is the half-plane Ψ with three interior points (x,y), (x,z) and (y,z) removed. The Borel–Moore homology group of this punctured half-plane is concentrated in dimension 1 and is generated by three rays connecting these three removed points to infinity, say, by rays shown in Figure 5. The complex $\Lambda(3,2)$ consists of one three-dimensional open face. The bundle $S(3,2) \to C(3,2)$ is orientable; thus statement 2 also follows from the Thom isomorphism.

(3) The space $C(4) \sim B(\mathbb{R}^1, 4)$ is obviously diffeomorphic to a 4-cell. Therefore, $S((4))$ is homeomorphic to the direct product $B(\mathbb{R}^1, 4) \times \mathbb{R}^{\omega-3n} \times \Delta^2(4)$ (where $\Delta^2(4)$ denotes the 5-dimensional simplex $\Delta(4)$ with all not two-connected faces removed), and statement (3) follows immediately from Theorem 2. For the realization of generators of the group $\bar{H}_4(\Delta^2(4)) \simeq \bar{H}_{\omega-3n+8}(\Phi_2 \setminus \Phi_1)$, see Example 2 in Section 2. □

REMARK. To specify the latter isomorphism, we need to choose an orientation of $B(\mathbb{R}^1, 4)$ and a (co)orientation of the fiber $\mathbb{R}^{\omega-3n}$. We do it as follows.

The standard orientation of $B(\mathbb{R}^1, 4)$ is defined by local coordinate systems, whose coordinate functions are coordinates of four points taken in increasing order.

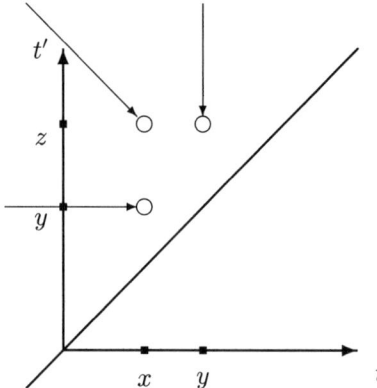

FIGURE 5. Basic cycles in the space $C(3,2)$

If $\xi_1, \ldots, \xi_n$ are coordinates in $\mathbb{R}^n$, so that any noncompact knot is given parametrically by n real functions $\xi_k(t)$, $k = 1, \ldots, n$, then the fiber $\mathbb{R}^{\omega-3n}$ over the configuration $(r < s < t < u)$ is distinguished by the equations $\xi_1(r) = \xi_1(s) = \xi_1(t) = \xi_1(u)$, $\ldots$, $\xi_n(r) = \xi_n(s) = \xi_n(t) = \xi_n(u)$. Then the transversal orientation of this fiber in the space $\mathcal{K} \sim \mathbb{R}^\omega$ will be specified by the skew form

$$\begin{aligned}(19) \quad & d(\xi_1(s) - \xi_1(r)) \wedge d(\xi_2(s) - \xi_2(r)) \wedge \cdots \wedge d(\xi_n(s) - \xi_n(r)) \\ & \wedge\ d(\xi_1(t) - \xi_1(s)) \wedge d(\xi_2(t) - \xi_2(s)) \wedge \cdots \wedge d(\xi_n(t) - \xi_n(s)) \\ & \wedge\ d(\xi_1(u) - \xi_1(t)) \wedge d(\xi_2(u) - \xi_2(t)) \wedge \cdots \wedge d(\xi_n(u) - \xi_n(t)).\end{aligned}$$

We have proved that the groups $\bar{H}_j(\sigma_3 \setminus \sigma_2)$ are trivial for all j other than $\omega - 3n + 8$, $\omega - 3n + 7$ and $\omega - 3n + 6$, and the calculation of these groups in these three dimensions is reduced to the calculation of a certain complex of the form $0 \to \mathbb{Z}^2 \to \mathbb{Z}^3 \to \mathbb{Z} \to 0$, whose boundary operator is the (horizontal) differential d_1 of our reversed spectral sequence.

6.2. The differential $d^1 : \bar{H}_{\omega-3n+8}(\Phi_2 \setminus \Phi_1) \to \bar{H}_{\omega-3n+7}(\Phi_1 \setminus \Phi_0)$. First we calculate this operator for the generator of the group $\bar{H}_{\omega-3n+8}(\Phi_2 \setminus \Phi_1)$, corresponding to the generator

(20)

of $\bar{H}_4(\Delta^2(4))$, where numbers of vertices correspond to the order of four points in $\mathbb{R}^1$. The boundary of any of these two graphs is the sum of 5 graphs with 4 edges each. One of them (obtained by removing the diagonal edge) appears in both sums and vanishes. Remaining boundary graphs (and the signs with which they participate in the algebraic boundary of the cycle (20)) are

(21)
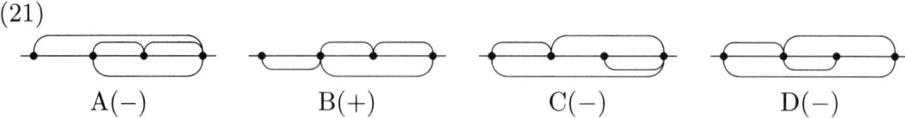

for the first graph in (20) and

(22)

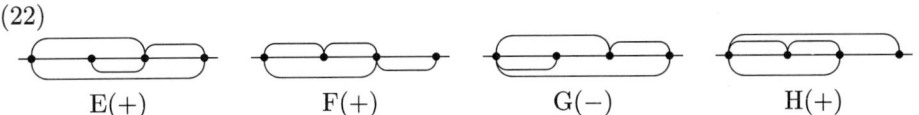

for the second. The term $\Phi_1 \setminus \Phi_0$ (in which the boundary lies) can be considered as a fiber bundle (we will call it the *former bundle*) whose base is the set of triples $\{x < y < z\} \subset \mathbb{R}^1$ and the fiber over such a point is the space of the *latter fiber bundle* with the base being the set of all pairs $\{v \leq w\} \in \Psi$ not coinciding with (x,y), (x,z) or (y,z) and the fiber being the direct product of an open tetrahedron with vertices called

(23) $\qquad\qquad (x,y), (x,z), (y,z), (v,w)$

and an affine subspace of codimension $3n$ in $\mathcal{K}$ (consisting of all maps $\phi : \mathbb{R}^1 \to \mathbb{R}^n$ such that $\phi(x) = \phi(y) = \phi(z)$ and $\phi(v) = \phi(w)$).

For an arbitrary basepoint (x,y,z) of the former bundle, consider the intersection set of the fiber over it with the parts of the boundary of the cycle (20), corresponding to eight graphs (21), (22). It is easy to calculate that these intersections are complete preimages of the latter bundle over the open segments in $\Psi \setminus ((x,y) \cup (x,z) \cup (y,z))$ labeled in Figure 6 by characters $A, \ldots, H$, corresponding to the notation of these graphs in pictures (21) and (22).

For instance, for any (4)-configuration $(r < s < t < u) \in C((4)) = B(\mathbb{R}^1, 4)$ consider the fiber in $\Phi_2 \setminus \Phi_1$ over this configuration and the part of the boundary of this fiber in $\Phi_1 \setminus \Phi_0$ corresponding to the graph E in (22). This part belongs to the fiber of the former bundle in $\Phi_1 \setminus \Phi_0$ over the point $(x < y < z) = (r < t < u)$, and coordinates (v,w) in the base of the latter bundle satisfy the relations $w = t \equiv y$, $v = s \in (x,y)$.

The part (E) of the boundary of the generator (20) of $\bar{H}_{\omega-3n+8}(\Phi_2 \setminus \Phi_1)$ coincides thus with the complete preimage in $\Phi_1 \setminus \Phi_0$ of the 4-dimensional submanifold

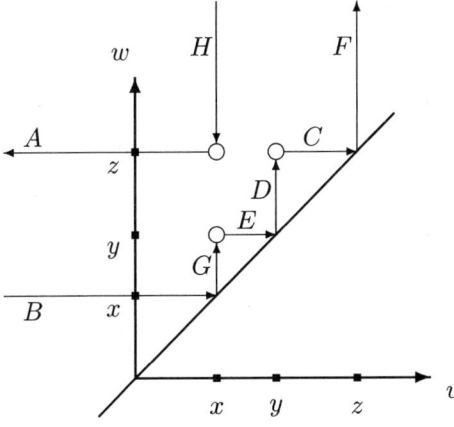

FIGURE 6. Boundary of the chain (20).

in $C(3,2)$ formed by *all* configurations $((x < y < z), (v, w))$ such that $w = y$, $x < v < y$.

To calculate the coefficients with which this and other similar preimages corresponding to other letters $A, \ldots, H$ appear in the *algebraic* boundary, we need to fix their orientations. Again, any such orientation consists of orientations of three objects:

(a) the 4-dimensional base manifold in the space of all $(3,2)$-sets $((x, y, z), (v, w))$;
(b) the tetrahedron (23); and
(c) for any point of this 4-dimensional manifold, the subspace in $\mathbb{R}^\omega$ consisting of maps $\phi: \mathbb{R}^1 \to \mathbb{R}^n$, such that $\phi(x) = \phi(y) = \phi(z)$ and $\phi(v) = \phi(w)$.

We fix these orientations as follows.

The manifold in the configuration space will be oriented by the pair of orientations, the first of which is lifted from the orientation $dx \wedge dy \wedge dz$ of the base of the former bundle, and the second is given by the increase of the unique "free" number v or w; i.e., all horizontal segments in Figure 6 should be oriented to the right, and all vertical segments should be oriented up.

The tetrahedron (23) over any point $((x < y < z), (v < w))$ of the space $C(3,2)$ will be oriented by the order (23) of its vertices.

The (co)orientation of the fiber $\mathbb{R}^{\omega - 3n}$ over the same point $((x < y < z), (v < w))$ will be specified by the skew form

$$\begin{aligned}(24) \quad & d(\xi_1(y) - \xi_1(x)) \wedge d(\xi_2(y) - \xi_2(x)) \wedge \cdots \wedge d(\xi_n(y) - \xi_n(x)) \\ & \wedge \; d(\xi_1(z) - \xi_1(y)) \wedge d(\xi_2(z) - \xi_2(y)) \wedge \cdots \wedge d(\xi_n(z) - \xi_n(y)) \\ & \wedge \; d(\xi_1(w) - \xi_1(v)) \wedge d(\xi_2(w) - \xi_2(v)) \wedge \cdots \wedge d(\xi_n(w) - \xi_n(v))\end{aligned}$$

on its normal bundle.

Now, for any of 8 components (21), (22) of the boundary of the generator (20), we need to compare these orientations with those induced from the canonical orientations of similar objects in $\Phi_2 \setminus \Phi_1$. Here we present these comparisons only for one component (E).

Compare orientations of configurations. In the base $C((4))$ of $\Phi_2 \setminus \Phi_1$ the orientation is given by $dr \wedge ds \wedge dt \wedge du$, and in $\Phi_1 \setminus \Phi_0$ the orientation of the corresponding component of the boundary is given by $dx \wedge dy \wedge dz \wedge dv$. For the component E we have $x = r$, $y = t$, $z = u$, $v = s$; hence these orientations coincide.

Compare orientations of tetrahedra. The lexicographic order of vertices of the graph E in (22) is as follows: $((13), (14), (23), (34))$, or, in notation used in (23), $((x, y), (x, z), (v, w), (y, z))$. This ordering is opposite to (23).

Compare coorientations of subspaces $\mathbb{R}^{\omega - 3n}$. If n is even, then all our coorientations of all such fibers coincide canonically. For odd n this is not the case anymore. For example, it is easy to check that for the component E of the boundary orientations (19) and (24) are opposite.

Doing the same calculations for all other components $A, \ldots, H$, we obtain Table 1, in which the second line is the sign indicated in (21) or (22), and the next three lines are the results of comparisons or orientations of tetrahedra, configuration spaces and subspaces in $\mathcal{K}$ (in the case of odd n), respectively. The sixth line contains the final sign, with which the corresponding oriented component depicted by a segment in Figure 6, participates in the boundary of the generator (20) in the case of odd n; this sign is just the product of previous four signs. The last line

TABLE 1

	A	B	C	D	E	F	G	H
Coefficient	−	+	−	−	+	+	−	+
Compare Δ	−	−	+	−	−	+	−	−
Configurations	−	−	−	−	+	+	+	+
Subspaces	+	+	+	−	−	+	+	+
Odd n	−	+	+	+	+	+	+	−
Even n	−	+	+	−	−	+	+	−

TABLE 2

	A	B	C	D	E	F	G	H	A	B	C	D	E	F	G	H
Coefficient	−	+	+	−	+	−	−	+	+	−	+	−	+	−	+	−
Compare Δ	−	+	−	+	−	−	+	+	−	+	+	−	−	−	−	−
Configs	+	−	+	−	−	−	+	+	+	+	−	+	−	−	+	+
Subspaces	−	+	+	−	+	+	+	+	−	+	−	+	+	+	−	−
Odd n	−	−	−	−	+	−	−	+	−	−	+	+	+	−	+	−
Even n	+	−	−	+	+	−	−	+	+	−	−	+	+	−	−	+

indicates a similar sign for even n; it is equal to the product of three first signs in the column.

It follows from these calculations that in the case of odd n the boundary in $\Phi_1 \setminus \Phi_0$ of the element (20) is homologous to zero. Indeed, we can make all signs in the sixth line equal to + if defining the orientation of the 4-submanifold in $C(3,2)$ we orient segments A and H as shown in Figure 6. Therefore, our boundary chain coincides with the boundary of the complete preimage under the projection $S(3,2) \to C(3,2)$ of the domain, consisting of all configurations $((x < y < z), (v < w))$, such that the point (v,w) belongs to the domain in $\mathbb{R}^2$ bounded from left and from above by segments H and A (see Figure 6), and from right and from below by the union of segments B, G, E, D, C and F, depending on x, y and z.

In the case of even n the boundary of the element (20) is not a cycle in $\Phi_1 \setminus \Phi_0$. However, making the same calculations for two other generators $\begin{smallmatrix}1\\2\end{smallmatrix}\boxtimes\begin{smallmatrix}4\\3\end{smallmatrix} - \begin{smallmatrix}1\\2\end{smallmatrix}\boxtimes\begin{smallmatrix}4\\3\end{smallmatrix}$ and $\begin{smallmatrix}1\\2\end{smallmatrix}\boxtimes\begin{smallmatrix}4\\3\end{smallmatrix} - \begin{smallmatrix}1\\2\end{smallmatrix}\boxtimes\begin{smallmatrix}4\\3\end{smallmatrix}$ of the group $\bar{H}_4(\Delta^2(4))$, we obtain that the *difference* of corresponding chains in $\Phi_2 \setminus \Phi_1$ has the boundary in $\Phi_1 \setminus \Phi_0$, which is homologous to zero. These calculations are represented in Figure 7 and Table 2.

Finally, we get the following statement.

PROPOSITION 9. *For any n, the group $\bar{H}_{\omega-3n+8}(\sigma_3 \setminus \sigma_2)$ is isomorphic to $\mathbb{Z}$. Its basis element coincides in $\Phi_2 \setminus \Phi_1$ with the sum (if n is odd) or difference (for even n) of elements, corresponding to generators* $\begin{smallmatrix}1\\2\end{smallmatrix}\boxtimes\begin{smallmatrix}4\\3\end{smallmatrix} - \begin{smallmatrix}1\\2\end{smallmatrix}\boxtimes\begin{smallmatrix}4\\3\end{smallmatrix}$ *and* $\begin{smallmatrix}1\\2\end{smallmatrix}\boxtimes\begin{smallmatrix}4\\3\end{smallmatrix} - \begin{smallmatrix}1\\2\end{smallmatrix}\boxtimes\begin{smallmatrix}4\\3\end{smallmatrix}$ *of the group $\bar{H}_4(\Delta^2(4))$.*

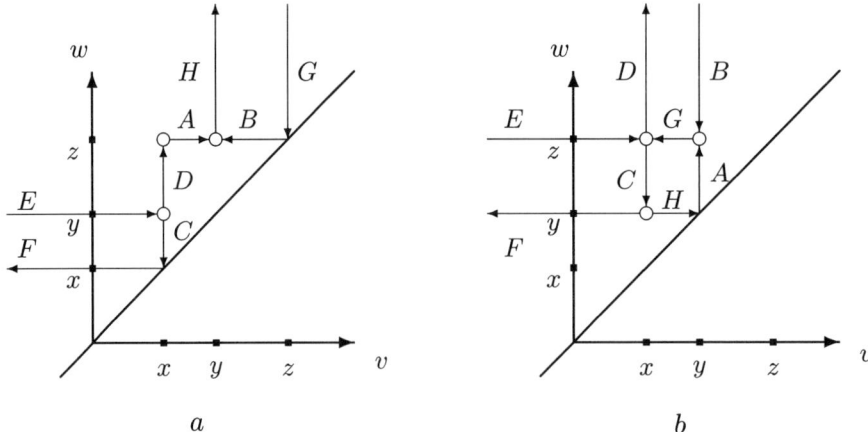

FIGURE 7. Boundaries of noninvariant 4-chains

For any n, the image of the operator $d^1 : \bar{H}_{\omega-3n+8}(\Phi_2 \setminus \Phi_1) \to \bar{H}_{\omega-3n+7}(\Phi_1 \setminus \Phi_0)$ is isomorphic to $\mathbb{Z}$ and the quotient group $\bar{H}_{\omega-3n+7}(\Phi_1 \setminus \Phi_0)/d^1(\bar{H}_{\omega-3n+8}(\Phi_2 \setminus \Phi_1))$ is isomorphic to $\mathbb{Z}^2$. □

6.3. The differential $d^1 : \bar{H}_{\omega-3n+7}(\Phi_1 \setminus \Phi_0) \to \bar{H}_{\omega-3n+6}(\Phi_0)$.

PROPOSITION 10. *The boundary operator d^1 of our spectral sequence maps any of three generators of the group $\bar{H}_{\omega-3n+7}(\Phi_1 \setminus \Phi_0)$ (indicated by rays in Figure 5) into a generator of the group $\bar{H}_{\omega-3n+6}(\Phi_0)$.*

We prove this statement for the boundary of one generator of the former group, depicted in Figure 5 by the horizontal ray, consisting of points (λ, y) with $\lambda < x$. The chain representing this generator is the fiber bundle whose base is the configuration space $B(\mathbb{R}^1, 4)$, and the fiber over its point $\{\lambda < x < y < z\}$ is the direct product of a tetrahedron (whose vertices are called (λ, y), (x, y), (x, z) and (y, z); see Figure 8a) and an affine space of dimension $\omega - 3n$ (consisting of all maps $\phi : \mathbb{R}^1 \to \mathbb{R}^n$ such that $\phi(\lambda) = \phi(x) = \phi(y) = \phi(z)$).

One face of the simplex in Figure 8a (opposite to the vertex (λ, y)) is a non-connected graph and the corresponding stratum belongs to σ_2. The strata swept out by three other faces (opposite to vertices (x, y), (x, z) and (y, z), respectively), belong to the term $\Phi_0 \subset \sigma_3 \setminus \sigma_2$; namely, they are complete preimages in $S(2, 2, 2)$

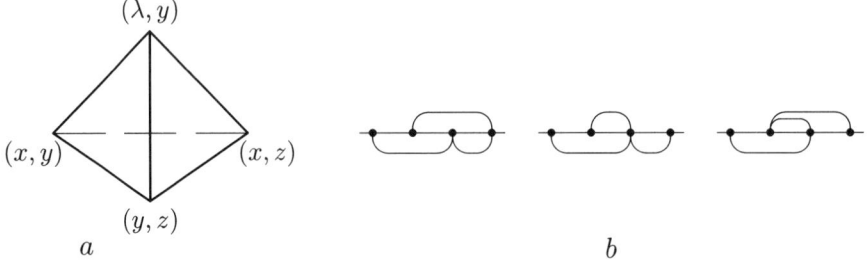

FIGURE 8. The boundary of the first generator of $E^1_{1, \omega-3n+6}$

of the 4-dimensional cycles in $C(2,2,2)$ consisting of all triples of the form

(25) $\quad \{(\lambda,y),(x,z),(y,z)\}, \quad \{(\lambda,y),(x,y),(y,z)\}, \quad \{(\lambda,y),(x,y),(x,z)\},$

respectively, with all possible $\lambda < x < y < z$; see Figure 8b. By the Thom isomorphism for the bundle $S(2,2,2) \to C(2,2,2)$, it remains to prove that the union of these three cycles in $C(2,2,2)$ defines a generator of the group $\bar{H}_4(C(2,2,2)) \equiv \bar{H}_4(B(\Psi,3),\mathbf{T})$ (if n is odd) or $\bar{H}_4(C(2,2,2),\pm\mathbb{Z}) \equiv \bar{H}_4(B(\Psi,3),\mathbf{T};\pm\mathbb{Z})$ (if n is even); or, equivalently, that its boundary in $\bar{H}_3(\mathbf{T})$ is the fundamental cycle of $\mathbf{T} \sim B(\mathbb{R}^1,3)$. But among three manifolds (25) only the first approaches $\mathbf{T}$ and its boundary obviously coincides with $\mathbf{T}$.

The calculation for two other generators of the group $\bar{H}_{\omega-3n+7}(\Phi_1 \setminus \Phi_0)$ is exactly the same. $\square$

Theorem 5 is a direct corollary of Lemma 6 and Propositions 9 and 10. $\square$

7. Cohomology classes of compact knots in $\mathbb{R}^n$

7.1. Classes of order 1.
It is well known that there are no first-order knot invariants in $\mathbb{R}^3$; see [**V2**]. However, the subgroup $F^*_{1,\mathbb{Z}_2} \subset H^*(\mathcal{K} \setminus \Sigma, \mathbb{Z}_2)$ of all $\mathbb{Z}_2$-valued first-order cohomology classes of the space of compact knots is nontrivial. For instance, the generator of the group $F^1_{1,\mathbb{Z}_2} \subset H^1(\mathcal{K} \setminus \Sigma, \mathbb{Z}_2)$ (which we describe below) proves that already the component of unknots in $\mathbb{R}^3$ is not simply connected.

THEOREM 6. A. *For any $n \geq 3$, the subgroup $F^*_{1,\mathbb{Z}_2} \subset H^*(\mathcal{K} \setminus \Sigma, \mathbb{Z}_2)$ of first-order cohomology classes of the space of knots in $\mathbb{R}^n$ contains exactly two nontrivial homogeneous components $F^{n-2}_{1,\mathbb{Z}_2} \sim F^{n-1}_{1,\mathbb{Z}_2} \sim \mathbb{Z}_2$.*

B. *If n is even, then both of these cohomology classes give rise to integer cohomology classes, i.e., $F^{n-2}_{1,\mathbb{Z}} \sim F^{n-1}_{1,\mathbb{Z}} \sim \mathbb{Z}$, and there are no other nontrivial integer cohomology groups $F^d_{1,\mathbb{Z}}$, $d \neq n-2, n-1$.*

C. *If n is odd, then the generator of the group $F^{n-1}_{1,\mathbb{Z}_2}$ is equal to the first Steenrod operation applied to the generator of $F^{n-2}_{1,\mathbb{Z}_2}$.*

D. *The generator of the group $F^{n-2}_{1,\mathbb{Z}_2}$ can be defined as the linking number with the $\mathbb{Z}_2$-fundamental cycle of the variety $\Gamma \subset \Sigma$, formed by all maps $\phi : S^1 \to \mathbb{R}^n$, gluing together some two opposite points of S^1; the generator of the group $F^{n-1}_{1,\mathbb{Z}_2}$ is the linking number with the $\mathbb{Z}_2$-fundamental cycle of the subvariety $\Gamma! \subset \Gamma$, formed by all maps $\phi : S^1 \to \mathbb{R}^n$, gluing together some two fixed opposite points, say, the points 0 and π. Moreover, if n is even, then these two varieties are orientable and the groups $F^{n-2}_{1,\mathbb{Z}}$, $F^{n-1}_{1,\mathbb{Z}}$ are generated by the linking numbers with the corresponding $\mathbb{Z}$-fundamental cycles.*

E. *If $n = 3$, then the cycles generating the groups $F^1_{1,\mathbb{Z}_2}$ and $F^2_{1,\mathbb{Z}_2}$ are nontrivial already in the restriction to the component of the unknot in $\mathbb{R}^3$.*

Proof. The term σ_1 of the main filtration of σ is the space of an affine fiber bundle of dimension $\omega - n$, whose base is the configuration space Ψ; in the case of compact knots this base space is diffeomorphic to the closed Möbius band. It is easy to check that this affine bundle is orientable if and only if n is even. Thus the term $E_1^{-1,q}(\mathbb{Z}_2)$ of the main spectral sequence with coefficients in $\mathbb{Z}_2$ is isomorphic to $\mathbb{Z}_2$ for $q = n-1$ or n and is trivial for all other q. Moreover, if n is even, then the terms $E_1^{-1,q}(\mathbb{Z})$ are isomorphic to $\mathbb{Z}$ if $q = n-1$ or n and are trivial for all other q. The basic cycles in σ_1 generating these groups are the manifolds

$\tilde\Gamma$ (respectively, $\tilde\Gamma!$) consisting of all pairs $(\phi,(x,y))$ such that $\phi(x) = \phi(y)$ and $y = x + \pi$ (respectively, $x = 0$, $y = \pi$). Thus the direct images of these cycles in Σ are exactly the fundamental classes of varieties Γ, $\Gamma!$, mentioned in statement D of Theorem 6. We need to prove that the homology classes in Σ of these cycles are nontrivial, i.e., the linking numbers with them (which we denote by $\{\Gamma\}$ and $\{\Gamma!\}$, respectively) are nontrivial cohomology classes in $\mathcal{K} \setminus \Sigma$.

For $n > 4$ this follows from the dimension arguments. Indeed, for any $n > 3$ our spectral sequence converges to the entire group $H^*(\mathcal{K} \setminus \Sigma)$, and by (3), (5) all groups $E_{\geq 1}^{p,q}$, $p < -1$, with $p + q < n - 1$ are trivial, hence all differentials d^r, acting into $E_r^{-1,q}$, also are trivial. For $n = 4$ there is a unique nontrivial group $E_1^{-2,4}$, from which, in principle, there could act a nontrivial homomorphism $d^1 : E_1^{-2,4} \to E_1^{-1,4} \sim \mathbb{Z}$. We will see in the next subsection that the group $E_1^{-2,4}(G)$ is isomorphic to the coefficient group G. Thus for $G = \mathbb{Z}$ and $n = 4$ the differential $d^1(E_1^{-2,4})$ is trivial by the Kontsevich realization theorem (6), and for $G = \mathbb{Z}_2$ the same follows from the functoriality of spectral sequences under the coefficient homomorphisms. Thus statements A, B, and D of Theorem 6 are proved for all $n > 3$.

The statement C for odd $n > 3$ will be proved as follows. First, we prove the equality $Sq^1(\{\Gamma\}) = \{\Gamma!\}$ in the group $H^*(\mathcal{K} \setminus \Gamma)$. Then the desired similar equality in the cohomology group of the subspace $\mathcal{K} \setminus \Sigma \subset \mathcal{K} \setminus \Gamma$ will follow from the naturality of Steenrod operations.

For any $n \geq 3$, the group $H^*(\mathcal{K} \setminus \Gamma)$ can be calculated by a cohomological spectral sequence similar to (but much easier than) our main spectral sequence. Namely, it is "Alexander dual" (in the sense of formula (4)) to the homological sequence associated to the standard filtration of the simplicial resolution γ of Γ. The term $\gamma_i \setminus \gamma_{i-1}$ of this filtration is the space of a fiber bundle with base $B(S^1, i)$ (where S^1 is the space of all diametral pairs of points of the original circle) and the fiber equal to the direct product of the open simplex $\dot\Delta^{i-1}$ and the space $\mathbb{R}^{\omega - ni}$. In particular $E_1^{p,q} = 0$ for $p > 0$ or $q + p(n-1) < 0$. For any $n > 3$, the group $H^{n-1}(\mathcal{K} \setminus \Gamma, \mathbb{Z}_2)$, containing the element $Sq^1(\{\Gamma\})$, is one dimensional and is generated by the linking number with the cycle $\Gamma!$. Consider any loop $l \subset \Gamma$ in the set of nonsingular points of Γ and such that going along l we permute two points $x, y = x + \pi \in S^1$ glued together by the corresponding maps. (Such l exists for any $n \geq 3$ because the codimension in the manifold $\tilde\Gamma \simeq \gamma_1$ of the preimage of the set of singular points of Γ is equal to $n - 1$ and we can realize l as the projection into Γ of almost any loop in $\tilde\Gamma$ permuting the points x and y.) Let L be the "tube" around l in $\mathcal{K} \setminus \Gamma$, i.e., the union of boundaries of small $(n-1)$-dimensional discs transversal to Γ with centers at the points of l. The fibration $L \to l$ with fiber S^{n-2} is nonorientable if n is odd; thus already in the restriction to L, $Sq^1(\{\Gamma\})$ is nontrivial and is equal to the $\mathbb{Z}_2$-fundamental cocycle of L. The union of these transversal discs spans L in $\mathcal{K}$, and its $\mathbb{Z}_2$-intersection number with $\Gamma!$ is equal to 1. In particular, the fundamental cycle of L is nontrivial in $H_{n-1}(\mathcal{K} \setminus \Gamma, \mathbb{Z}_2)$ for any $n \geq 3$ and generates this group for $n > 3$. Also, we get that for any odd $n \geq 3$ in restriction to L, $Sq^1(\{\Gamma\}) = \{\Gamma!\}$. Since for $n > 3$ the inclusion homomorphism $H^{n-1}(\mathcal{K} \setminus \Gamma, \mathbb{Z}_2) \to H^{n-1}(L, \mathbb{Z}_2)$ is an isomorphism, statement C of Theorem 6 is proved for all odd $n > 3$.

Finally, let be $n = 3$. Consider the loop $\Lambda : S^1 \to (\mathcal{K} \setminus \Sigma)$, with certain eight points as shown in Figure 9. Note that any two (un)knots of this family placed

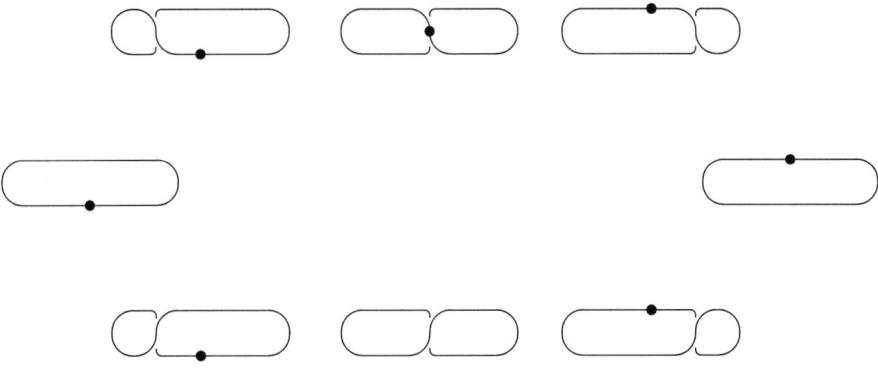

FIGURE 9. Nontrivial 1-cycle in the space of unknots

in this picture one over the other have the same projection to $\mathbb{R}^2$. Let us connect any two such unknots by a segment in $\mathcal{K}$ along which the projection to $\mathbb{R}^2$ also is preserved. The union of these segments is a disc in $\mathcal{K}$ spanning the loop Λ; it is obvious that the $\mathbb{Z}_2$-intersection number of this disc with the variety Γ is equal to 1. In particular the class $\{\Gamma\} \in H^1(\mathcal{K} \setminus \Sigma, \mathbb{Z}_2)$ is nontrivial.

It is easy to see that this loop Λ is homotopic to the loop Λ', consisting of knots obtained from the standard embedding $\phi : S^1 \to \mathbb{R}^2 \subset \mathbb{R}^3$ by rotations by all angles $\alpha \in [0, 2\pi]$ around any diagonal of $\phi(S^1)$, and also to the loop Λ'', consisting of all knots ϕ_τ, $\tau \in [0, 2\pi]$, having the same image as ϕ and given by the formula $\phi_\tau(\alpha) = \phi(\alpha + \tau)$.

Let us fix a sphere $S^2 \subset \mathbb{R}^3$ and consider the space GC of all naturally parametrized great circles in it. This space is obviously diffeomorphic to $SO(3) \sim \mathbb{R}P^3$, and its group $H_1(GC, \mathbb{Z}_2)$ is generated by our loop $\Lambda' \sim \Lambda$. To complete the proof of statements A, C, and D for $n = 3$ and to prove statement E, it remains to show that the linking number of the generator of $H_2(GC, \mathbb{Z}_2)$ with the variety Γ! equals 1.

If $n = 3$, then the group $H^2(\mathcal{K} \setminus \Gamma, \mathbb{Z}_2)$ containing the element $Sq^1(\{\Gamma\})$, is two dimensional: besides the class $\{\Gamma!\}$ it contains a basic element of second order coming from $E^{-2,4}$ of the canonical spectral sequence calculating $H^*(\mathcal{K} \setminus \Gamma, \mathbb{Z}_2)$. The reduction mod γ_1 of its dual class in $\bar{H}_{\omega-3}(\Gamma)$ is the fundamental cycle of $\gamma_2 \setminus \gamma_1$. As we will see in the next subsection, this fundamental cycle is homologous to zero in the space $\sigma_2 \setminus \sigma_1 \supset \gamma_2 \setminus \gamma_1$. Thus the element $Sq^1(\{\Gamma\}) \in H^2(\mathcal{K} \setminus \Sigma, \mathbb{Z}_2)$ belongs to the group of elements of order 1, which is generated by the class $\{\Gamma!\}$. On the other hand, $Sq^1(\{\Gamma\}) \equiv \{\Gamma\}^2$ is nontrivial already in restriction to the submanifold $GC \subset \mathcal{K} \setminus \Sigma$ (because $\{\Gamma\}$ is), thus it coincides with $\{\Gamma!\}$. Theorem 6 is completely proved. $\square$

7.2. Classes of order 2.

THEOREM 7. *For any $n \geq 3$, the group $F^*_{2,\mathbb{Z}}/F^*_{1,\mathbb{Z}}$ of order 2 integer cohomology classes of the space of compact knots reduced modulo classes of order 1, contains at most two nontrivial homogeneous components in dimensions $2n - 6$ and $2n - 3$. The component in dimension $2n - 6$ is always isomorphic to $\mathbb{Z}$, the component in dimension $2n - 3$ is isomorphic to $\mathbb{Z}$ if $n > 3$ and is a cyclic (maybe trivial) group* if $n = 3$. The generator of the group in dimension $2n - 3$ has bi-order $(2,0)$, and the generator in dimension $2n - 6$ has bi-order $(2,1)$.

Proof. The auxiliary filtration of the space $\sigma_2 \setminus \sigma_1$ consists of two terms $\Phi_0 \subset \Phi_1$. The term Φ_0 is the space of a fiber bundle over $B(\Psi, 2)$ whose fiber is the direct product of a vector subspace of codimension $2n$ in $\mathcal{K}$ and an open interval. It is easy to calculate that the Borel–Moore homology group $\bar{H}_*(\Phi_0)$ of the space of this bundle coincides with that of its restriction on the subset $B(S^1, 2) \subset B(\Psi, 2)$, where S^1 is the *equator* of the Möbius band Ψ consisting of all pairs of the form $(\alpha, \alpha+\pi)$. This subset $B(S^1, 2)$ is homeomorphic to the open Möbius band. The generator of its fundamental group $\pi_1(B(S^1, 2)) \simeq \mathbb{Z}$ preserves the orientation of the bundle of $(\omega - 2n)$-dimensional subspaces in $\mathcal{K}$ and reverses the orientation of both the bundle of open intervals and the base $B(S^1, 2)$ itself. Therefore, $\bar{H}_*(\Phi_0)$ coincides with the homology group of a circle up to the shift of dimension by $\omega - 2n + 2$, so that it has only two nontrivial homology groups $\bar{H}_{\omega-2n+3}$ and $\bar{H}_{\omega-2n+2}$, both isomorphic to $\mathbb{Z}$. The first (respectively, second) of them is generated by the fundamental cycle of the preimage of the set $B(S^1, 2) \subset B(\Psi, 2)$ (respectively, by that of the circle in $B(S^1, 2)$ formed by all pairs of pairs of the form $((\alpha, \alpha + \pi), (\alpha + \pi/2, \alpha + 3\pi/2))$, where α is defined up to the addition of a multiple of $\pi/2$).

The term $\Phi_1 \setminus \Phi_0$ is the space of a fiber bundle over $B(S^1, 3)$ whose fiber is the direct product of a vector subspace of codimension $2n$ in $\mathcal{K}$ and an open triangle. Both these bundles (of vector subspaces and triangles) over $B(S^1, 3)$ are orientable, thus $\bar{H}_*(\Phi_1 \setminus \Phi_0) \simeq \bar{H}_{*-(\omega-2n+2)}(B(S^1, 3))$; i.e., this group is equal to $\mathbb{Z}$ in dimensions $\omega - 2n + 5$ and $\omega - 2n + 4$ and is trivial in all other dimensions. The $(\omega - 2n + 5)$-dimensional component is generated by the fundamental cycle of $\Phi_1 \setminus \Phi_0$, and the $(\omega - 2n + 4)$-dimensional one by the preimage of the cycle in $B(S^1, 3)$ formed by all triples $(\alpha, \beta, \gamma) \subset S^1$ such that $\alpha + \beta + \gamma \equiv 0 \pmod{2\pi}$.

The unique suspicious differential $d^1 : \mathcal{E}^1_{1,*} \to \mathcal{E}^1_{0,*}$ of the reversed auxiliary spectral sequence sends the $(\omega - 2n + 4)$-dimensional generator of $\bar{H}_*(\Phi_1 \setminus \Phi_0)$ to a multiple of the $(\omega - 2n + 3)$-dimensional generator of $\bar{H}_*(\Phi_0)$. Now we prove that the coefficient in this multiple is equal to ± 1.

By construction, the image of this differential is the fundamental cycle of the preimage in Φ_0 of a certain 2-dimensional cycle $\Delta \subset B(\Psi, 2)$. Namely, this cycle is the space of a fiber bundle over S^1 whose fiber over the point $\alpha \in S^1$ is homeomorphic to an open interval and consists of all unordered pairs $\{(\alpha, \beta); (\alpha, \gamma)\}$ such that $\beta \neq \gamma$ and $\alpha + \beta + \gamma \equiv 0 \pmod{2\pi}$ (although either β or γ can coincide with α).

LEMMA 7. *The submanifolds Δ and $B(S^1, 2)$ of $B(\Psi, 2)$ are homeomorphic and there is a proper homotopy $\Delta \times [0, 1] \to B(\Psi, 2)$ between this homeomorphism $\Delta \to B(S^1, 2)$ and the identical embedding of Δ into $B(\Psi, 2)$.*

Proof. We will consider all points α, β, etc., as points of the unit circle in $\mathbb{R}^2$. Any point $\{(\alpha, \beta); (\alpha, \gamma)\} \in \Delta$ can be deformed to a point of $B(S^1, 2)$ by a homotopy h_t, $t \in [0, 1]$ (see Figure 10), along which the chords $[\alpha'(t), \beta(t)]$ and $[\alpha''(t), \gamma(t)]$ preserve their directions. The easiest possible "direct and uniform" realization of such a deformation depends continuously on the initial point of Δ; it is easy to calculate that any point of $B(S^1, 2)$ can be obtained by such a homotopy from exactly one point of Δ. □

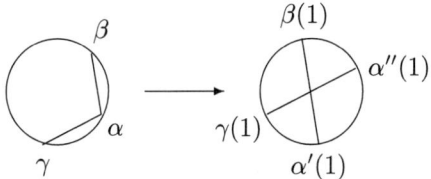

FIGURE 10. The homotopy $\Delta \to B(S^1, 2)$.

So, the differential $d^1 : \mathcal{E}^1_{1,*} \to \mathcal{E}^1_{0,*}$ kills both groups $\bar{H}_{\omega-2n+4}(\Phi_1 \setminus \Phi_0)$ and $\bar{H}_{\omega-2n+3}(\Phi_0)$, and the column $E_1^{-2,*}$ of the main cohomological spectral sequence contains exactly two nontrivial terms $E_1^{-2,2n-4} \sim E_1^{-2,2n-1} \sim \mathbb{Z}$. By the Kontsevich realization theorem (see subsection 4.2) they survive after the action of all differentials of the spectral sequence and define certain cohomology classes of $\mathcal{K} \setminus \Sigma$ in dimensions $2n - 6$ and $2n - 3$, respectively. However, in the case $n = 3$ we cannot be sure that the $(2n-3)$-dimensional class or its multiple will not be killed by some unstable cycle in Σ not counted by our spectral sequence. (An example of a situation of this sort is described in [**V8**] or Section VII.2 of [**V6**].) The $(2n-6)$-dimensional class in this case is a well-known knot invariant, and Theorem 7 is proved.

Note however that for $n = 4$, its statement concerning the group in dimension $2n - 3$ depends on an unpublished theorem of Kontsevich. (For $n > 5$ the fact that nothing acts into the corresponding cell $E^{-2,2n-1}$ follows from the dimensional reasons (see (5)) and for $n = 5$ from the simplest version (6) of the Kontsevich theorem, whose proof essentially coincides with that for the classical case $n = 3$, see [**K1**], [**BN2**].) □

8. Three problems

PROBLEM 1. To establish a direct correspondence between the calculus of two-connected graphs and that of Chinese Character Diagrams, see [**BN2**].[3] Perhaps this correspondence will give us the most natural proof of Theorem 2. The starting point of our construction was the space of all smooth maps $M^1 \to \mathbb{R}^n$. Is there some analog of it behind the Chinese Character Diagrams?

PROBLEM 2. To present a precise (and economical) description of generators of groups $\bar{H}_*(\Delta^2(k))$.

PROBLEM 3. The multiplication conjecture in subsection 5.5.2.

References

[A1]   V. I. Arnold, *On some topological invariants of algebraic functions*, Trudy Moskov. Mat. Obshch. **21** (1970), 27-46; English transl., Trans. Moscow Math. Soc., **21** (1970), 30–52.

[A2]   _____, *The spaces of functions with mild singularities*, Funktsional. Anal. i Prilozhen., **23** (1989), no. 3, 1–10; English transl. in Functional Anal. Appl. **23** (1989), no. 3.

[BBLSW] E. Babson, A. Björner, S. Linusson, J. Shareshian, and V. Welker, *The complexes of not i-connected graphs*, preprint, 1997.

---

[3]Note in proof. Recently this problem was solved by V. Turchin, see his article "Homology isomorphism of the complex of 2-connected graphs and the graph-complex of trees", in the book "Mathematics at the Independent Moscow University", Amer. Math. Soc., 1998.

[BN1]   D. Bar-Natan, *Some computations related to Vassiliev invariants,* preprint, 1994.
[BN2]   _____, *On the Vassiliev knot invariants,* Topology, **34** (1995), 423–472.
[BL]   J. S. Birman and X.-S. Lin, *Knot polynomials and Vassiliev's invariants,* Invent. Math. **111** (1993), 225–270.
[B]   A. Björner, *Topological methods, Handbook of combinatorics,* R. Graham, M. Grötschel, and L. Lovász (eds.) North–Holland, Amsterdam, 1995, pp. 1819–1872.
[Fo]   J. Folkman, *The homology group of a lattice,* J. Math. Mech. **15** (1966), 631–636.
[Fu]   D. B. Fuchs, *Cohomology of the braid group* mod 2, Funktsional. Anal. i Prilozhen., **4** (1970), no. 2, 62–73. English transl. in Functional Anal. Appl., **4** (1970), no. 2.
[K1]   M. Kontsevich, *Vassiliev's knot invariants,* I. M. Gelfand Seminar, Advances in Soviet Math., vol.16, Part 2, Amer. Math. Soc., Providence, RI, 1993, p. 137–150.
[K2]   M. Kontsevich, *Private communication, April 1994, Texel Island.*
[T]   V. E. Turchin, *Homology of complexes of two-connected graphs,* Uspekhi Mat. Nauk, **52** (1997), no. 2, 189–190; English transl., Russian Math. Surveys, **52** (1997), no. 2, 426–427.
[V1]   V. A. Vassiliev, *Stable cohomology of complements of discriminants of singularities of smooth functions,* Sovremennye Problemy Matematiki. vol 33, Moscow, VINITI, 1988, p. 3–29; English transl., J. Soviet Math. **52** (1990), no. 4, p. 3217–3230.
[V2]   _____, "Cohomology of knot spaces," *Theory of Singularities and its Applications* (V. I. Arnold, ed.), Advances in Soviet Math., vol. 1, Amer. Math. Soc., Providence, R.I., 1990, pp. 23–69.
[V3]   _____, "Complexes of connected graphs", *The Gelfand mathematical seminars 1990–1992* (L. Corwin, I. Gelfand, and J. Lepovsky, eds.), Birkhäuser, Boston, 1993, 223–235.
[V4]   _____, *Invariants of ornaments,* Advances in Soviet Math., vol. 21, Amer. Math. Soc., Providence, R.I., 1994, pp. 225–262.
[V5]   _____, *Complements of discriminants of smooth maps: topology and applications,* revised edition, Transl. Math. Monographs, vol 98, Amer. Math. Soc., Providence, RI, 1994.
[V6]   _____, *Topology of complements of discriminants,* Phasis, Moscow, 1997 (Russian).
[V7]   _____, *On invariants and homology of spaces of knots in arbitrary manifolds,* Mathematics at the Independent University of Moscow (B. L. Feigin and V. A. Vassiliev, eds.), Amer. Math. Soc., Providence RI, 1998, pp. 155–182.
[V8]   _____, *Homology of spaces of homogeneous polynomials without multiple zeros in* $\mathbb{R}^2$. Trudy Mat. Inst. Steklov., 1998, to appear.

INDEPENDENT MOSCOW UNIVERSITY, B. VLAS'EVSKII PER., 11, MOSCOW, 121002, RUSSIA
E-mail address: vassil@vassil.mccme.rssi.ru